SEVEN DAY LOAN

This book is to be returned on
or before the date stamped

Phosphorus Management Strategies for Lakes

Proceedings of the 1979 Conference sponsored by:

The New York State College of Agriculture
and Life Sciences, a Statutory College of
the State University at Cornell University

The International Joint Commission (U.S. and
Canada)

Edited By

RAYMOND C. LOEHR, Director
Environmental Studies Program
Cornell University
Ithaca, New York

COLLEEN S. MARTIN, Administrative Aide
Environmental Studies Program
Cornell University
Ithaca, New York

WALTER RAST, Environmental Scientist
International Joint Commission
Washington, D.C.

ANN ARBOR SCIENCE
PUBLISHERS INC
P.O. BOX 1425 • ANN ARBOR, MICH. 48106

PREFACE

The purpose of these proceedings is to summarize available information dealing with the technical, economic and institutional aspects of phosphorus management strategies in lakes. The papers indicate: (1) the relative magnitude of point and nonpoint phosphorus inputs to lakes, (2) mathematical models of phosphorus dynamics and interactions in lakes, (3) availability of various forms of phosphorus to aquatic life, (4) effectiveness of phosphorus management strategies that have been and should be tried, and (5) considerations concerning the implementation of possible strategies.

Eutrophication in lakes and reservoirs has been recognized as a challenging water quality problem for many decades and continues to be of concern in many regions and countries. Phosphorus is a common cause of such problems, and agencies, corporations and communities spend considerable effort developing and implementing appropriate phosphorus management strategies. Thus, the information in this volume can relate to lakes in many locations and should be useful for regulatory and planning personnel, consulting engineers, individuals from academic and other institutions, and environmental leaders.

To provide realistic information and approaches, the Great Lakes were used as a specific case study, and each paper addresses specific aspects of phosphorus management strategies in the Great Lakes. The Great Lakes are a very relevant example to be used in considering phosphorus management strategies for lakes. In 1978 the governments of Canada and the United States reaffirmed their determination to restore and enhance the water quality in the Great Lakes. To achieve this goal, it was agreed that there should be a maximum effort to develop programs, practices and technology necessary for a better understanding of the Great Lakes ecosystem and to eliminate or reduce, to the maximum extent practicable, the discharge of pollutants into the Great Lakes system. One of the pollutants requiring control is phosphorus.

Eutrophication of Lake Erie and Lake Ontario remains a major problem. Phosphorus loadings to these and the other Great Lakes continue to be larger than those established in earlier agreements. Further reductions of phosphorus from point and nonpoint sources are to be achieved. Therefore, the phosphorus management strategies will be a combination of point and nonpoint source controls.

Appropriate phosphorus control approaches need to be determined carefully and critically, since they can have greatly different economic impacts on different regions, and ultimately the costs will be borne by the public. Pertinent information on the costs of various management strategies is included in many of the papers.

The New York State College of Agriculture and Life Sciences, a Statutory College of the State University at Cornell University, was pleased to have the International Joint Commission (IJC) co-sponsor the conference. The program committee consisted of members of the Phosphorus Management Strategies Task Force of the IJC. This task force has the responsibility of assessing the phosphorus management strategies applicable to the Great Lakes and advising the IJC of the pros and cons of appropriate strategies. These proceedings will be of considerable value to the task force and the IJC in its deliberations.

The assistance of Ms. Doreen Kirchgraber, who played an important role in preparing for the conference, of the moderators for each conference session, of the task force for their insight and knowledge, and of those who asked questions and discussed the papers after each presentation is greatly appreciated.

Any opinions, findings, conclusions or recommendations expressed in this publication are those of the authors and do not necessarily reflect the view of the College of Agriculture and Life Sciences or the International Joint Commission.

Raymond C. Loehr, Director
Environmental Studies Program
New York State College of Agriculture
 and Life Sciences
Cornell University
Conference Chairman and Member IJC
 Phosphorus Management Strategies
 Task Force

CONTENTS

INTRODUCTION AND CONFERENCE OBJECTIVES

PHOSPHORUS INPUTS, TRENDS AND CURRENT MANAGEMENT STRATEGIES FOR THE GREAT LAKES

MODELING OF PHOSPHORUS DYNAMICS IN THE GREAT LAKES

AVAILABILITY OF PHOSPHORUS TO AQUATIC LIFE

POINT AND NONPOINT SOURCE
PHOSPHORUS MANAGEMENT STRATEGIES

INTRODUCTION
and
CONFERENCE OBJECTIVES

INTRODUCTION

D. L. Call

Dean, College of Agriculture and Life Sciences
Cornell University
Ithaca, New York

It is a privilege to open this conference on Phosphorus Management Strategies for the Great Lakes and to welcome the participants. The College of Agriculture and Life Sciences is very pleased to have the International Joint Commission as a co-sponsor of the conference. Decisions on the appropriate phosphorus management strategies that should be applied to the Great Lakes will have significant environmental, economic and institutional implications. Obviously, these decisions will not be made lightly, and this conference will be a very important opportunity to assist with the detailed discussion and evaluation that this topic deserves.

The College of Agriculture and Life Sciences has been concerned with environmental problems since its origin 75 years ago. This 11th in a series of conferences to provide a focus on important state and national environmental management concerns is but one of the extensive, environmentally related activities in which the College is involved. Several topics may serve to illustrate the breadth of environmental interests of the College and its faculty:

- integrated pest management;
- fertilizer management;
- impact of air pollutants;
- land as a waste management alternative;
- agricultural waste management;
- water resources management;
- control of pollutants from rural land; and
- ecosystem control and protection.

Many of these activities relate to the topic of this conference.

3

The 1978 Agreement [1] between Canada and the United States on Great Lakes Water Quality contained the objective that the waters of the Great Lakes be "free from nutrients directly or indirectly entering the waters as a result of human activity in amounts that create growths of aquatic life that interfere with beneficial uses." One of the programs to achieve this objective has been measures for the reduction and control of inputs of phosphorus and other nutrients into the Great Lakes.

These measures will require control of the phosphorus in both point (municipal and industrial) sources and nonpoint (agricultural, urban and rural) sources. Control of the phosphorus in point sources has received the major emphasis thus far. Limits have been imposed on the amount of phosphorus in the effluent from municipal waste treatment facilities discharging more than one million gallons per day. Any needed reductions in effluent phosphorus concentrations will depend on the target input loads to the lakes that will meet the desired water quality objectives. The approaches used to establish the target loads and the resultant accuracy of the target loads are essential to the determination of reasonable phosphorus management strategies and undoubtedly will receive considerable evaluation and discussion at this conference.

Point sources, however, are not the only sources that may need control. The lakes most affected by phosphorus are Lakes Erie and Ontario, and agricultural sources have been identified as the major nonpoint source contributor of phosphorus to these lakes. In developing appropriate management strategies, some control of nonpoint sources probably will occur.

In considering possible phosphorus management strategies, care should be taken to avoid the trap of uniform approaches for all lakes or for specific types of sources. While uniform treatment can be an administratively convenient legal and political approach, it may not equitably achieve the desired water quality goals. The Great Lakes should be viewed as an interconnected ecosystem with many factors affecting the internal ecosystems.

Phosphorus management strategies should be applied selectively for each of the Lakes to reduce total costs. Identical policies applied to different Lakes will not produce water of similar quality.

The costs of reducing phosphorus loads vary depending upon the source. A major study by faculty in the College of Agriculture and Life Sciences [2] evaluated the possibilities of reducing phosphorus inputs to Cayuga Lake, the largest of the Finger Lakes in New York. The results, shown on page 5, indicated a significant difference in costs of reducing phosphorus.

One may debate the accuracy of these costs, but the most important inference is that the costs to reduce the dissolved phosphorus load from different sources can be orders of magnitude apart. Thus a decision to uniformly reduce the phosphorus loads from all sources by a certain

Method of Phosphorus Reduction	Cost of Reducing the Dissolved Phosphorus Loading ($/kg)
Tertiary treatment of municipal sewage	5-12
Collection and treatment of septic tank waste and other unsewered wastes	15-20
Control through reduction in corn acreage	400
Avoid winter spreading of manure	570-1030

percentage or amount can have greatly different economic impacts on different regions and different parts of the economy. Ultimately these costs will be borne by the public.

Preliminary estimates [3] of the cost to achieve the phosphorus target loads are about $69 million for Lake Erie and $22 million for Lake Ontario. It must be ascertained that such an investment is actually needed, if needed is spent wisely, and if spent will achieve the desired objectives.

The need to spend wisely always has been important. However, the increasing public recognition of limited resources and of valid competition for available funds demands that increasingly careful attention be paid to the spending of public funds. As an example, the Great Lakes Water Quality Agreement [1] contains objectives for control of toxic substances such as pesticides and heavy metals and of pollution from shipping sources, and objectives for continued surveillance and monitoring. Achievement of these objectives also will require adequate support and, in turn, may compete with resources needed for phosphorus control.

These comments should not be interpreted as negative or as suggesting that there is no need to continue to control the amount of phosphorus entering the Great Lakes. Rather the comments were intended to reinforce the need to put phosphorus control in perspective with other control measures required for the Great Lakes, the need to evaluate critically the best approaches to achieve the phosphorus target loads in the Great Lakes, and the need to identify critically the economic, social and institutional implications of implementing various technical alternatives to control the phosphorus inputs to the Great Lakes.

It is obvious from the program that the conference was developed to provide detailed information needed for critical evaluation and to permit the comprehensive evaluation that will identify the most appropriate phosphorus management strategies for the Great Lakes. It is encouraging to note the considerable amount of time that has been provided for discussion of the various topics and presentations. It is important that all the pertinent information and insights be provided on this important topic. It is your questions, observations, comments, and discussion that will provide a wide perspective. I trust that each of you will actively participate in the conference.

Phosphorus control for the Great Lakes will have important ramifications for the states and provinces draining into the Great Lakes. In addition, the management strategies that result from the information and discussion at this Conference can have an impact on decisions in other locations where phosphorus control is needed for lakes. As I indicated earlier, the College of Agriculture and Life Sciences is very pleased to work with the International Joint Commission in developing and sponsoring this very relevant and important conference.

I am looking forward to a very productive and exciting conference.

REFERENCES

1. International Joint Commission "Great Lakes Quality Agreement of 1978," Washington, DC and Ottawa, Ontario, November 22, 1978.
2. Porter, K. S. (Ed.) *Nitrogen and Phosphorus–Food Production, Waste and the Environment* (Ann Arbor, MI: Ann Arbor Science Publishers Inc., 1975).
3. Berg, N. A., and M. G. Johnson "Final Report to the International Joint Commission from the International Reference Group on Great Lakes Pollution from Land Use Activities–PLUARG," International Joint Commission, Windsor, Ontario, 1978.

CONFERENCE OBJECTIVES

R. J. Sugarman
Chairman, U.S. Section
International Joint Commission
Washington, DC

I am honored to open this conference.

This is an exceptionally important event in the effort to control and reverse eutrophication of the lower Great Lakes, and to protect the upper lakes from similar problems.

In our lower lakes report of 1970, the commission advised the Governments of Canada and the United States to enter into an agreement on an integrated program of phosphorus control to include: (1) a reduction of phosphorus in detergents; (2) reductions of phosphorus in municipal and industrial waste effluents, and (3) reductions in phosphorus discharges from agriculture. By April 1972 the nations had formulated the Great Lakes Water Quality Agreement, including target loadings for the lower lakes.

By that agreement, IJC was given the job of reporting the trends in phosphorus loadings to the lakes and advising on measures to be taken. It was thought that the target phosphorus loads set in the agreement could be met largely through reductions in phosphorus from municipal wastewater treatment plant effluents. I defer for the moment the question as to the adequacy of the target loads to restore the health of the lower lakes. Reductions were made, but target loads were not met. The obvious question is what further measures are required, and how, if at all, can we implement them?

These target phosphorus loads had been set on the basis of the information available at the time. They were designed to represent the loads that would exist if the apparently achievable goal of 1 mg/l phosphorus in effluents of treatments plants discharging in excess of 1 mgd were met.

In 1976, the IJC told Governments that Lake Erie had begun to show signs

7

of recovery. However, Lake Ontario had begun to exhibit some of the deterioration indicators of Lake Erie. As better information has been obtained it is clear that the target loads are probably not achievable by the present programs. Nor did the nearshore areas recover as fast as we had hoped.

We were then four years into the extensive and expensive sewage treatment plants construction grants program, and the massive National Pollutant Discharge Elimination System of permits had begun under the U. S. Federal Water Pollution Control Act–however, progress was slow.

In Canada, treatment plant construction and phosphorus removal facilities planned to meet conditions stipulated under the agreement were nearly 100% completed in 1977, but loadings still were not being met. We do not yet know whether the quick-fix approach had inherent operations problems or whether O&M methods were to blame. In both cases, clearly, better approaches are needed.

One way to reduce loads is by reducing phosphorus in wastewater influents. For years the IJC has been recommending a phosphate detergent ban. Reports from areas where such a ban has been enacted indicate that reduced phosphorus loadings do result. These bans or limits could be extended to more products and more jurisdictions. The reductions achievable, however, are but a small fraction of the present loadings, and industry questions the cost-effectiveness of further such reductions.

The commission has noted that land treatment, hitherto given little consideration, routinely achieves concentrations as low as 0.1 mg/l at installations like Muskegon, Michigan, a 33 mgd facility, and has recommended its consideration. But resistance remains to this method. Hopefully, the land treatment issue will be explored here.

Other problems persist as well. The compatibility and comparability of data from one jurisdiction to another is still far from ideal, though greatly improved since 1972. Many questions as to sampling adequacy, consideration of climatic variables, data comparability, monitoring reports, diffuse source estimates, including atmospheric inputs, and the effects of dilution on meaningfulness of phosphorus concentrations challenge and trouble the interpreter of the data. Today there are such wide divergencies in lake-response models and predictions that efforts to focus on programs are severely hampered. No agreement is possible on a treatment goal of 0.5 mg/l in wastewater treatment plant effluents because scientists do not agree as to whether it will improve lake conditions.

Our Upper Lakes Reference Group and PLUARG both identified some of the hitherto neglected variables and reduced the amount of uncertainty as to their importance. Improved monitoring techniques and data analysis techniques were developed. Better predictive models were constructed and further calibration was possible. The models now apparently include at least some consideration of loadings from diffuse land sources (both urban and rural) and air impacts. But they seem to indicate that municipal sewage treatment plant phosphorus effluent levels must approach zero in Lake Erie, and possibly Lake Ontario, before satisfactory lakewide phosphorus and oxygen conditions can be achieved. But these models are based on the assumption that all phosphorus is equally causative. It has been forcefully argued that much of the phosphorus in each source is benign. Thus, our Science Advisory Board advised us in 1978 that the problems of phosphorus were becoming more, instead of less, complex.

It is argued that we should wait until we see the effects of the present programs before attempting improvements. It is argued that we should not proceed to develop and implement a definitive phosphorus management strategy for the Great Lakes when we do not know if the phosphorus we will remove is the phosphorus of concern. It is argued we should wait until the answer is clearer before spending additional monies to abate pollution from hard-to-control diffuse sources. Indeed, total phosphorus does seem a poor, indeed fictional, surrogate for available phosphorus, and present programs may bring about more improvements, and money *is* in too short supply to be spent without clear reason. But we cannot stand still. Billions of dollars are being committed to municipal wastewater treatment plants which may not achieve 1.0 mg/l effluent phosphorus levels, much less the lower levels indicated for Lakes Erie and Ontario and for local areas of the Upper Lakes. Decisions are being made all the time. And more phosphorus is accumulating all the time.

In recognition of these inadequacies and uncertainties, the 1978 Great Lakes Water Quality Agreement includes tentative phosphorus loadings for each lake, based on municipal treatment plants discharging in excess of 1 mgd throughout the Great Lakes Basin achieving an effluent concentration of 1 mg/l of phosphorus, except for the lower levels believed to be required for plants in the Lake Erie and Ontario basins, and in Saginaw Bay.

Maximum practicable diffuse source reductions are called for in Lakes Superior, Huron and Michigan, while a 30% cutback is agreed to for the lower lakes.

By mid-1980, the governments are to confirm these tentative targets and, based upon them, establish phosphorus load allocations and compliance schedules. The governments agree to consider our recommendation in this process. We must therefore address these complex questions now.

To provide help to us, the Science Advisory Board formed a committee of experts to advise it on phosphorus management strategies. Our Water Quality Board has joined the effort, and the planning of phosphorus management strategies is now a joint activity. The commission, our two boards under the new agreement, and the Phosphorus Management Strategies Task Force are hopeful that the people here today can begin to help us formulate implementable, cost-effective phosphorus management strategies for the Great Lakes Basin, strategies we can recommend to the Governments with confidence.

We recognize that scientific certainty is not around the corner. But events cannot wait. We simply must have your best judgment and your best thinking based on the available data as to: (1) what levels of phosphorus should be achieved to restore and protect the Great Lakes; (2) which sources and types of phosphorus should be of most concern; and (3) what measures are indicated to remove this phosphorus as safely and economically as possible.

That is why IJC is co-sponsoring this conference and why I am here. We look to you to advance the state-of-the-art in phosphorus management and for guidance in development of phosphorus management strategies for the Great Lakes.

SECTION I
PHOSPHORUS INPUTS, TRENDS AND CURRENT
MANAGEMENT STRATEGIES FOR THE GREAT LAKES

CHAPTER 1

ACTION TAKEN TO CONTROL PHOSPHORUS IN THE GREAT LAKES

R. W. Slater and G. E. Bangay

Environmental Protection Service
Environment Canada
Toronto, Ontario

BACKGROUND

Since the time of Champlain, the Great Lakes/St. Lawrence River system has exerted a compelling influence on the development of that region. Unlike the colonies of the eastern seaboard, which were built on the solid foundation of agriculture and the sea, the French settlements on the St. Lawrence, capitalizing on the accessibility that this drainage system offered to the remainder of the continent, turned instead to the vagaries of the fur trade. By 1765 the British had assumed control of this vast trading empire giving them complete control of the continent. For 20 years, until 1785, the British continued to exploit the commercial advantages offered by the Great Lakes system. However, the signing of the Treaty of 1783 which ended the War of Independence between Great Britain and the United States arbitrarily divided the Great Lakes system for the first time by a boundary devoid of geographic and historical meaning. In an effort to minimize the disruption to the region, the two countries signed the Jay's Treaty of 1794. It essentially guaranteed the right of passage throughout the region to all inhabitants and the right to free trade.

Since the Jay's Treaty, the governments of the two countries have entered into a succession of treaties and agreements to minimize disruptions in trade and commerce on the lakes. In 1909 these finally culminated in the signing of

13

the Boundary Waters Treaty. This Agreement applied to the 2000 miles of the international boundary which are marked by navigable and nonnavigable waters.

The principal achievement of this treaty was the establishment of an international commission which would effect the aims of the treaty and whose conclusions were to be justified by public opinion. Under the terms of the treaty, the commission could perform judicial, investigative, administrative and arbitrational functions.

Although the primary issue at the time of signing of the treaty was the maintenance of levels and flows in the boundary waters for purposes of navigation,* other concerns were also expressed. Article IV of the Boundary Waters Treaty, 1909, contains the following provision:

> "It is further agreed that the waters herein defined as boundary waters and waters flowing across the boundary shall not be polluted on either side to the injury of health or property on the other" [1].

During the same period, when jurisdictional differences and approaches to management of this shared resource were evolving, important changes in man's activities in the basin were also taking place. Ultimately, these changes would hold serious implications for the lakes and would determine the shape of institutions yet to come.

After 1783, the arrival of the United Empire Loyalists in the British portion of the basin brought agriculture into the middle of a fur trading region. This immigration signaled the beginning of permanent, large-scale settlements based on the new staples of timber and wheat. The rapid advancement of this settlement though the fertile lowlands of the basin resulted in the destruction of the Great Lakes forest. With this large-scale land disturbance, the level of upland soil erosion accelerated, releasing greater quantities of eroded soil particles for transport down the tributaries to the lakes.

Wheat was the first agricultural commodity of importance to the region. By 1900, however, the importance of livestock to the agricultural economy was firmly established. Between 1931 and 1971 receipts from livestock and livestock-related products accounted for an average of 72% of the total cash receipts from farming operations [2].

Until 1941 the area of land actively farmed continued to increase in the basin. Since that time, however, the acreage has decreased, but important

---

*In its Preliminary Article, the Treaty defines boundary waters as "the waters from main shore to main shore of the lakes and rivers and connecting waterways, or the portions thereof, along which the international boundary between the United States and the Dominion of Canada passes, including all bays, arms, and inlets thereof, but not including tributary waters which in their natural channels would flow into such lakes, rivers, and waterways, or waters flowing from such lakes, rivers, and waterways, or the waters of rivers flowing across the boundary".

changes have occurred in the way that this land is now used. The acreage of land devoted to more erosion-prone row cropping as opposed to the less hazardous, close-grown crops, has grown rapidly. Existing projections do indicate that this land base used by agriculture is likely to continue to decrease [3]; however, the intensity of use will increase to meet the growing national and international demand for food. All of these changes have had important implications for water quality.

In the initial stages of settlement, the essentially agrarian population was thinly dispersed about the fertile lowlands of the basin. Later, increasingly large settlements developed to service these agricultural interests. The combination of rich land and high-quality water resources and an easy access to markets encouraged the rapid growth of these centres. In 1851, only 14% of Ontario's population could be classified as urban; by the time of the signing of the Boundary Waters Treaty, this had risen to 50% and, in 1971, 80% of the population lived in urban centres [4]. In just over a century, settlement in the basin has proceeded from a thinly scattered, rural population to the rapidly evolving Great Lakes megalopolis. In 1971 the population of the Great Lakes Basin was approximately 35 million; by the year 2020 this may rise as high as 60 million [3].

During this same period, the land-based economic activities of agriculture, forestry and fisheries are expected to show a decline in their share of the total growth in economic activity. In contrast, the share of secondary industries is projected to increase in size.

As these changes in the basin's economic base and demographic patterns have occurred, so too have their impacts on water quality. It wasn't until 1912, however, that the two governments expressed their joint concern about the deteriorating state of the boundary waters. In a Reference to the International Joint Commission (IJC), the two governments asked the Commission to investigate and report on the following questions:

1.   To what extent and by what causes and in what localities have the boundary waters between the United States and Canada been polluted as to be injurious to the public health and unfit for domestic or other uses?

2.   In what way or manner, whether by the construction and operation of suitable drainage canals or plants at convenient points or otherwise, is it possible and advisable to remedy or prevent the pollution of these waters, and by what means or arrangements can the proper construction or operation of remedial or preventive works, or a system or method of rendering these waters sanitary and suitable for domestic and other uses, be best secured and maintained in order to insure the adequate protection and development of all interests involved on both sides of the boundary and to fulfill the obligations undertaken in Article IV of the Waterways Treaty of January 11, 1909 between the United States and Great Britain, in

which it is agreed that the waters therein defined as boundary
waters and waters flowing across the boundary shall not be polluted on either side to the injury of health or property on the
other?

In 1916 the experts enlisted by the IJC made their final report. They concluded that in the Great Lakes, with the exception of small areas near large
cities and in shipping lanes, the water was in its natural state of purity. The
Detroit and Niagara Rivers were identified as being the most grossly polluted,
and thirty miles of the St. Clair River were also noted as being unfit for
drinking. Discharges of raw sewage by cities and towns along these rivers were
identified as the major sources of pollution. The investigators also concluded
that these aforementioned conditions were undoubtedly responsible for the
abnormal level of typhoid fever being experienced at the time.

It was not until 1946 that a further reference on pollution of boundary
waters was submitted to the IJC by the two governments. This reference,
including later extensions, was specific to an investigation of water quality
problems in the connecting channels of Lakes Ontario, Erie, Huron and
Superior. In its final report to the two governments, the IJC made the following observations:

1. Despite improvements in sewage treatment, bacterial concentrations
   in the connecting channels were, in places, 3-4 times greater on
   average than in 1912.
2. Industrial wastes, which were of little concern in 1912, were in 1951
   seen as a major problem from both the perspective of exerting an unfavorable oxygen demand and creating toxicity problems.
3. Vessel wastes, which were identified as a problem in 1916, were still
   viewed as having an impact on water quality.

Both of these investigations into the water quality of the Great Lakes were
restricted either by the number of parameters considered or the areas studied.
No attempt had been made to begin the development of a comprehensive
data base through which trends and changes in overall water quality could be
observed. It was not until 1957 that the first efforts were made in this direction [5].

The pollution of the Great Lakes from municipal and industrial sources
continued to be a problem during the 1950s. The pace of new developments
had outstripped the ability of regulatory agencies to deal with them. Numerous U.S. cities, including Rochester, Buffalo, Cleveland and Milwaukee, had
abandoned their beaches to bacterial pollution. The technology of industrial
production had surpassed the technology of residuals management. The
opening of the St. Lawrence Seaway in 1959 raised new concerns about raw
sewage and garbage disposal from ocean-going ships. The development of
synthetic detergents after World War II which incorporated phosphates as
builders raised the spectre of increased nutrient enrichment of boundary

waters. Finally, in 1964 the two governments jointly informed the IJC that they had reason to believe the waters of Lake Erie, Lake Ontario and the international section of the St. Lawrence were being polluted by sewage and industrial wastes. The two governments had "agreed upon a joint reference of the matter" to the commission pursuant to the provisions of Article IX of the Boundary Waters Treaty of 1909.

The Commission was requested to inquire into and report to the two governments as soon as practicable upon the following questions:

1. Are the waters of Lake Erie, Lake Ontario, and the international section of the St. Lawrence River being polluted on either side of the boundary to an extent which is causing or is likely to cause injury to health or property on the other side of the boundary?
2. If the foregoing question is answered in the affirmative, to what extent, by what causes, and in what localities is such pollution taking place?
3. If the Commission should find that pollution of the character just referred to is taking place, what remedial measures would, in its judgment, be most practicable from the economic, sanitary and other points of view, and what would be the probable cost thereof?

To make the necessary investigations and studies to form the basis for its report to the Governments of the United States and Canada, the Commission established two advisory boards: (1) The International Lake Erie Water Pollution Board, and (2) The International Lake Ontario and St. Lawrence River Water Pollution Board. Their studies, which took place between 1963 and 1968, represented the first comprehensive look at water quality in the Great Lakes and, as a result, has proved to be the foundation for all subsequent action taken to manage Great Lakes water quality.

In 1969 the International Joint Commission made its report to the two governments. This report, which was based on the earlier reports of its advisory boards, concluded that transboundary water pollution was occurring to an extent that was causing injury to health and property on the other side of the boundary. The report also identified a number of specific problems which required remedial action:

1. accelerated eutrophication of Lake Ontario and a condition of advanced eutrophication in the western basin of Lake Erie, largely as a result of increased nutrient inputs, especially phosphorus;
2. a disruption of beneficial uses of water as a result of changing trophic conditions;
3. dramatic changes in the species composition of commercial fish catches;
4. pollution problems related to dredge spoil disposal;
5. bacterial pollution of nearshore waters; and
6. pollution from organic contaminants, spills of oil and other hazardous substances, radioactive materials, viral contamination and thermal pollution.

Although many of the problems described by this reference simply reconfirmed the continued existence of problems noted in earlier references, the identification of water quality problems associated with nutrient inputs added new emphasis to the need for improved management of the Great Lakes. A number of specific recommendations were made by this reference group to deal with the problem of eutrophication, including:

1. immediate reductions in the phosphorus content in detergents;
2. implementation of programs for the reduction of phosphorus from municipal and industrial waste effluents, including a timetable for reductions;
3. the development of programs for the control of phosphorus from agricultural operations;
4. regulation of any new uses of phosphorus which could result in appreciable additions to the lakes.

In recognition of the importance of these findings, the two governments began negotiations which culminated in a new strategy for managing Great Lakes water quality. On April 15, 1972, the Prime Minister of Canada and the President of the United States signed the Great Lakes Water Quality Agreement. This agreement, together with annexes, texts and terms of reference charted a course of action for the two governments in dealing with Great Lakes water quality problems. It also established an unprecedented opportunity for continued cooperation between the two central governments and the key state and provincial administrations in dealing with water quality problems of mutual concern. Certainly, the degree of cooperative effort envisioned by the agreement had not been duplicated elsewhere in the world.

With respect to the control of phosphorus, the agreement committed the governments to a scheduled reduction in phosphorus discharges from municipal and industrial sources to the Lower Lakes. The Agreement also directed the IJC to undertake two new references. The Upper Lakes Reference was directed to provide much needed information on water quality in the upper lakes, and the Pollution from Land Use Activities Reference was asked to provide information on the quantities and control of pollution from diffuse or nonpoint sources of pollution.

The Agreement also called for the creation of new institutions to assist in the implementation of the agreement and the continued management of the lakes. The Great Lakes Water Quality Board was established to assist the Commission in the exercise of the powers and responsibilities assigned to it under the agreement, and the Great Lakes Research Advisory Board was directed to review and assess the adequacy of existing and proposed research in contributing to the achievement of the terms of the Agreement.

Since the report of the Lower Lakes Reference Group, a number of important steps have been taken to reduce phosphorus input to the Great Lakes. In Canada, immediate steps were taken to reduce the phosphorus content in laundry detergents. In 1970, levels were reduced to 8.5% P by weight, and in 1972 this was further reduced to 2.2% P. In the United States, the existence of eight separate state jurisdictions has resulted in a somewhat less rapid and uniform reduction in phosphorus in detergents; however, only Ohio and Pennsylvania still do not control this source. Important reductions have also been achieved at municipal sewage treatment plants. In Lake Erie, the point source load has been reduced from approximately 10,000 metric ton/yr in 1972-73 to 5700 metric ton/yr in 1977 and is expected to be 2100 metric ton/yr when all controls are in place to meet the 1 mg/l phosphorus effluent limit. In Lake Ontario, direct point source discharges have been reduced from approximately 6300 metric ton/yr in 1972-73 to 2600 metric ton/yr in 1977 and are expected to be 1500 metric ton/yr when all controls are in place to meet the 1 mg/l phosphorus effluent limit.

In order to accomplish these changes, the governments of Canada and the United States have committed approximately $4.5 billion for sewerage construction since the signing of the Agreement. In Lake Ontario some improvements in whole lake conditions which may be attributed to this investment have been observed [6]. Since 1973, Dobson has reported a decline in dissolved inorganic phosphorus over the winter period and an increase in the summer mean secchi depth for Lake Ontario. Trends in conditions in Lake Erie are far less clearly discernible. An analysis of nearshore water quality data collected by the Ontario Ministry of Environment has provided evidence of a declining trend in total phosphorus concentrations in nearshore waters of the western basin of Lake Erie [7]. Other nearshore waters in Lake Erie did not exhibit this clearly defined trend. In Lake Ontario, nearshore areas west and east of Toronto have also shown a decline in total phosphorus concentrations between 1976 and 1978.

Public perceptions of changes in water quality are also interesting to note. In a survey of public perceptions, water users stated that Lake Erie is improving while changes in other lakes were less evident. Nonusers generally considered water quality to be deteriorating [8].

As a part of the fifth year review of the progress made under this agreement, the parties decided to strengthen the existing terms of the agreement and to address their mutual concerns about the growing problem of persistent toxic substances and the atmospheric transport of pollutants. These actions lead to the signing of a second Great Lakes Water Quality Agreement in November 1978. Unlike the earlier agreement, which was signed during a period when "environment" was a more prominent and favored issue, this agreement has been promulgated at a time when the competing issues of economics and energy are receiving precedence.

Before signing this new agreement, the two parties were made aware of a number of important issues which required resolution before any firm commitments to further phosphorus load reductions could be made. These concerns related specifically to the different phosphorus loading estimates which were made by PLUARG and the Surveillance Subcommittee and the dissimilar target loading estimates made by Task Group III* and PLUARG.*

For these reasons, Annex 3, Section 3 of the Agreement states that, within 18 months after the date of entry into force of this agreement, the parties, in cooperation with the state and provincial governments, confirm the future phosphorus loads (as stated in the agreement) and, based on these, establish load allocations and compliance schedules.

Thus, despite the existence of a wealth of new information generated on the Great Lakes since 1964, those responsible for managing Great Lakes water quality are still left with a number of unresolved problems. It is of critical importance to the implementation of any future phosphorus management strategy that these problems be dealt with now. This conference can play a key role. Clear statements on the following five points are required.

1. What are our present best estimates of phosphorus loadings in the Great Lakes?
2. What confidence do we have in the accuracy of phosphorus water quality objectives and the models which translate this into loadings? How sensitive are these models when compared with the current loading information and the attainable load reduction from additional remedial programs?
3. What is the "state of the art" concerning phosphorus availability and how should this be used in formulating future phosphorus management strategies?
4. What is the potential for control of the remaining sources of phosphorus?
5. What will be the social and economic impacts of the continued management of phosphorus inputs to the Great Lakes?

Each of these questions poses a separate problem for scientists and managers alike. Each must be dealt with separately. However, the answers provided through subsequent discussions will ultimately come together to shape the direction of our future phosphorus management strategies. Your individual and collective wisdom, based on the best available information, while bearing in mind the significance of unknowns, could provide the basis for actions over the coming decades.

---

*Task Group III was the bilateral technical group which developed the target loads for the 1978 Water Quality Agreement.

*Pollution from Land Use Activities Reference Group was charged with determining the extent and sources of pollution of the Great Lakes from land drainage and the nature of remedial action to correct the identified problems.

ESTIMATES OF CURRENT LOADS

The existence of at least 3 separate estimates of the total load of phosphorus to the Great Lakes requires clarification. The compatibility of these estimates has yet to be demonstrated, although attempts are now under way to determine this. Important differences also exist with respect to the calculation of tributary loads and the variations attributable to natural variations in the hydrologic cycle. These variations may be important, especially when viewed in the context of the loading reductions achievable through further remedial programs. Thus, these variations must be explained adequately before commitments for further reductions can be expected. If it is possible to settle on a best estimate of current loads, what should be done to maintain our ability to generate such estimates?

GOALS & OBJECTIVES

The target loads for the lower lakes contained in the 1972 agreement are based upon those which could be accomplished by controlling sewage treatment plant effluents with plants larger than 1 mgd to the level of 1 mg/l phosphorus. As such, the achievement of these loads was not associated with meeting a specific, predicted water quality in the lakes. Over the past few years, significant advances have been made in our knowledge of the process of eutrophication and, consequently, it is possible to predict with greater precision than hitherto possible the relationship between phosphorus loads and resulting water quality. This was the route followed by the parties in their review of the agreement and is a significant departure from the approach employed in the current agreement.

The job of setting target loads in the revised agreement was assigned to Task Group III during the course of the Fifth Year Review. These proposed target loads have been incorporated in the revised agreement as "tentative target loads" and are subject to confirmation by the parties. These proposed targets were further refined by PLUARG, primarily for the upper lakes on the basis of more recent information. It is important to note that the loadings in the 1972 agreement still prevail and that the parties have yet to conduct the formal examination of the tentative target loads required under the 1978 agreement.

In the lower lakes, the "tentative target loads" were developed using computer models. These targets are directed towards the maintenance of aerobic conditions in hypolimnetic waters of Lake Erie and the restriction of nuisance growths of algae in both Lakes Erie and Ontario. For the upper lakes, the loads were calculated on the basis of all municipal STPs (with capacities greater than 1 mgd) achieving 1 mg/l in their effluent. There does not appear to have been a rigorous peer group evaluation of either the water quality objectives or the models. As a consequence, there is a degree of uncertainty with respect to the derived target loads.

Managers require a clear view of the confidence modelers have in predicting water quality effects given estimated phosphorus loadings. Further information is also needed on the translation of water quality objectives into realistic statements of social and economic benefits. Until this is accomplished, we will continue to flounder in a sea of physical science jargon which makes only a partial connection with the needs of managers. Further remedial action can only be taken if there is a recognition of the importance of translating often obscure statements of scientific fact into terms meaningful to the public. Human needs remain the first priority in any discussion of water quality.

PHOSPHORUS AVAILABILITY

Existing phosphorus management strategies have been based on the reduction of total phosphorus loads. Recent information made available through PLUARG and others raises the question of the importance of available phosphorus versus the unavailable forms. A basic factor in designing both the scope and timing of future remedial programs for phosphorus loading reductions will be the significance of available versus nonavailable forms.

The modeling work conducted by Task Group III used measurements of total phosphorus inputs to calibrate their predictive models (the only exception being the exclusion of apatite phosphorus forms contributed in eroded lakeshore material). Since the signing of the 1972 agreement, all of our efforts to control phosphorus have concentrated on those forms of phosphorus generally considered to be most available—phosphorus in laundry detergents and in municipal sewage treatment plant effluents. Will new programs directed toward reducing phosphorus inputs from nonpoint sources yield the same level of lake response given the situation where only 40% or less of the particulate phosphorus in tributary discharges was estimated to be available [9]?

REMEDIAL PROGRAMS:

It is self-evident that remedial measures must be designed to meet environmental objectives in a cost-effective and practical manner. A variety of options for obtaining further reductions in phosphorus at point and nonpoint sources has been proposed.

In 1977, 35% of the total reported flow from those municipal sewage treatment plants reporting flow and phosphorus loading discharge data in the Great Lakes Basin had mean annual phosphorus concentrations below 1 mg/l. Only 4% of this flow contained mean annual phosphorus concentrations of 0.5 mg/l or less [10]. Some serious reservations do exist about the proposals to achieve 0.5 mg/l P in the effluent from sewage treatment plants larger than 1 mgd. Given existing technology, is it practical to expect that sewage treatment plant operators can consistently obtain this level, especially at the larger plants and in the absence of additional stages of treatment? A related concern is the existing level of expertise held by plant operators. This is especially important in the smaller sewage treatment plants where the initial requirements for plant operators are not as stringent as those in the large plants. The human as well as the technical element will therefore be critical in determining the success of future phosphorus reductions at municipal sewage treatment plants.

The existing means for underwriting the costs of sewage treatment will also contribute to the level of phosphorus treatment achieved. In many cases, local governments receive partial grants or bans from other levels of government to enable them to construct and expand sewage treatment facilities. The provision of these capital resources, however, does not assist municipalities in meeting the subsequent high operating costs of the facilities. The availability of operating funds will also be important in determining the success of any efforts to further reduce phosphorus concentrations. Further information must be provided to allow for an accurate determination of the efficiency of proceeding with this option.

The new approaches and measures proposed for the control of nonpoint sources require further study. Many of these remain relatively untried in terms of their efficiency in reducing phosphorus movement to the lakes. In particular, only a few types of controls requiring little expenditure have been demonstrated to be cost-effective.

BENEFITS & COSTS OF FURTHER ACTION

Since the report of the Lower Lakes Reference in 1969, those involved in the study of the lakes and in their management have spoken generally of the benefits to society of controlling phosphorus. Decaying masses of algae fouling bathing beaches, plugged water intakes and reduced property values are only a few of the reasons which have been presented as a rationale for encouraging governments to act to reduce P inputs. Objective evaluation of these claims was not undertaken and was hardly necessary given the immediacy of the problem in the late 1960s.

Today the situation is different. Information which will permit an assessment of the costs and benefits of achieving different levels of phosphorus management must be forthcoming. You can be assured that those competing for the same tax dollars will be presenting their case in this way. So must we.

CONCLUSION

From the time of the French settlements on the St. Lawrence, the Great Lakes have continued to shape the development of the region. Successive stages of this economic development have modified the jurisdictional and institutional arrangements found in the region. Closely intertwined with this changing economic base has been the transformation in the role of the lakes. Each step in this continuous process of development has increased the complexity of managing the often competing and conflicting activities carried out there.

Only recently has there been an appreciation of the broader implications of the links which exist between economic development in the basin and the quality of water in the lakes. The study of these linkages has brought attention to a growing number of problems of mutual concern to the two countries. The complexity of these problems has required a divison of effort on the part of the scientific community who study the problems, and the managers who ultimately must design and implement remedial programs.

This situation has resulted in the fragmentation of Great Lakes' problems and solutions. Nowhere do we find the necessary integration of effort in an institutional sense to deal with this situation. The time has come for managers and scientists alike to develop answers which possess a high degree of certainty, and which are significant in providing a means for effectively managing the lakes.

REFERENCES

1. Bloomfield, L. M. and G. F. Fitzgerald. *Boundary Waters Problems of Canada and the United States—The IJC 1912-1958* (Toronto: Carswell Co. Ltd., 1958), p. 76.
2. Bangay, G. E. "Livestock & Poultry Wastes in the Great Lakes Basin, Environmental Concerns and Management Issues," Social Science Series No. 15, Burlington, (1976) p. 5.
3. "Land Use and Land Use Practices in the Great Lakes Basin: Joint Summary Report—Task B, United States and Canada," International Reference Group on Great Lakes Pollution from Land Use Activities, Windsor, Ontario, (September 1977).
4. Stone, L. D. "Urban Development in Canada," Dominion Bureau of Statistics, (1967) p. 29.

5. "Great Lakes Basin," A symposium presented at the Chicago Meeting of the American Association for the Advancement of Science, 29-30 December, 1959.

6. Dobson, H. "Lake Ontario is Changing: A Progress Report on the Study of Eutrophication Trends," unpublished.

7. "Great Lakes Water Quality 1977: Appendix B", Annual Report of the Surveillance Subcommittee to the Implementation Committee, Great Lakes Water Quality Board, July 1978, p. 24.

8. "Sixth Annual Report to the IJC on Great Lakes Water Quality," Great Lakes Quality Board, Windsor, Ontario, July 1978, p. 34.

9. "Environmental Management Strategy for the Great Lakes System," International Reference Group on Great Lakes Pollution from Land Use Activities, Windsor, Ontario, July 1978, p. 23.

10. "Great Lakes Water Quality 1977: Appendix C," Annual Report of the Remedial Measures Subcommittee to the Implementation Committee, Great Lakes Water Quality Board, July 1978.

CHAPTER 2

POINT SOURCE LOADS OF PHOSPHORUS TO THE GREAT LAKES

H. Zar

Great Lakes National Program Office
U. S. Environmental Protection Agency
Chicago, Illinois

INTRODUCTION

The purpose of this chapter is to discuss the progress in reduction of phosphorus loads from point source dischargers to the Great Lakes Basin since the original U.S.-Canada Great Lakes Water Quality Agreement was adopted in 1972. The loading estimates presented are derived from those developed anually by the Remedial Programs Subcommittee of the Great Lakes Water Quality Board (Board) [1-6]. Reference will also be made to critical revisions of the Board's estimates that have recently been made, those of the Pollution of Land Use Activities Reference Group (PLUARG) [7], Task Group III (a U.S.-Canadian review group established during development of the 1978 Great Lakes Water Quality Agreement [8], and the U.S. Army Corps of Engineers Lake Erie Wastewater Management Study (Corps) [9].

BACKGROUND

Estimates of individual point source loads made by the Board are based primarily on self-monitoring data supplied to the pollution control agencies of the eight states surrounding the Great Lakes and Ontario by dischargers as part of their normal reporting requirements. The frequency of measurement and type of sample now in use vary somewhat among facilities, with large municipal plants generally obtaining daily composite samples. At the beginning of each year the agencies derive estimates for average total phosphorus load and flow for the preceding calendar year or water year, relying on their own inspections and compliance monitoring data where necessary.

These data, together with load estimates for other contaminants and compliance information, are supplied to the Windsor Regional Office of the International Joint Commission (IJC) for assembly. Loading and compliance data for 376 municipalities and 448 industries were reported for 1977 [6]. The Remedial Programs Subcommittee then develops a loading report which is reviewed by the board and submitted to the IJC.

The extent and quality of data are largely determined by the regulatory purposed for which they were originally gathered. Data for earlier years are spotty or absent in places because effluent and monitoring requirements for phosphorus were often not in place. Current estimates have the advantage of a more sizable data base but omit municipal plants with flow less than 3800 m^3/day (1 mgd) and small industrial dischargers of phosphorus which do not have phosphorus requirements. Data for larger plants have also been unavailable in some years.

LOADING ESTIMATES

The consequence of the above omissions is that the Board estimates of "reported" loads tend to be underestimates. In an effort to improve consistency, in recent years some estimates for loads from nonreporting municipalities under 1 mgd have been included. The intent has been to achieve a reasonably accurate estimate of point source loads for plants above 1 mgd, subject to time constraints and without overburdening the jurisdictions. The quality of the estimates has been improving as the amount of effluent data increases and the jurisdictions begin to employ automated data-handling systems.

For the year 1976, Task Group III and PLUARG performed separate efforts to improve the Board estimates, by further consideration of available data. Rast and Gregor [10] compared these efforts along with those of the Corps of Engineers for direct phosphorus discharges. Direct discharges are limited to those which discharge directly to the lake or to a tributary point downstream of the monitoring station established by the Surveillance Subcommittee. As can be seen in Table I, the estimates are quite close except for the Lake Erie value of Task Group III which is, in turn, based on the Corps study. The data set used by the Corps is different in two major ways from that used by the Board [11]. First, the Corps performed effluent

Table I. Point Source Direct 1976 Phosphorus Loads (kg/day)

	PLUARG	TASK GROUP III	WATER QUALITY BOARD
Lake Superior	463	495	441
Lake Michigan	2936	2978	2937
Lake Huron	422	482	422
Lake Erie	16,255	19,732[a]	16,455
Lake Ontario	5827	6090	5833
Totals	25,903	29,777	26,088

[a]Corps of Engineers estimate, includes plants below 1 mgd and more 1975 data.

analyses at a number of municipal plants with flow less than 1 mgd of small industrial plants and was able to estimate their contribution. Second, the Corps obtained a significantly higher estimate for Detroit using surveillance data from the end of 1975.

The Corps direct point source load value for 1976 for all plants is some 20% higher than that obtained by the Board for the same year for plants over 1 mgd. The Corps reports [11] values for 1977 which are also higher than the Board estimate. The Corps 1977 estimate for total municipal point source discharge to the Lake Erie Basin is 20,279 kg/day or about 10% more than the Board's value for plants over 1 mgd.

While the loading estimates assembled by the Board are short of being perfect, they do provide a useful estimate of progress in controlling phosphorus, particularly for plants over 1 mgd to which the 1 mg/l Great Lakes Agreement target applies. Table II, derived from the 1977 Board Report, is a summary of all municipal phosphorus loads to the Great Lakes Basin above 1 mgd. As may be noted, average municipal phosphorus effluent concentration has dropped to 2.2 mg/l with plants in the Canadian Lake Erie and Ontario Basins and the U. S. Huron and Michigan Basins doing particularly well. Total loads dropped between 1975 and 1977 with the greatest reduction in the Canadian portion of the Lake Ontario Basin and the U. S. portion of the Lake Erie Basin. The reported total load for 1976 is abnormally low because of some reporting deficiencies, primarily underestimates of loading from New York State. Concentrations are considerably higher than the 1 mg/l target except for the Canadian portion of Lake Erie but represent considerable progress from earlier years.

Improvements in concentration may be traced somewhat further back by reference to the 1972 and 1974 reports of the Board. As indicated in Table III, average phosphorus effluent concentration has dropped from 6.2 mg/l in

Table II. Municipal Phosphorus Loads in the Great Lakes Basin[a]

Lake Basin	Population Served	Loadings (kg/day)			1977 Average Concentration (mg/l)
		1975	1976	1977	
Superior					
United States	259,100	447	511	315	3.3
Canada	121,900	155	178	279	4.5
Huron					
United States	732,000	427	326	477	1.4
Canada	446,000	470	479	489	2.1
Michigan					
United States	3,901,000	5768	6548	4703	1.4
Erie					
United States	6,400,000	21,180	17,880	17,827	2.6
Canada	300,000	600	690	686	1.02
Ontario					
United States	1,700,000	5000	4210	6149	3.0
Canada	3,800,000	6780	3620	3130	1.3
Total	17,660,000	40,827	34,442	34,055	2.2

[a]Actual phosphorus loadings for 1975, 1976, and 1977 were measured at all sewage treatment plants over 3800 m^3/day (1 mgd).

Table III. Average Municipal Phosphorus Effluent Concentration (mg/l)

Basin	1972 Direct Discharge Only	1974 Direct & Indirect	1977 Direct & Indirect
Lake Superior	11.2	4.7	3.8
Lake Huron	4.3	2.0	1.7
Lake Michigan	—	2.8	1.4
Lake Erie	5.6	3.2	2.5
Lake Ontario	7.3	2.2	2.1
Aggregate	6.2	2.9	2.2

1972 to 2.9 mg/l in 1974 to 2.2 ppm in 1977. The 1972 value is suspect and
it exceeds a concentration of 4.5 ppm obtained by D. J. Appleby for a group
of plants in Ontario for 1972 [12]. Appleby's value for 1974 is 2.2 and for
1976 is 1.5. Thus the trend is similar even if the values are lower than Table
III. Detailed data for 1978 were not available for inclusion in this chapter, but
early 1978 results indicate load reductions at Detroit, further reductions at
Canadian plants, and other results which suggest the overall effluent concentration may drop below 2 mg/l.

Another indication of trends can be found by tabulating direct loads for
all the Water Quality Board reports between 1972 and 1977 as in Table IV.

Table IV. Reported Direct Point Source Phosphorus Loads (kg/day), by Year

Lake Basin	1972	1973	1974	1975	1976	1977
Superior	1252	1400	566[a]	438	441	470
Huron	412	493	386	683	424	939
Michigan	–	1253[b]	3103	3090	2936	1946
Erie	24,671	29,519	19,461	18,355	16,455	15,978
Ontario	18,240	6830[c]	5682[c]	9424	6034[c]	7519
Total	–	39,495	29,198	31,990	26,290	26,852

[a]Duluth no longer included.
[b]Sparse data.
[c]Missing New York data.

Early data are sparse and there is some difficulty with changes in monitoring
stations affecting the definition of "direct"discharger; nevertheless, a relatively impressive load reduction is seen to occur. A slightly better impression of
the 1972-1977 period may be obtained by separating out the industrial from
municipal estimates, as in Table V. The reported industrial loads are quite
irregular but also small when compared to municipal loads, except in Lakes
Superior and Huron. Again, these values omit municipalities under 1 mgd.

The identity of the municipal dischargers of most concern can be seen in
the ranking of phosphorus loads taken from Appendix C of the 1977 Board
Report [6] as shown in Table VI. Detroit appears at the top with the cities
of Buffalo, Toledo, Syracuse, Niagara Falls and Cleveland below it. Programs
are in place in all of these dischargers with loading reduction expected for
many of them in 1978. Construction completion is expected for the Buffalo
and Syracuse, New York facilities this year. Correction of the problems at
Niagara Falls should occur this year also. Construction at Cleveland Southerly
will not be complete until 1982, but repairs to the parts of the plant that
were damaged by explosion in 1977 have been made. Detroit is under a court

Table V. Reported Direct Point Source Phosphorus Loads (kg/day), by Type of Source

Lake Basin	Industrial						Municipal					
	1972	1973	1974	1975	1976	1977	1972	1973	1974	1975	1976	1977
Superior		354	255	267	280	295	1252	1046	311[a]	171	161	175
Huron	72	47	0	353	86	496	340	446	386	330	338	443
Michigan		35	122	168	88	138		1218[b]	2981	2922	2848	1808
Erie	1149	771	346	185	753	370	23,522	28,748	19,115	18,170	15,702	15,608
Ontario	818	598	324	511	220	415	17,422	6232[c]	5358[c]	8913	5814[c]	7104
Total	INDUSTRIAL REPORTING IRREGULAR							37,690	28,151	30,506	24,863	25,138

[a] Duluth no longer included.
[b] Sparse data.
[c] New York data incomplete.

Table VI. Ranked Municipal Phosphorus Dischargers to the Great Lakes

Rank	Name of Plant and Location[a]	Lake Basin	Flow (10^3 m^3/day)	Total Phosphorus Loading (kg/day)	Average Annual P Concentration	Ranked Loading[b]
1	Detroit, MI	Erie	3040	10,336	3.4	7296
2	Buffalo, NY	Ontario	655	1796	2.7	1114
3	Toledo, OH	Erie	377	955	2.5	578
4	Syracuse Metro, NY	Ontario	239	813	3.4	574
5	Niagara Falls, NY	Ontario	247	815	3.3	568
6	Cleveland Southerly, OH	Erie	360	898	2.5	538
7	Grand Rapids, MI	Michigan	169	689	4.1	520
8	Akron, OH	Erie	298	712	2.4	414
9	Kalamazoo, MI	Michigan	127	420	3.3	293
10	Milwaukee South, WI	Michigan	249	534	2.1	285
11	Cleveland Westerly, OH	Erie	123	383	3.1	260
12	Tonawanda, NY (S.D. #2)	Ontario	57	309	5.4	252
13	Lorain, OH	Erie	53	271	5.1	218
14	Erie, PA	Erie	243	438	1.8	194
15	Cornwall, Ontario	Ontario (St. Lawrence River)	57	247	4.4	192
16	Wayne Co., Wyandotte, MI	Erie	272	435	1.6	163
17	Duluth Main, MN	Superior	66	224	3.4	159
18	Mississauga Lakeview, Ontario	Ontario	173	327	1.9	154
19	Sault Ste. Marie, Ontario	Huron	42	186	4.4	144
20	Chili, NY	Ontario	45	159	3.5	114
21	Bay City, MI	Huron	67	158	2.3	91
22	Appleton, WI	Michigan	46	126	2.7	80

Table VI, continued

Rank	Name of Plant and Location[a]	Lake Basin	Flow (10^3 m^3/day)	Total Phosphorus Loading (kg/day)	Average Annual P Concentration	Ranked Loading[b]
23	Green Bay, WI	Michigan	111	188	1.7	77
24	Hamilton, Ontario	Ontario	254	329	1.3	75
25	Wyoming, MI	Michigan	46	119	2.6	73
26	Rochester, NY (FVL)	Ontario	244	317	1.3	73
27	Toronto Highland, Ontario	Ontario	133	200	1.5	67
28	Racine, WI	Michigan	80	143	1.8	63
29	Lakewood, OH	Erie	43	100	2.3	57
30	Sudbury, Ontario	Huron	50	100	2.0	50
31	Mississauga Clarkson, Ontario	Ontario	45	77	1.7	32
32	Brantford, Ontario	Erie	44	75	1.7	31
33	Battle Creek, MI	Michigan	48	78	1.6	30
34	Oshawa, Ontario	Ontario	49	73	1.5	24
35	Kitchener, Ontario	Erie	65	84	1.3	19
36	Niagara Falls, Ontario	Ontario	46	60	1.3	14

[a]Municipality flows are equal to or greater than 3.8 x 10^4 m^3/day (10 mgd).

[b]Ranking equation = actual loading (kg/day) – loading at 1 mg/l.

order which mandates full phosphorus removal at the end of 1981. Phosphorus loads have dropped from 1977 levels because of the implementation of a phosphorus detergent ban in Michigan.

SUMMARY AND RECOMMENDATIONS

In reviewing the Water Quality Board reports between 1972 and 1977, it was clear that the methods of calculating point source phosphorus loads had changed somewhat from year to year. It was also clear that estimates of earlier years might be improved with the advantage of hindsight. Because the Windsor Regional Office has now automated the load calculation process, it might be useful to insert earlier data in that system and to try to improve the load estimates that have been discussed. Since the jurisdictions now also have better information systems to handle this data, and the Great Lakes community continues to be interested in phosphorus load estimates, it seems appropriate for greater effort to be expended in assessing quality and statistical accuracy of the data. Efforts in this area have been increased.

Despite some of the difficulties in the data which have been described, these load estimates provide a good instrument for tracking the progress in reducing point source phosphorus loads to the Great Lakes Basin in accordance with the Great Lakes Agreement and Jurisdictional requirements. Substantial reductions have occurred, although not as yet to the 1 mg/l concentration target. Progress is well underway to continue to bring effluent concentration down to the target levels.

REFERENCES

1. "Great Lakes Water Quality 1972," 1st Annual Report of the Great Lakes Water Quality Board to the International Joint Commission, Windsor, Ontario, (1973) 315 pp.
2. "Great Lakes Water Quality 1973," 2nd Annual Report of the Great Lakes Water Quality Board to the International Joint Commission, Windsor, Ontario, (1974) 113 pp.
3. "Great Lakes Water Quality 1974," 3rd Annual Report of the Great Lakes Water Quality Board to the International Joint Commission, Windsor, Ontario, (1975) 170 pp.
4. "Great Lakes Water Quality 1975," 4th Annual Report of the Great Lakes Water Quality Board to the International Joint Commission, Windsor, Ontario (1976) 72 pp.
5. "Great Lakes Water Quality 1976," 5th Annual Report of the Great Lakes Water Quality Board to the International Joint Commission, Windsor, Ontario, (1977) 72 pp.
6. "Great Lakes Water Quality 1977," 6th Annual Report of the Great Lakes Water Quality Board to the International Joint Commission, Windsor, Ontario, (1978) 89 pp.

7. "Environmental Management Strategy for the Great Lakes System," Final Report of Task Group III to the Governments of the United States and Canada, Windsor, Ontario, (1978) 121 pp.

8. "Fifth Year Review of Canada-United States Water Quality Agreement," Report of Task Group III to the Government of the United States and Canada, Windsor, Ontario, (1978) 86 pp.

9. "Lake Erie Wastewater Management Study," U. S. Army Corps of Engineers (as reported in references 7 and 8).

10. Rast, W. and D. Gregor. "Report on Differences in Great Lakes Phosphorus Load Estimates," International Joint Commission Memorandum, (January 1979).

11. Yaksich, S., U.S. Army Corpos of Engineers, Buffalo. Personal communication.

12. Appleby, D. J. "The Impact of Phosphorus Control Activities: The Experience in Ontario," ERCO Industries Ltd., Islington, Ontario, (1977) 37 pp.

CHAPTER 3

NONPOINT SOURCE PHOSPHORUS INPUTS TO THE GREAT LAKES

D. J. Gregor and M. G. Johnson

Canada Centre for Inland Waters
Burlington, Ontario

INTRODUCTION

Studies requested by the International Joint Commission (IJC) concerning water quality in Lakes Erie and Ontario, submitted to the IJC in 1969 by the International Lake Erie, Lake Ontario-St. Lawrence River Water Pollution Board [1], indicated that land drainage sources of pollutants were significant. Specifically, the total tributary loads of phosphorus to Lakes Erie and Ontario were estimated to be 30 and 21%, respectively, of the total lake load. Subsequent reduction in municipal sewage treatment plant or direct point source loads to these lakes would increase the relative importance of land drainage and other nonpoint sources of phosphorus loads. In addition, land drainage sources of phosphorus, as well as other major nonpoint sources, are highly variable and were poorly quantified in the late 1960s. Thus, if a rational approach to phosphorus management was to be attained, it was necessary to better understand and further quantify the nonpoint sources of phosphorus to the Great Lakes. This would require identification of the factors which determine nonpoint land drainage loads and identification and assessment of the relative importance of generalized source areas. Direct to the lake atmospheric and nonpoint urban drainage, as well as loads derived from shorebluff erosion and other sources, also required investigation.

In November 1972, the IJC appointed an International Reference Group on Great Lakes Pollution from Land Use Activities (PLUARG) to investigate nonpoint source pollutants. PLUARG submitted its final report to the IJC in July 1978. This report and its supportive documents represent essentially the "state-of-the-art" for nonpoint or diffuse sources of pollutants to the Great Lakes. Consequently, the majority of material herein is based upon results of the PLUARG study [2]. Two other major studies in the Great Lakes Basin, concurrent with the PLUARG study, provided useful information on more specific topics. The first, the Lake Erie Waste Water Management Study conducted by the United States Army Corps of Engineers, was concerned only with Lake Erie. The second, a report prepared by Task Group III [3] as part of the fifth year review of the Canada-U. S. Great Lakes Water Quality Agreement, provided a basis for and specified target phosphorus loads to each lake.

The purpose of the PLUARG study was to:

1. determine and evaluate the causes, extent and locality of pollution from land use activities;
2. gain an understanding of the relative importance of various land uses in terms of their diffuse pollutant loads to the Great Lakes;
3. examine the effects of the diffuse pollutant loads on Great Lakes water quality; and
4. determine the most practicable remedial measures for decreasing the diffuse pollutant loads to an acceptable level and the estimated costs of these measures.

Based upon the results of the PLUARG study, the magnitudes of phosphorus inputs to the Great Lakes from agricultural, forestry, urban, atmospheric and other nonpoint sources will be summarized. The factors controlling the relative importance of rural and urban nonpoint sources and some conditions affecting quantification of phosphorus loads from these sources will be considered. The major nonpoint source areas within the Great Lakes basin will be identified. Finally, rural and urban nonpoint loadings will be placed in perspective with point source phosphorus loadings (Chapter 2, this volume) for the purpose of illustrating remedial management alternatives outlined by PLUARG.

MAGNITUDES OF NONPOINT PHOSPHORUS LOADS

The distribution of the major land uses within each of the lake basins is illustrated in Figure 1. The preponderance of forested lands in the basins of Lakes Superior (95%), Michigan (50%), Huron (66%) and Ontario (56%) and the large proportion of agricultural land in the Lake Erie basin (59%) provides a necessary perspective for this discussion of nonpoint source loads.

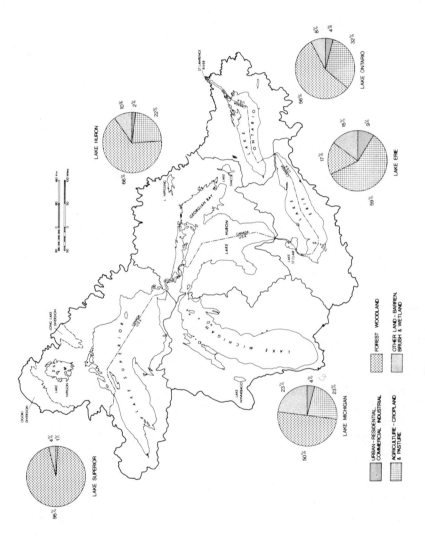

Figure 1. Major land uses for each lake basin.

The magnitudes of the total phosphorus (TP) loads for 1976 from the major sources (including direct and indirect point sources) to the Great Lakes are summarized in Figure 2.*

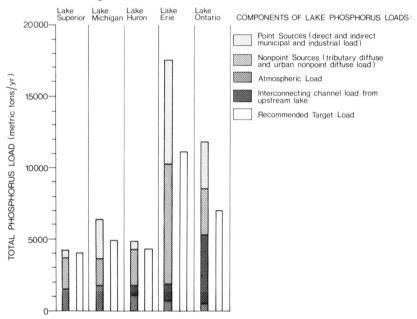

Figure 2. 1976 phosphorus loads and recommended target loads for the Great Lakes.

The relative importance of TP loads from atmospheric and tributary sources is indicated in Table I. As a percentage of the total estimated 1976 load, the atmospheric input accounts for about 4% in Lakes Erie and Ontario, and 23, 26 and 37% in Lakes Huron, Michigan and Superior, respectively. The total estimated tributary TP load ranges from a low of 34% of the total Lake Ontario load to 55-60% of the total load of the other four lakes. Assuming that 100% of the municipal and industrial plant source TP loads discharging to these tributaries reaches the lakes, the total diffuse tributary load can be approximated by difference. As Indicated in Table I, the total diffuse tributary load accounts for only 28 and 30% of the total load to Lakes Ontario and Michigan, respectively, but 48, 50 and 53% for Lakes Erie, Huron and Superior, respectively. A generalized model was used by PLUARG [2] and Drynan and Davis [4] to estimate the contributions of

---

*Further discussion of the relative importance of the major sources to achieving the target loads, also indicated in Figure 2, will be undertaken in the final section of the chapter.

Table I. Relative Importance of Nonpoint Source Total Phosphorus Loads [2]

Lake	Total Load (metric tons/yr)	Atmospheric Load (% of total load)	Estimated Nonpoint Proportion of Total Tributary Load (% of total load)	Estimated Contributions of Major Land Uses to Estimated Nonpoint Tributary Loads (% of total load)		
				Agriculture	Urban	Forest & Other
Superior	4,200	37	53	4	4	45
Michigan	6,350	57	30	21	4	5
Huron	4,850	60	50	34	6	10
Erie	17,450	4	55	32	10	6
Ontario	11,750	34	28	19	5	4

major land uses to the total diffuse tributary load. Evidently forests contribute the greatest proportion of the diffuse load to Lake Superior (about 85%), with agriculture a major source (approximately 65-70%) in the other four lakes (Table I). Diffuse urban loads contribute about 20% of the total diffuse tributary load to Lakes Ontario and Erie.

Two other nonpoint sources of TP to the lakes have not been considered explicitly in this discussion. The first, TP loadings due to shore bluff erosion, will be excluded from the discussion entirely, because although the magnitude of this TP load is potentially high, the biological availability of phosphorus derived from this source is small [2,5].

The second, most pertinent to Lake Erie, is the recycling of phosphorus within the lake, including the regeneration of phosphorus from lake bottom sediments. This source is not well quantified at the present time and will be considered only briefly below.

DETERMINANTS AND SOURCES OF NONPOINT PHOSPHORUS LOADS

Atmospheric Sources

The total atmospheric phosphorus loads, as discussed above, contribute a relatively large proportion of the lake load of each of the upper Great Lakes. This is in part due to the larger size of Lakes Superior, Michigan and Huron, as indicated by the absolute loadings, and to the much larger total lake loads of Lakes Ontario and Erie (Figure 2). However, phosphorus loading rates (load per unit area of lake surface) to the lakes generally decrease from south to north [6]. The estimated atmospheric phosphorus loads to Lakes Superior, Michigan, Huron, Erie and Ontario are 1566, 1682, 1129, 774 and 438 metric tons/yr, respectively [2].

More than 75% of the total atmospheric phosphorus load to Lake Superior is estimated to be derived from U. S. sources, especially Chicago [7]. As would be expected, Lake Michigan loadings are dominated by the large industrialized metropolises to the west and south. An estimated 70% of the total Lake Michigan atmospheric phosphorus load is derived from Chicago, Indianapolis, Cincinnati, St. Louis, Milwaukee, and Green Bay as well as generally from the states of Illinois and Iowa [6]. American sources contribute the greatest proportion (more than 80%) of the total atmospheric phosphorus loads to Lake Huron [7].

American sources contribute 90% of the total atmospheric loading to Lake Erie and 80% to Lake Ontario with Cleveland the source for the largest proportion of the Lake Erie load. In comparison, Lake Ontario loadings result, to a much larger extent, from the general industrial activity within the general area of the Great Lakes, and consequently, a dominant single source is not identified. Major Canadian sources are Toronto (which includes industrialized centres such as Hamilton and Oshawa), and Montreal [6].

Urban and Rural Nonpoint Sources

Although nonpoint TP loads derived from land runoff separate logically into urban and rural land uses, land use is not the only factor influencing land derived nonpoint loadings. Other factors, the most important of which, in the rural context, are physiography, soils and other natural characteristics of the land, land use intensity and materials usage (e.g., amount of fertilizer applied), must also be considered. In urban areas, the type and degree of industrialization proportion of impervious surfaces (paved or otherwise) and sewer systems are dominant. Meteorological and climatological conditions cause annual and seasonal variations, respectively, in land-derived phosphorus loads. Individual land management practices are also important but are the subject of another chapter in this proceedings, and therefore will not be addressed here.

Probable rural unit area TP loads for combinations of land use, land characteristics, and intensity of use can be generalized from site-specific studies. PLUARG developed and used this information in an "overflow model", designed to categorize contributing area locations and provide the means to determine cost-effective management strategies and associated reductions in nonpoint source loads [2,8].

The TP unit area loads for urban areas are summarized in Table II. The

Table II. Generalized Urban Total Phosphorus Unit Area Loads (kg/ha/yr) [8]

Land Use/Characteristic		Total P Unit Area Load (kg/ha/yr)
Areas of Combined Sewer Systems	High Industry	11
	Medium Industry	10
	Low Industry	9
Areas of Separated Sewer Systems	High Industry	3
	Medium Industry	2.5
	Low Industry	1.25
Unsewered Areas		1.25
Towns (sewer system not differentiated) 1000-10,000 population		2.5
Developing Urban		25

controlling factors are the presence of combined sewer systems and to a lesser extent, the degree of industrialization. The highest unit area load from any land use, whether rural or urban, is 25 kg/ha/yr from urbanizing areas. However, the urbanizing area within the basin at any time is relatively small and each development remains in this aggravated state for only a short time.

In addition, the biological availability of the phosphorus derived from this source, although highly variable among development sites, is likely relatively low because of the natural characteristics of the subsoil which has been exposed (i.e., analogous to the biological availability of the natural material comprising the shore bluffs [5].

Predicted TP unit area loads for nonurban or rural portions of the Canadian drainage basin are summarized in Table III. The land use cate-

Table III. Generalized Nonurban Total Phosphorus Unit Area Loads (kg/ha/yr) [8]

Land Form Land Use/Intensity	Fine Textured		Medium Textured		Coarse Textured		Misc. Types
	Level	Sloping	Level	Sloping	Level	Sloping	
>50% Rowcrop Low Animal Density	1.5		0.63		0.23		0.87 (sand on clay)
>50% Rowcrop Med. Animal Density					0.27		
25-50% Rowcrop Med. Animal Density	0.58	0.69	0.25	0.36	0.15	0.19	
25-50% Rowcrop High Animal Density	0.64	0.79	0.30	0.44	0.20	0.27	0.70 (sand on clay)
<25% Rowcrop Med. Animal Density	0.35	0.42	0.10	0.15	0.10	0.12	0.45 (sand on clay)
<25% Rowcrop High Animal Density	0.40	0.51	0.14	0.20	0.14	0.16	
>60% Forest	0.20	0.20	0.10	0.10	0.10	0.10	0.09 (shield)

gories are not identical between the U. S. and Canada, and thus slightly different unit area loads must be used [8] but the results are comparable. As indicated in Table III, unit area loads tend to increase as the proportion of rowcrops, animal density and slope increase and the soil texture becomes finer. Forested areas are characterized by relatively low unit area loads. Consequently, although the magnitude of forest-derived loads to some lakes is very large, especially Lake Superior, where a large proportion of the basin is forested, the load from this source can be considered essentially background (except where locally aggravated for short periods of time by harvesting operations) [9].

Identification of relatively homogeneous watershed subbasins on the basis of land use, use intensity and land characteristics, as well as further characterization of these subbasins according to a number of variables (e.g., appropriate unit area loads, population, growth rate, population density, sewerage system, point source loadings and in-stream transmission factors), allows the "overview model" to estimate tributary mouth loads. The model schematic is illustrated in Figure 3. Tributary mouth TP loads predicted by the

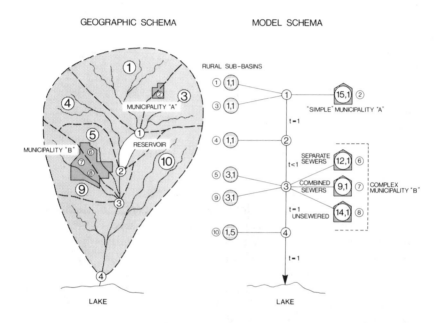

Figure 3. Schematic representation of watershed for overview modeling.

model compare favorably with measured loads for Canada and U. S. as indicated in Table IV. The data base used in the model load predictions included site-specific studies conducted by PLUARG, water quality data for upstream subwatersheds collected by provincial and state agencies, as well as other investigations, and consequently was essentially independent of the data used to calculate the tributary mouth loads. In this way the "overview model" delineated the most important nonpoint TP source area within the Great Lakes Basin.

The relative importance of various rural source areas for southern Lake Huron and Lakes Erie and Ontario is illustrated in Figures 4, 5 and 6, respectively. Major rural nonpoint source areas in southern Lake Huron include the highly agriculturalized, essentially fine-textured areas south of Saginaw Bay in the U. S. and along the southeastern shore of Lake Huron

Table IV. Summary of Monitored and Predicted Total Phosphorus Loads
for Selected Lakes (metric tons)

	1975 Model Prediction	1975 Monitored Load
Lake Erie		
Canada	1,378	1,234
U.S.	16,434	16,128
Total	17,812	17,362
Lake Ontario		
Canada	718	722
U.S.	1,875	2,057
Total	2,593	2,779
Lake Huron		
Canada	618	511
U.S.	2,204	1,764
Total	2,822	2,275

in Canada. In the Lake Erie basin, major source areas include much of the Maumee River basin, discharging into the western basin, and essentially all of the U. S. drainage basin of the central and eastern basins of Lake Erie. In Ontario, the land between Lake St. Clair and Lake Erie typified by fine-grained soils and widespaced rowcrops, stands out as a relatively compact but important rural nonpoint source area to Lake Erie. In the Lake Ontario basin (Figure 6), the major rural nonpoint source areas occur within Ontario at the western end of the lake. In comparison to Lake Erie, however, the unit area loads in the Lake Ontario basin are generally low.

NECESSARY CONSIDERATIONS IN THE MANAGEMENT OF NONPOINT SOURCES

The six years since the 1972 Water Quality Agreement have evidenced a major advance in nonpoint source loadings information. Partially as a result of the work of PLUARG, as well as other stimuli, a proliferation of research and planning for the purpose of identifying sources, loads and management practices relative to phosphorus and other pollutants derived from other than

---

*Expected load reductions and estimated costs of various remedial options are also part of the "overview model", allowing determination of cost-effective TP management strategies within the basin. This portion of the model is discussed briefly in another chapter in this volume and is detailed in the PLUARG final report [2] and the supporting PLUARG Technical Report [8].

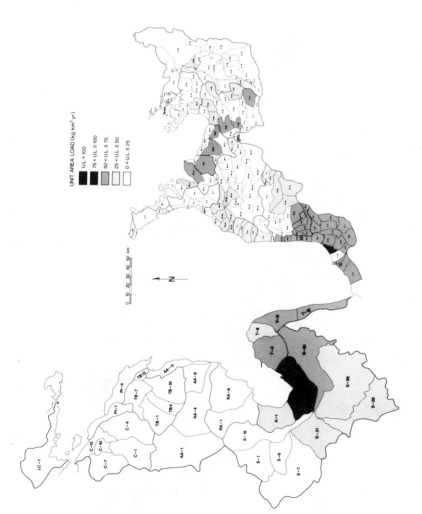

Figure 4. Total phosphorus delivered to lake—Lake Huron basin (kg/km² /yr x 0.01 = kg/ha/yr).

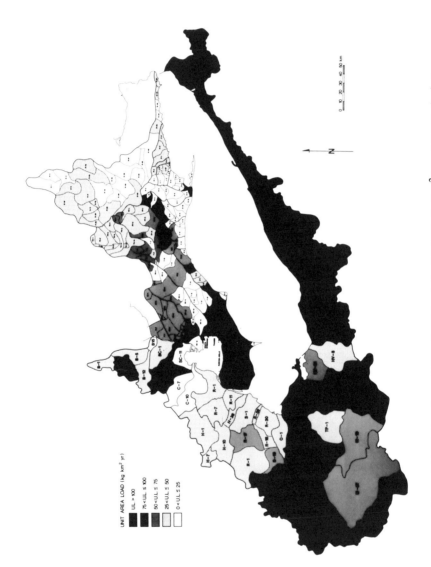

Figure 5. Total phosphorus delivered to lake–Lake Erie basin (kg/km^2/yr x 0.01 = kg/ha/yr).

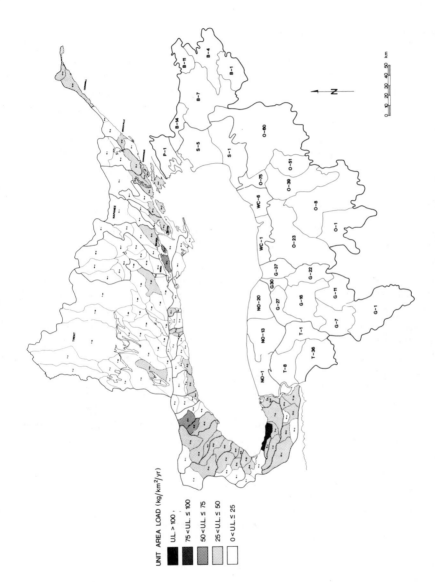

Figure 6. Total phosphorus delivered to lake—Lake Ontario basin (kg/km² /yr x 0.01 = kg/ha/yr).

point sources is being witnessed within the basin. There are, however, a number of factors which have a bearing on nonpoint source management strategies. The most important of these are discussed below.

Biological Availability of Phosphorus

The biological availability of phosphorus varies among sources. As mentioned above, the available fraction of the TP load from shoreline erosion is so small in comparison to other sources that it can generally be ignored. The availability of tributary phosphorus loads, direct point source loads and atmospheric loads needs to be considered to provide cost-effective management strategies. The importance of biological availability to phosphorus management of the Great Lakes is such that this topic is considered elsewhere in this book.

Accurate Load Estimates

Although a great deal of progress has been made in the monitoring of tributary water quality and the estimation of pollutant loadings, there are still a number of problems which are encountered. Probably one of the greatest concerns with these data is the possibility of not sampling major storm or spring melt events which can contribute a very large proportion of the total tributary load. This problem can only be minimized through improved sampling following research into the "event response" nature of each tributary.

Other considerations related to nonpoint source loads include:

1. The use of a variety of tributary and calculation methodologies and the addition of loading results from these methodologies to provide a total load with no appreciation of the comparability of the loads.

2. Although there is a tendency toward standardization of loading methodologies there has been little attempt to assess the accuracy of any method in various types of streams under various sampling strategies.

3. Recent unpublished studies suggest that sampling frequency is extremely critical to the accuracy of any loading calculations and the inaccuracies resulting from a limited data base (i.e., a sampling frequency of about 30 days) likely exceed inaccuracies resulting from any loading calculation methodology.

Sediment-associated loads (important for assessing the biologically available portions of total phosphorus) have generally been ignored in the tributary surveillance program. The magnitude of sediment-associated tributary loads to the Great Lakes and partitioning of sediment phosphorus among particle size fractions warrants greater emphasis to provide useful information in the development of management strategies. Although the atmospheric loads reported here are considered the best available at this time, there is continued extensive effort to better quantify loadings from this source. As the results of this research become available, a better understanding of the effects of atmospheric source phosphorus and management alternatives should be possible.

In-lake resuspension and diffusion/convection of soluble phosphorus from sediments (i.e., internal loading) needs to be better understood, especially with respect to the impact on shallow lakes wherein nutrient-rich sediment may greatly delay water quality improvements expected from reduced external loadings.

In-Stream Transmission

The in-stream transmission factor or stream delivery ratio remains an area of limited understanding within the Great Lakes Basin, but is probably critical to the development of management strategies. If the transmission factor for a certain point on a stream is less than one (i.e., not everything entering the stream reaches the lake), it may be unjustifiable, in the Great Lakes context, to manage the upper portion of some basins as stringently as the lower portion. PLUARG generally assumed a transmission factor of one for the long term, except where large lakes or impoundments in the watershed function as permanent nutrient sinks for a portion of the tributary load (e.g., Finger Lakes in New York, Kawartha Lakes in Ontario, and some of the larger reservoirs).

Meteorological and Climatological Effects

Natural meteorological fluctuations can directly affect overland delivery to a stream, transmission within the stream and volume of the stream, and consequently are responsible for large temporal variations in tributary loads to the lakes. Specifically, wet, dry and normal years, and individual but very severe storms can affect nonpoint source loads. For example, maximum tributary flow variability between 1962 and 1976 for Canadian and U. S. tributaries, respectively, were ± 39 and ± 45% relative to the mean flow to Lake Erie (Table V). Assuming a direct relationship between flow and tributary TP load yields an estimated annual mean and maximum expected range for Canadian and U. S. tributary TP loads of 2400 ± 1000 metric ton/yr and 8600 ± 4100 metric ton/yr for Lakes Ontario and Erie, respectively (Table V). In comparison the 1976 tributary loads to Lakes Ontario and Erie were about 4100 and 9600 metric ton/yr, respectively. Obviously, effective mananagement designs and assessment of the effects of

Table V. Flow and Estimated Load Variability Due to Climatic and Meteorologic Fluctuations

	Lake Ontario			Lake Erie		
	Canadian Tributaries[a]	U.S. Tributaries	Niagara River	Canadian Tributaries[a]	U.S. Tributaries	Detroit River
15 Year Annual Mean Flow (m³/sec)	223	396	5862	83	304	5494
Minimum Mean Annual Flow (m³/sec) [year]	129 [1964]	263 [1965]	4588 [1964]	50 [1963]	160 [1963]	4390 [1964]
Maximum Mean Annual Flow (m³/sec) [year]	301 [1967]	619 [1972]	7023 [1973]	113 [1976]	457 [1972]	6315 [1973]
Flow Variability Relative to Annual Mean Flow C–B/2A x 100 (%)	± 39	± 45	± 21	± 38	± 49	± 18
1976 Mean Annual Flow (m³/sec)	259	597	6712	113	316	5947
Relationship of 1976 Mean Annual Flow to Annual Mean Flow E–A/A x 100 (%)	± 16	± 51	± 15	± 36	± 4	± 8
1976 Load as Reported by PLUARG [metric tons/ (metric tons/yr)]	1247	2800	4769[b] (5562)[c]	1911	7732	1080[b]
Estimated Annual Mean Load (metric tons/yr)[d]	1000	1400	4100[b] (4700)[c]	1200	7400	1000[b]
Estimated Annual Load Variability due to Flow [D x H] (metric tons/yr)[d]	±400	±600	±900[b]	±500	±3600	±200[b]
Estimated Total 1976 Lake Load Using Estimated Annual Mean Load and Variability (metric tons/yr)[e]	9400 ±1900			16,400 ±4300		

[a]Canadian flow data are back calculated because flow data were not always available for all rivers for all years.
[b]Does not include sewage treatment plants discharging directly to the interconnecting channel as these are considered to be direct dischargers to the downstream lake.
[c]Represents the load measured at the mouth of the Niagara River and therefore includes the point source excluded in (b) above.
[d]Loads are rounded to the nearest 100 metric tons.
[e]Loads from other sources are identical to those reported by PLUARG [2].

management strategies (i.e., whether or not the recommended target loads are being achieved) requires consideration of natural tributary load variability. Natural variability in lake levels and subsequently flow and thus load of interconnecting channels is also important.

The presence of the spring melt is a major climatic control of tributary loads. The importance of spring melt-related loads to total lake load is such that greater monitoring of this phenomenon is warranted.

Atmospheric phosphorus loads also vary considerably from year to year and season to season. Loading rates to the Lower Great Lakes for the years 1973 and 1974 were maximum during the summer, generally June through September, and minimum during the winter season, essentially November through March [7].

Continued research should provide more information regarding the magnitude of seasonal and annual nonpoint source TP load variations and their controlling factors which will be necessary for the development of rational nonpoint source phosphorus management strategies. In addition, the number of years (out of 10 or 100, for example) during which the load objective will not be met due to natural load variability must be considered. The consequences which will result that year and in following years from not meeting the objective will also require consideration in the development of phosphorus management strategies.

Potential Contributing Areas and Hydrologically Active Areas

Major source areas of TP, or potential contributing areas (PCA), have been identified (Figure 4) and discussed briefly. As indicated above, soil type, morphology, land use intensity and materials usage are important factors in determining nonpoint source loads. The most critical problem areas are rowcrops on fine textured soils, some concentrated livestock operations, developing urban areas, and highly impervious portions of major urban centres. It should be noted that identification of PCA does not necessarily reflect the presence of water quality problems, since the way these areas are managed is very important to phosphorus loadings. Furthermore, not all the land within a PCA contributes to the load of that area. Additional site specific investigations are required to prioritize portions of the PCA with respect to their total load contributions. A first step in setting priorities on land areas for management purposes is the identification of areas contributing large proportions of pollutants directly to surface waters. These areas are normally located close to rivers, streams, lakes and impounds and have been termed hydrologically active areas (HAA) [2].

Setting priorities in this manner identifies major nonpoint source areas requiring some form of treatment or management to reduce nonpoint source loads and thus contributes to basinwide phosphorus management strategies.

Other Nonpoint Sources

A variety of specialized land uses including the nonsewered waste disposal of septic systems, sanitary landfills, streambank erosion, groundwater inputs, land disposal of mine tailings, sludge disposal on land and recreational activities have also been considered. These specialized land uses may affect local water quality conditions but apparently do not generally contribute to Great Lakes problems due to nonpoint source loadings of phosphorus [2].

Resuspension and in-lake recycling of phosphorus is essentially a nonpoint source. The processes of recycling and resuspension in Lake Erie have been discussed in Project Hypo [10] and by Lam and Jaquet [11]. However, the magnitude of these sources relative to total loadings has not generally been quantified.

NONPOINT AND POINT SOURCE PHOSPHORUS
LOADINGS IN PERSPECTIVE

It is not practical to consider phosphorus management strategies for either point or nonpoint sources independently of each other. Although the emphasis in this chapter has been on nonpoint sources, it is necessary, for perspective, to return to the total TP loadings situation (i.e., combining the nonpoint source TP loadings information with the point source loadings information illustrated in Figure 1). On the basis of these 1976 total TP loadings data, the effects of various phosphorus management strategies can be investigated for each of the lakes relative to target load recommendations.

The result of several scenarios involving various sewage treatment plant phosphorus effluent standards (i.e., 1.0 mg P/l, 0.5 mg P/l and 0.3 mg P/l) upon total lake loadings, relative to the 1976 loadings, is illustrated in Table VI and Figure 7. Relative to the 1976 phosphorus loads, point source controls to 1 mg P/l can evidently attain the recommended target loads for Lakes Superior, Michigan and Huron.*

Additional point source loading reductions as well as nonpoint source controls would evidently be necessary (Table VI) to achieve the Lakes Erie and Ontario targets of 11,000 and 7000 metric ton/yr, respectively, based on 1976 data [2]. Even the strictest effluent control considered in these scenarios (0.3 mg P/l) was unable to achieve the target in 1976 (although other factors, such as natural variability of tributary and atmospheric phosphorus loads, ultimately need to be considered).

A simple population-dependent point-source projection has been used to predict the TP load to each lake in the year 2020 (Table VI), assuming that

---

*Although the Lake Huron target load is essentially attained, additional phosphorus load reductions are necessary within Saginaw Bay to meet the objective established for this embayment [2].

Table VI. Present and Future Phosphorus Loads to Lakes Erie and Ontario Under Several Phosphorus Reduction Scenarios (metric tons/yr)

	Lake Erie		Lake Ontario	
	Present	Future[d]	Present	Future[d]
Existing 1976 Total Load[a]	17,474		11,755	
Existing 1976 Nonpoint Load[b]	8,445		3,581	
Recommended Target Loads	11,000		7,000	
Reduction Scenarios[c]				
Scenario 1: (STP[e] at 1 mg/l) Total Load	13,400	14,700	9,400	11,000
Additional Reduction Required to Meet Target Load	2,400	3,700	2,400	4,000
Percent of Existing Nonpoint Load	28	44	67	112
Scenario 2: (STP at 0.5 mg/l) Total Load	12,000	12,600	8,200	9,000
Additional Reduction Required to Meet Target Load	1,000	1,600	1,200	2,000
Percent of Existing Nonpoint Load	12	19	34	56
Scenario 3: (STP at 0.3 mg/l) Total Load	11,500[f]	11,900[f]	7,800[g]	8,300[g]
Additional Reduction Required to Meet Target Load	500	900	800	1,300
Percent of Existing Nonpoint Load	6	11	22	36

[a]Total load taken from PLUARG [2].

[b]Includes tributary diffuse and municipal nonpoint direct phosphorus loads; does not include direct atmospheric and upstream lake loads.

[c]Only sewage treatment plants with flows ⩾ 1 mgd are reduced to the indicated effluent standards.

[d]Sewage treatment plants and upstream lake loads have been projected on the basis of population trends. All other lake inputs were kept constant in these scenarios.

[e]STP = sewage treatment plant.

[f]Based on assumption that P concentrations in Lake Huron sewage treatment plant effluent (> 1 mgd) are reduced to 0.5 mg/l.

[g]Based on assumption that P concentrations in Lake Erie sewage treatment plant effluents (> 1 mgd) reduced to 0.3 mg/l.

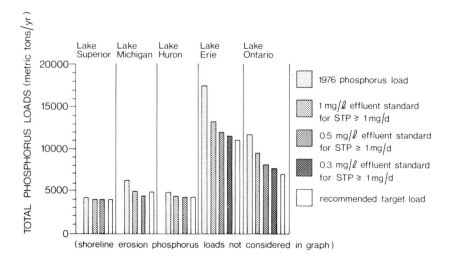

Figure 7. Total phosphorus loads (metric tons/yr)

nonpoint source loads remain constant (relative to 1976). Evidently, additional point source controls (below the 1.0 mg P/l municipal sewage treatment plant effluent standard) will eventually be necessary in the Upper Lakes (see PLUARG, [2] for additional discussion of Lake Huron) to achieve the presently ascribed loading objectives. Further reductions will also be necessary within Lake Erie and Ontario.

Continuing to use 1976 as an example, it is evident that some combination of point and nonpoint source controls is necessary to achieve the loading targets for Lakes Erie and Ontario. The "overview model" developed by PLUARG [2,8] provides information on some "cost-effective" management alternatives to reduce TP loadings appropriately. Scenario examples which would achieve the target loads for Lakes Erie and Ontario (based upon the residual loads after the 1 mg P/l effluent standard is hypothetically attained for the 1976 data, from Table VI) are summarized in Table VII and VIII, with respect to estimated costs and phosphorus load conditions. As indicated in Table VIII, half of the reduction required for Lake Ontario will likely be a natural result of achieving the target load for Lake Erie.

Details of the operation and development of the "overview model" and of the specifics of the various remedial management alternatives are outlined in Johnson et al. [8]. The point source alternatives are self-explanatory. Rural nonpoint remedial alternatives are identified here at two levels. Level 1 consists of voluntary sound management or conservation practices, including a variety of management options, all of which are considered to be in the landowner's self-interest, through reduction, of nutrient and soil losses from

Table VII. Scenario Example for Cost-Effective Phosphorus Load Reductions
For Lake Erie[a] [8]

Remedial Program	Estimated Load Reductions (metric tons)	Estimated Annual Cost (million $)
Voluntary Sound Management on all Agricultural Land	450	minimal
0.5 mg P/l Effluent Residual at all Municipal Sewage Treatment Plants	1305	10.5[b]
Second Level of Effort on all Croplands in Fine Textured Soil Areas	350	22.5
First Level of Effort on Nonpoint Inputs from Urban Areas	445	36.5
Totals	2550	69.5

[a]Estimated necessary load reduction based on 1976 load is 2400 metric tons.
[b]New program costs over and above costs of achieving effluent residual of 1.0 mg P/l.

Table VIII. Scenario Example of Cost-Effectiveness
Phosphorus Load Reductions for Lake Ontario[a] [8]

Remedial Program	Estimated Load Reductions (metric tons)	Estimated Annual Cost (million $)
Voluntary Sound Management on all Agricultural Land	80	minimal
0.5 mg P/l Effluent Residual at all Municipal Sewage Treatment Plants	1000	7.5[b]
First Level of Effort on Nonpoint Inputs from Urban Areas	140	14.0
Total	1220	21.5
Estimated Reductions from Lake Erie Program	1220	
Total Reduction	2440	

[a]Estimated necessary load reduction based on 1976 load is 2400 metric tons.
[b]New program costs over and above costs of achieving effluent residual of 1.0 mg P/l.

the land. Level 2 rural includes conservation tillage, contour strip cropping, cover crops and improved municipal drainage. The first level of urban nonpoint remedial alternatives, identified in the scenario example for Lake Erie, is essentially a program designed to reduce pollutants at the source through the use of, for example, more efficient street sweeping and sediment control in developing urban areas or construction sites.

CONCLUSIONS

It should be stressed that the phosphorus load reductions ilustrated here and derived through a modeling exercise undertaken by PLUARG are not intended to represent a rigid scheme or recommended sequence of controls for managing phosphorus in the Great Lakes. Rather, the information should be viewed as one means of quantitatively comparing various management alternatives. Only in this way can the cost-effectiveness of management alternatives be assessed to better determine least cost, socially and politically acceptable point and nonpoint source phosphorus management strategies.

There are, however, a variety of factors which will have to be kept in mind in the development of nonpoint-source phosphorus management alternatives. These include the factors discussed above, specifically, the philosophy of "potential contributing areas", and "hydrologically active areas", accurate load estimates, in-stream transmission, natural load variations, design criteria, surveillance strategies, other nonpoint sources, and continued growth. These factors may or may not be adequately quantified or understood or may only be relevant to local problems as opposed to eutrophication of the Great Lakes. In addition, new information will become available, and land uses and practices will change, all of which may affect management decisions, while changing social and economic factors may result in the need to reassess, for example, the goals presently considered to be desirable for the Great Lakes. Some of these considerations will be addressed and perhaps satisfactorily resolved at this conference or in the near future. However, a final answer or even a final formula for Great Lakes phosphorus management should not be viewed as the ultimate goal. Rather, a process is necessary whereby remedial alternatives can be assessed relative to environmental, social and economic factors to identify acceptable phosphorus management strategies.

REFERENCES

1. International Lake Erie, Lake Ontario-St. Lawrence River Water Pollution Board. "Pollution of Lake Erie, Lake Ontario, and the International Section of the St. Lawrence River, Vol. 1, Summary," Report to the International Joint Commission, Windsor, Ontario, (1969) 150 pp.

2. International Reference Group on Great Lakes Pollution from Land Use Activities (PLUARG). "Environmental Management Strategy for the Great Lakes System," International Joint Commission, Windsor, Ontario, (1978) 115 pp.

3. Task Group III Report to the Government of the United States and Canada (J. R. Vallentyne and N. A. Thomas, co-chairman), a part of the Fifth Year Review of the Canada-United States Great Lakes Water Quality Agreement—unpublished report, Windsor, Ontario. 86 pp.

4. Drynan, W. R. and M. J. Davis. "Application of the Universal Soil Loss Equation to the Estimation of Nonpoint Sources of Pollutant Loadings to the Great Lakes," Technical Report prepared for PLUARG, Windsor, Ontario, (1978) 38 pp.

5. Thomas, R. L. and W. S. Haras. 1978. "Contribution of Sediment and Associated Elements to the Great Lakes from Erosion of the Canadian Shoreline." Technical Report to PLUARG, Windsor, Ontario. 57 pp.

6. Acres Consulting Services Ltd. "Atmospheric Loading of the Lower Great Lakes and the Great Lakes Drainage Basin," Technical Report prepared for PLUARG, Windsor, Ontario, (1977) 70 pp.

7. Acres Consulting Services Ltd. "Atmospheric Loading of the Upper Great Lakes, Vol. 1, Summary," (1975) 20 pp.

8. Johnson, M. G., J. Comeau, W. C. Sonzogni, T. Heidtke and B. Stahlbaum. "Management Information Base and Overview Modelling," Technical Report submitted to PLUARG, Windsor, Ontario, (1978) 90 pp.

9. Nicolson, J. A. "Forested Watershed Studies; Summary Technical Report," Technical Report submitted to PLUARG, Windsor, Ontario, (1978) 45 pp.

10. Burns, N. M. and C. Ross. "Project Hypo, an Intensive Study of the Lake Erie Central Basin Hypolimnion and Related Surface Water Phenomena," Canada Centre for Inland Waters, Paper No. 6, Burlington, Ontario, (1972) 182 pp.

11. Lam, D. C. L. and J. M. Jaquet. "Computations of Physical Transport and Regeneration of Phosphorus in Lake Erie, Fall 1970," *J. Fish. Res. Bd. Can.* 33:550-563.

CHAPTER 4

REVIEW OF CONTROL OBJECTIVES:
NEW TARGET LOADS AND INPUT CONTROLS

N. A. Thomas

Large Lakes Research Station
U. S. Environmental Protection Agency
Grosse Ile, Michigan

A. Robertson

National Oceanic and Atmospheric Administration
Great Lakes Environmental Research Laboratory
Ann Arbor, Michigan

W. C. Sonzogni

Water Resources Division
Great Lakes Basin Commission
Ann Arbor, Michigan

INTRODUCTION

In accordance with Article IX of the 1972 Great Lakes Water Quality Agreement between the U. S. and Canada, a comprehensive review of accomplishments was made after the fifth year of its implementation. Because of the complicated issues and data review required concerning the control of phosphorus, a bilateral technical group (Task Group III) was formed. The charge to the task group included establishment of water quality objectives related to eutrophication, estimation of current phosphorus loadings, and establishment of target loads required to meet the desired water quality objectives.

The development of this paper has followed the same format as the charges to the task group. Water quality objectives for phosphorus used by the task group were drawn heavily from other groups of the International Joint

61

Commission (IJC). In certain cases, these objectives were modified where attainment did not appear possible. Estimates of present Great Lakes phosphorus loads and water quality conditions were required to provide a starting point from which to begin the development of target loads. Target loads to achieve these desired water quality conditions were then developed. Various phosphorus management scenarios were developed to achieve the target loads. Finally, some perspectives were developed for the target loads and input controls.

The 1972 Great Lakes Water Quality Agreement [1] between Canada and the United States mandates objectives for certain water quality properties of the Great Lakes and specifies that additional objectives should be developed where deemed necessary. The IJC Great Lakes Water Quality Board assigned the responsibility for developing these objectives to its Water Quality Objectives Subcommittee (WQOS). This group asked for assistance in this task from the IJC Research Advisory Board (now known as the Scientific Advisory Board), and the Scientific Basis for Water Quality Criteria Committee (SBWQC) was set up by that board to fill this role. WQOS and SBWQC met numerous times and attempted to identify the present and potential water quality problems in the Great Lakes and to develop objectives for concentrations of water quality parameters that, if met, would eliminate or greatly reduce these problems.

The 1972 Great Lakes Water Quality Agreement recognizes that eutrophication is a major problem on the Great Lakes, and specifies upper limits for loadings and effluent concentrations for phosphorus in the lower lakes (i.e., Lakes Erie and Ontario) to help alleviate this problem. Thus, when WQOS and SBWQC initiated their work, they concentrated primarily on developing objectives for toxic substances, and did not feel an immediate need to consider nutrients. However, during the course of their deliberations it became increasingly apparent that the treatment of nutrients in the 1972 agreement should be reviewed for several reasons. For one thing, the objectives were only for the lower lakes, and yet, eutrophication was a potential problem for all the lakes, and was a present problem, not only in the lower lakes, but also in certain areas of the the upper lakes (i.e., Lakes Superior, Michigan and Huron). Further, the loading and concentration limitations in the 1972 agreement were based largely on what was considered to be feasible in an engineering sense. SBWQC and WQOS had been directed by the Water Quality Board to develop objectives to protect all beneficial uses of the Great Lakes. There was a substantial question as to whether the objectives in the 1972 agreement, being defined by feasibility, would provide such protection. Thus, a work group of SBWQC was set up to look further into the development of modified nutrient objectives for the Lakes. This chapter presents a summary of the conclusions of this subcommittee and the modifications made by the Task Group, including the objectives and accompanying rationale that were developed.

Philosophy of Approach

As its first step the subcommittee considered what general philosophy of approach should be used to develop nutrient objectives. It was decided that attempts should be made to determine the type of ecosystem desired in each of the major subdivisions of the Great Lakes. Progressing from these determinations, estimates of nutrient concentrations required to allow such conditions to exist in these subdivisions would then be made. The 1972 Agreement was used as a guide for identifying the ecosystem conditions desired. Annex 2 of this agreement, entitled, "Control of Phosphorus," lists four general objectives for ecosystem quality. These objectives formed the starting point for subcommittee deliberations. Through a combination of objective analysis and expert opinion, the upper limits for nutrient concentrations which would allow these conditions to be met in each subdivision were estimated. These specific objectives, and the rationale for their development, are presented in a later section of this chapter.

Nutrients to be Considered

Organisms need a large number of chemical elements for growth and reproduction. Many of these elements are required only in trace quantities. However, certain elements, especially carbon, nitrogen, oxygen, hydrogen, sulfur and phosphorus, are needed in larger amounts because they are the basic building blocks for organic matter. If any of the required elements is in short supply relative to needs in a biological community, growth and reproduction will be limited, as will be the biomass or yield of the community.

In the Great Lakes, three elements (phosphorus, nitrogen and silicon) have been implicated, at least to some extent, in the limitation of biological production. Of these three, phosphorus is by far the most common limiting factor. However, under certain conditions, nitrogen may become the limiting factor, especially when man's activities add large amounts of phosphorus to the lake. Silicon, although not required by most types of phytoplankton, is essential for the growth of one very important freshwater group, the diatoms. In the Great Lakes, the amounts of silicon that are available can become so low that the growth and yield of the diatoms are limited. Examination of the roles of these elements in the Great Lakes ecosystem provides convincing evidence that only one of these, phosphorus, should be controlled by setting objectives for concentration limitations. The reasoning for this decision is explained for each of the limiting elements in the three following subsections.

Phosphorus. It is now well established that phosphorus most commonly limits yield in freshwater phytoplankton communities, although nitrogen, or more rarely some other vital element, may play this role in certain environments [2-4]. Substantial addition of phosphorus to a body of water, whether intentional or inadvertent, usually causes increases in photosynthesis and algal biomass. Such fertilization also causes many other changes in the water body. The species of phytoplankton shift to types that are better adapted to nutrient-rich environments. The increased algal productivity and biomass result in increased decomposition of organic matter, which often causes depletion of oxygen. Such oxygen depletion occurs especially during the summer in the lower layers (hypolimnion) of those lakes that become thermally stratified, because stratification isolates these layers from exchange with atmospheric oxygen. The animals also undergo profound shifts in abundance and relative numbers of the different types. With varying species of algae present, and with the chemical conditions altered, animal communities that are entirely different from those present before fertilization are favored.

The relationships of phosphorus to observed water quality conditions, and resultant problems associated with increased phosphorus levels, are represented in Figure 1. These chemical and biological changes that

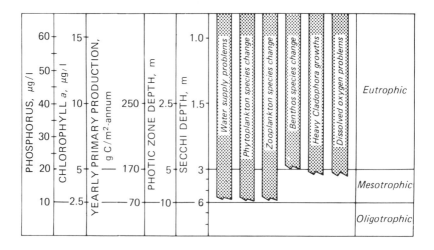

Figure 1. Enrichment problem relationships.

accompany phosphorus additions are usually, although not always, detrimental to man's interests. Water contact sports are often affected by the increased concentrations of algae, which may even float as green mats on the surface, accumulate and decay on the beaches. The thick growths of algae may also clog the intakes of water treatment plants and add unpleasant tastes and odors to the water. In addition, the types of fish may be altered, usually in the direction of increased numbers of fish not favored by sport and commercial fisherman.

Such changes often occur naturally in a lake, usually over a long period of time. However, in the Great Lakes, such changes (often collectively referred to as eutrophication) recently have been occurring at an accelerating rate [5-7]. This accelerated rate of eutrophication is believed to be due largely to the addition of phosphorus derived from human activities. Large amounts of phosphorus are added to the lakes from such sources as agricultural and urban runoff, discharges from industrial plants and effluents from municipal sources carrying human sewage, detergent residues and garbage. Thus, objectives for the upper limit of phosphorus concentration in various parts of the Great Lakes have been developed to aid in control of this very serious problem.

Nitrogen. The evidence presently available suggests that before man began adding his wastes in large quantities to the Great Lakes, phosphorus was the limiting factor almost everywhere in these lakes (i.e., the ratio of phosphorus before they exhausted the available nitrogen). However, the effluents from sewage plants and the runoff from agricultural and urban lands often contain much more phosphorus than nitrogen, relative to plant needs. Thus, in certain parts of the Great Lakes, especially Lake Erie and, to a lesser extent, Lake Ontario, nitrogen appears at times to be the factor limiting growth of most algal types [8,9]. This condition favors the nitrogen-fixing blue-green algae. These forms are especially likely to cause water quality problems, such as objectionable tastes and odors in drinking water, and the curtailment of recreational activities because of masses of algal material floating on the lake surface and accumulating on the beaches.

Because of the ability of blue-green algae to fix gaseous nitrogen, it is very difficult, if not impossible, to control eutrophication problems through limitations on nitrogen inputs to a water body. There is an inexhaustible supply of gaseous nitrogen in the atmosphere. As the gaseous nitrogen in the water is used up by the nitrogen-fixing algae, it is readily replaced from atmospheric sources. Further, the nitrogen-fixing forms, favored when nitrogen is limiting, are especially prone to cause deterioration of water quality. For these reasons then, it is impractical and, in fact, counterproductive to attempt to control eutrophication by restricting inputs of nitrogen, even in areas where it is currently limiting the growth of most algal forms. To rehabilitate such areas, phosphorus inputs must be lowered to the point where phosphorus replaces nitrogen as the limiting factor, and then further reduced so that the growth and yield of all algal forms is reduced.

Silicon. Algae belonging to the diatom group usually predominate throughout the year in phytoplankton in nutrient-poor (oligotrophic) lakes. Diatoms are generally considered desirable from man's point of view, because they serve as a good food source for zooplankton and because they usually do not increase to nuisance levels. An extremely high concentration can cause clogging problems at water treatment plants.

Diatoms have a special requirement for silicon that most other algal groups do not share. This element, in a form available to the algae, is required by diatoms so that they can form siliceous cell walls. If available silicon levels fall too low, the diatoms can no longer meet this requirement, and they are replaced by other forms, usually green and blue-green algae, that do not require silicon.

Schelske and Stoermer [7] have shown that phosphorus additions to Lake Michigan have stimulated diatom growth to such a degree that available silicon levels (measured as soluble, reactive silica) in the summer are approaching the point where they limit continued diatom growth. Yet, there is still adequate phosphorus for other types of algae to grow and succeed the diatomaceous communities. This process has probably already occurred in Lake Erie and Ontario as shown by the fact that the soluble, reactive silica levels are lower, and the diatoms are of much less relative importance, during the summer in these lakes than in the Upper Lakes [10].

Thus, silicon depletion seems to be an existing and increasing problem in the Great Lakes. However, the problem is the reverse of that of phosphorus. For phosphorus, the concentrations are increasing, allowing increased algal growth and shifts to less desirable types. For silicon the concentrations are decreasing, causing shifts away from desirable types. Therefore, as the problem is too little rather than too much silicon, there is no necessity to set an objective limiting its input. Further, the proposed phosphorus objectives should reduce algal growth, thus eliminating silicon deficiencies in most areas.

Time and Space Scales. Concentrations of phosphorus vary greatly, both within and among each of the five Great Lakes, and seasonally at any one place. Thus, the time and space scales on which to apply the phosphorus objectives must be considered.

The variations imposed by seasonal fluctuations may be largely ignored if objectives are developed which apply to a single time period, the early spring. At that time, before the algae start their yearly period of active growth, total nutrient concentrations are usually at their maxima [9]. In late spring and summer, when algal growth and cell concentrations are higher, sinking cells tend to carry nutrients to the bottom, thus reducing the total nutrient concentrations in the lake surface waters. Subsequently, the cells decompose and nutrients again become available at the start of the next growing season which largely determines the potential limits of growth and yield throughout the year.

With regard to spatial scale, objectives have only been developed for lake-wide averages or, in the cases of Lakes Huron and Erie, for two or three major subbasins. In Lake Huron, Saginaw Bay is so different and separate from the main lake that it was deemed necessary to develop a separate objective for it. Also, the three basins of Lake Erie differ sufficiently that separate objectives have been developed for each of them. Nearshore areas, where localized effects of rivers and direct discharges are present, are specifically excluded from this consideration of lakewide, or large area averages. The development of objectives for such large areas thus ignores localized problems and conditions. Therefore the objectives will not, even when met, cure all environmental deterioration due to nutrient enrichment. Local or regional problem areas that remain after the general objectives are achieved will require consideration on a case-by-case basis by the local jurisdictions involved.

Baseline Environmental Conditions. In order to develop the nutrient objectives, it also was felt necessary to agree on a definition of present environmental concentrations. These are presented in Table I, including spring concentrations of total phosphorus, inorganic nitrogen and total reactive silica, along with summer epilimnetic concentrations of two important indicators of surface water quality, total chlorophyll-a and Secchi disc depth. Since Secchi depth decreases as plankton biomass increases, inverse Secchi depth [18] is also included.

Several investigators [11,19,20] have suggested that a concentration of 20 μg/l total phosphorus in the spring may be used as an approximate lower limit of the condition where eutrophication is well advanced (eutrophy). These authors have also suggested that below 10 μg/l, few if any, of the effects of eutrophication are evident. Such waters are in a nutrient-poor condition (oligotrophy). The intermediate or transitional state (mesotrophy) is defined as being between 10 and 20 μg/l.

Using these limits as guidelines, the present trophic status of the Great Lakes is summarized in Figure 2. Only Lakes Superior and Huron are safely in the oligotrophic state. Lake Michigan is approaching the mesotrophic state, while the Lower Lakes and Saginaw Bay must all be classified as eutrophic, although Lake Ontario and the central and eastern basins of Lake Erie are not yet strongly so.

Table I. Average Values for Nutrients and Water Quality Variables in Great Lakes in the Late 1960s and Early 1970s

Lake	Spring			Summer		
	Total Phosphorus (μg P/l)	Inorganic Nitrogen (μg N/l)	Soluble Reactive Silica (mg SiO$_2$/l)	Chlorophyll a (μg/l)	Secchi Depth (m)	Inverse Secchi Depth (m^{-1})
Superior	4.6[a]	275[b]	2.25[b]	1.0[a]	8.8[a]	0.114
Michigan	9.0[c,d]	200[c]	1.5[c]	2.0[c,d,e]	~6.0[f,g]	0.127
Huron (main lake)	5.2[a]	259[b]	1.36[b]	1.2[a]	8.3[a]	0.120
Saginaw Bay	30.2[h]	304[h]	0.82[h]	18.2[h]	1.3[h]	0.79[h]
Western Erie	39.5[i]	631[b]	1.32[b]	11.1[b]	1.5[b]	0.667
Central Erie	21.2[i]	133[b]	0.33[b]	3.9[a]	4.4[a]	0.227
Eastern Erie	23.8[i,j]	180	0.30[b]	4.3[b]	4.5[b]	0.222
Ontario	24.0[i]	279[b]	0.42[b]	5.3[a]	2.5[a]	0.400

[a]Surface values only [11].
[b]Surface values only [9].
[c]Holland and Beeton [14].
[d]Rousar and Beeton [15].
[e]Holland [16].

[f]Only two cruises available [12].
[g]Only one cruise available [13].
[h]Nutrients are mean annual values [17].
[i]Great Lakes Water Quality Board [13a].
[j]Mean annual value.

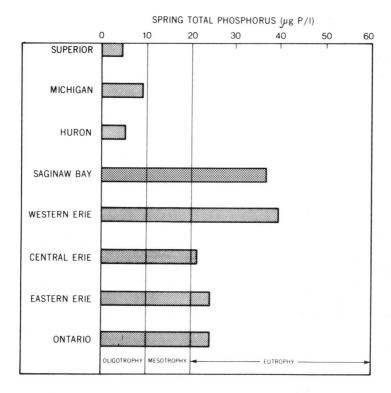

Figure 2. Current trophic status of the Great Lakes.

Phosphorus Objectives

With agreement on the philosophy and basic mode of approach, the Nutrient Work Group proceeded to develop specific objectives for phosphorus concentrations to meet the general environmental objectives of the 1972 Great Lakes Water Quality Agreement. These concentrations, with their accompanying rationale, are presented below.

Lake Superior and the Main Body of Lake Huron.

Annex 2 of the Water Quality Agreement [1] states that an objective of the phosphorus control program is the "... stabilization of Lake Superior and Lake Huron in their present oligotrophic state."

Thus, the Agreement calls for maintenance of the present water quality; and as both of these lakes seem to be phosphorus-limited [9], phosphorus concentrations should thus be maintained at no greater than their present levels. Therefore, based on the present conditions specified in Table I, it is recommended that the spring lakewide mean total phosphorus concentration in Lake Superior and the main body of Lake Huron should not exceed 5 μg/l.

Saginaw Bay

The Agreement does not deal with Saginaw Bay as an entity separate from Lake Huron. Yet this bay is so large that its overflow has a substantial effect on the water quality conditions in the boundary waters of the main lake [15]. Further, the main lake objective cannot be applied to this bay because the present phosphorus concentrations, as well as those that prevailed before large-scale human settlement in the area, are much higher than in the main lake [22]. Thus, it seems essential to develop a separate objective for this area.

Article II, section (e) of the agreement states that the waters of the Great Lakes System should be "free from nutrients entering the waters as a result of human activity in concentrations that create nuisance growth of aquatic weeds and algae."

Applying this general objective to an area such as Saginaw Bay requires that we know what part of the phosphorus comes from human activities. Chapra and Robertson [22] suggest that Saginaw Bay had natural total phosphorus concentrations somewhat above 10 μg/l, based on a mathematical model of the phosphorus inputs and concentrations to the Great Lakes. Further, recent observations by Thomas [23] suggest that the growth of *Cladophora* reaches nuisance levels in the Great Lakes at about 15 μg/l. Thus, it is recommended that the spring areawide mean total phosphorus concentrations in Saginaw Bay should not exceed 15 μg/l.

Western Lake Erie

The conditions in this part of Lake Erie bear a strong similarity to those in Saginaw Bay. Both areas presently have high total phosphorus concentrations and severe eutrophication problems. Both were also areas of naturally high phosphorus concentrations relative to most of the other parts of the Great Lakes [22]. The natural concentrations in western Erie were, however, probably somewhat lower than in Saginaw Bay, approximately 7 or 8 μg/l [24].

As pointed out for Saginaw Bay, the agreement calls for the prevention of nuisance growths of algae in the Great Lakes. More specifically, it establishes as one of the objectives of the phosphorus control program the "reduction in present levels of algal growth in Lake Erie."

Thus, to reduce these levels, and for basically the same reasons as for

Saginaw Bay, it is recommended that the spring basinwide average total phosphorus concentration in western Lake Erie should not exceed 15 μg/l.

Central Lake Erie

Annex 2 of the Agreement states than objective of the phosphorus control program is the "restoration of year-round aerobic conditions in the bottom waters of the central basin of Lake Erie".

The anaerobic conditions that now often exist during the summer result from the decomposition of large amounts of organic matter. This material is produced by the algae in the upper, lighted waters, which subsequently falls to the bottom and decomposes as the algae become senescent and die. Such decomposition requires oxygen. This gas is removed from the water as the process proceeds. If the waters where decomposition is occurring are in contact with the air, the dissolved oxygen that is removed is quickly replaced from the atmosphere, and anaerobic conditions do not develop. However, during the summer the central basin of Lake Erie is thermally stratified so that the hypolimnetic waters are not in contact with the atmosphere. Further, the central basin is fairly shallow, so the hypolimnion is only a few meters thick. Thus, any decomposition that occurs along the bottom withdraws oxygen from this limited hypolimnetic reservoir, so that substantial decomposition quickly leads to oxygen depletion.

To restore year-round aerobic conditions, phosphorus concentrations must be controlled so that the amounts of organic matter produced which settle to the bottom and decompose are not so large as to use up the oxygen supply of the bottom waters. Chapra [25] has developed a mathematical model that allows computation of this concentration. According to his calculations, a springtime average phosphorus concentration that does not exceed 10 μg/l will assure that, under normal meteorological and limnological conditions, no appreciable area of the central basin will become anaerobic during summer stratification. Thus, it is recommended that the spring, basinwide average concentration of total phosphorus should not exceed 10 μg/l. It is worth noting, however, that as emphasized by Chapra [25], his model is based on the normally occurring conditions. Because no detailed studies have looked at variability in this basin, it cannot be guaranteed that this objective will insure aerobic conditions each year.

Eastern Lake Erie

The agreement objectives of reduction of algal growths and prevention of nuisance conditions that apply to western Lake Erie also apply to the eastern basin. The environmental conditions here are quite different from those in the shallow western end, however. This basin is much deeper and is farther removed from most large sources of nutrient inputs. Thus, its present phosphorus concentrations (as well as those that occurred before man's intervention) are much lower than the concentrations in the western basin. This area is one of potentially oligotrophic conditions. As discussed earlier, 10 μg/l total phosphorus can be taken as the approximate upper limit of oligotrophy. Restoration to such a state should reduce the algal levels and restore high water quality. Thus, it is recommended that the spring basinwide average concentration of total phosphorus should not exceed 10 μg/l.

Lake Ontario

This lake has experienced substantial degradation of water quality and major alterations in biota due to man's activities. The recently completed International Field Year for the Great Lakes (IFYGL) program has revealed the severe deterioration that has occurred in its ecosystem. Detailed studies by Stoermer et al. [26] on the phytoplankton, by McNaught and Buzzard [27] and McNaught et al. [28] on the zooplankton, and by Nalepa and Thomas [29] on the benthos have shown that the lake is presently inhabited largely by enrichment-tolerant (eutrophic) forms. These authors believe that the lake's original biota were those characteristic of a much less eutrophic situation. In fact, Chapra's [24] model estimates the phosphorus concentrations in this lake originally fell within the oligotrophic range. Enrichment has led to large increases in phytoplankton production, and hence to nuisance growth of algae. The agreement objectives of reduction of algal growths and prevention of nuisance growths apply here. The lake should be returned to a state closer to its original oligotrophic condition. To meet this goal it is recommended that the lakewide average total phosphorus concentrations in the spring should not exceed 10 μg/l.

THE IMPACT OF THE PROPOSED PHOSPHORUS
OBJECTIVES ON WATER QUALITY

Although phosphorus is the major cause of accelerated eutrophication in the Great Lakes, it is difficult to interpret concentrations of this element directly as a measure of lake trophic status. A more effective interpretation can be obtained by relating the objectives proposed in this report to parameters that more directly measure algal biomass and water transparency. One way to do this is by the use of empirical correlations, such as have been used by Chapra and Dobson [30]. They have plotted measurements of chlorophyll-a concentrations in the Great Lakes against measurements of total phosphorus concentrations, using a linear regression with a zero intercept to estimate the relationship between these two variables. The zero intercept is based on the fact that, in the absence of the dependent variable, the independent variable would not be expected to be present. This results in an equation relating chlorophyll-a to total phosphorus. Chapra and Dobson [3] have also used the same type of procedure to establish the relation between chlorophyll-a and Secchi depth. However, in this case, they assume a hyperbolic relation between the variables, based on the Beer-Lambert law for extinction of light.

Using the equations of Chapra and Dobson, estimates of the maximum chlorophyll-a concentrations and minimum Secchi depths that would result if the phosphorus objectives are met have been developed. These estimates are presented in Table II. Using Chapra and Dobson's chlorophyll-phosphorus

Table II. Proposed Objectives for Total Phosphorus with Estimated Levels of Chlorophyll a, Secchi depth and Trophic State That Would Result.

	Total Phosphorus (μg P/l)	Total Chlorophyll (μg P/l)	Secchi Depth (m)	Trophic State
Superior	5	1.3	8.0	Oligotrophic
Michigan	7	1.8	6.7	Oligotrophic
Huron	5	1.3	8.0	Oligotrophic
Saginaw Bay	15	3.6	3.9	Mesotrophic
Western Erie	15	3.6	3.9	Mesotrophic
Central Erie	10	2.6	5.3	Oligomesotrophic
Eastern Erie	10	2.6	5.3	Oligomesotrophic
Ontario	10	2.6	5.3	Oligomesotrophic

equation, the maximum chlorophyll concentrations that correspond to the phosphorus objectives have been estimated. Using these chlorophyll values, Secchi depth values have then been estimated from the chlorophyll-Secchi depth equation for those systems in which phytoplankton are primarily responsible for light attenuation.

PRESENT PHOSPHORUS LOADS

As a result of both the intensive studies that were undertaken on the upper Great Lakes between 1974-1976 and the availability of tributary data, 1976 was selected for the establishment of current phosphorus loads. Phosphorus loadings were developed for direct industrial and municipal contributions, tributary inputs (both monitored and unmonitored), atmospheric deposition and shore erosion (however, the latter was not used in any of the associated calculations). Wherever possible, input loads were obtained from the 1976 Great Lakes Water Quality Board report [13]. Because the Task Group which was established to provide a technical basis for the renegotiation of the Water Quality Agreement had additional time to review the values established for many of the municipal and industrial sources to tributary streams, loading estimates for Lake Huron and Superior were altered to reflect current additions from these discharge sites. Previous estimates for Lake Huron were given as total loadings to the lake without consideration of the point of origin. The Task Group, following information that had been originally provided in the Upper Lakes Reference Group report [21], separated the loadings entering Lake Huron via the North Channel and Georgian Bay. Recent studies by the U. S. Environmental Protection Agency Large Lakes Research station provided estimates of the phosphorus loss through Saginaw Bay. The Task Group selected this technique as a method of providing a more realistic estimate for the phosphorus loading to the open waters of Lake Huron. Load estimates for Lake Ontario were derived both from the 1976 Great Lakes Water Quality Board Report [13] except for minor modifications, and from direct industrial and municipal inputs. The major differences in loading estimates were based on studies conducted by the U. S. Army Corps of Engineers on the U. S. tributaries to Lake Erie. The Corps study included more frequent sampling, thereby reducing the error in the estimation of the load between sampling periods.

TARGET LOADS

For the systems that were highly nutrient enriched, eutrophication models were required to calculate the loading necessary to meet the water quality objectives. To provide a common basis for simulation, base year loads were developed. Many tributary phosphorus loads were known to vary with flow. In order to establish a relationship between water quality and phosphorus loads, it was necessary to consider tributary flows. Base year input loads were developed by scaling the present data for tributaries and interlake exchange flows to correspond to the historical average tributary flows.

In preparing the target loads, the Task Group separated those which could be addressed by maintaining a nondegradation objective. This essentially required maintaining the inputs that have produced the present water quality conditions.

The Upper Lakes Reference Group [21] concluded that, " ...there was some degradation of water quality of both Lake Superior and Lake Huron caused by the input of nutrients... The degradation related to nutrients and organics affects both the open water and nearshore areas of the Upper

Lakes... ." The Upper Lakes Reference Group also concluded that the inplace and planned point source phosphorus control programs are presently adequate to protect the Upper Lakes, but that Lake Superior, Georgian Bay and the North Channel are at or near the target loads. The reference group recommended that a series of controls be implemented to provide a method of insuring that the target loads are not exceeded in future years.

During the negotiations of the 1972 agreement, one of the first attempts was made at establishing phosphorus loadings through a comparison of existing loads and related water quality conditions. Since that time, a fuller understanding of the Great Lakes system has been achieved, and the ability to numerically simulate biological phosphorus has improved. To provide the best target loads, five mathematical models were used. A minimum of three models was used in any one basin to describe existing conditions and to simulate changes as a result of load reductions. Numerical models were used on Saginaw Bay, the three basins of Lake Erie and Lake Ontario. Within the range for which simulations were made, approximately linear relationships between phosphorus loading and in-lake phosphorus concentrations were observed. While the slope of the regression curves were slightly different among the models, the results were not substantially different. The target loads were obtained from the arithmetic mean of projected results for the case of total phosphorus concentrations. The simulations did, however, suggest differences in secondary parameters (e.g., chlorophyll-a and dissolved oxygen). Refer to Bierman [31] for a detailed discussion of the model comparisons and the synthesis of model results.

Table III summarizes the target loads specified in the 1978 Water Quality Agreement, present loading and water quality conditions, and projected conditions resulting from existing loads. The water quality criteria for the recommended loads were based on total phosphorus concentrations; in the central basin of Lake Erie, dissolved oxygen was the criterion.

Lake Superior

The 1972 Water Quality Agreement, recognizing that data on the upper lakes was lacking, called for studies on these systems. During the period 1974-1976, lakewide, annual average phosphorus concentrations in Lake Superior averaged 4.6 μg/l (Table III), with a resultant chlorophyll-a value of 1.0 μg/l. The 1976 phosphorus load for this system was estimated at 3500 metric ton/yr. It was concluded by the Upper Lakes Reference Group (and adopted by Task Group III) that the Lake Superior loading should be equivalent to a 1-mg/l phosphorus effluent concentration for all municipal sewage treatment plants discharging in excess of 1 mgd. This value would provide an equivalent loading of 3400 metric ton/yr. The rationale for the required loading recommended by the Task Group involved several considerations:

Table III. Great Lakes Phosphorus Loading and Nutrient-Chlorophyll Concentrations

Lake	1976 Target Load from 1972 WQA (metric tons)	1974-1976 Annual Average Total Phosphorus (μg/l)	Summer Average Chlorophyll a (μg/l)	1976 Load (metric tons)	Base Year Load (metric tons)	Target Load (metric tons)	Total Phosphorus Concentration (μg/l)	Chlorophyll a (μg/l)
Superior	1796	4.6	1.0	3500	same as 1976	3400	4	Present
Michigan	5350	7.4	1.8	6700	same as 1976	5600	7	Present
Main Lake Huron	3088	5.2	1.2	3000	same as 1976	2800	5	Present
Saginaw Bay		40	23	1197	870	440	15	
Georgian Bay		4.5	1.2	630	same as 1976	600	4	Present
North Channel		5.5	1.7	550	same as 1976	520	5	Present
Lake Erie Whole Lake	14603			19676	20000	11000		
Lake Erie Western		44.9	12.4	12689	14499		20	5
Lake Erie Central		22.5	5.3	4382	4007		10	3.0
Lake Erie Eastern		23.9	3.8	2605	1463		8	2.5
Lake Ontario	9070	21	5.3	12795	11000	7000	10	2.5

1. A portion of the Upper Lakes Reference Group was not convinced that Lake Superior was in equilibrium with its loads.

2. The confined circulation pattern in the western tip provides increased phytoplankton growth.

3. A margin of safety in phosphorus loading would have to be made for future growth.

Lake Huron

The lake-wide, annual mean phosphorus concentration in Lake Huron was 5.2 μg/l, with a resultant chlorophyll-a concentration of 1.2 μg/l. The Task Group reaffirmed the Upper Lakes Reference Group's recommendation that all point sources > 1 mgd have an effluent phosphorus limitation of 1 mg/l. The loading to Lake Huron resulting from such limitation would be 2800 metric ton/yr to the open waters of the lake. This would minimize the future degradation of the southern basin of Lake Huron, as well as provide for future population growth.

Lake Michigan

Phosphorus loadings to Lake Michigan were established because of the input of this system to Lake Huron. Lake Michigan is known to have substantial algal enrichment problems (e.g., prolific *Cladophora* growth interfering with utilization of swim beaches, and adverse impacts on water supplies). Task Group III recommended a target load of 5600 metric ton/yr to this system. This load would result after all municipal sources > 1 mgd achieved 1 mg/l phosphorus in their effluent. The annual mean concentration of phosphorus was expected to decrease slightly, while associated changes in chlorophyll-a are expected to be undetectable.

North Channel-Georgian Bay

The Upper Lakes Reference Group recognized the problem of associating the phosphorus inputs entering the North Channel and Georgian Bay with Lake Huron water quality. Nutrient budgets were calculated for each of these water bodies. Losses to the sediments were estimated for each, thereby providing an appropriate net interbasin transport value. Target loads for the North Channel and Georgian Bay were based on achieving 1 mg/l phosphorus in effluents from municipal treated plants > 1 mgd. Most of the municipal inputs on Georgian Bay had phosphorus removal for the correction of localized problems. The small differences in the existing target loads were provided for future population growth.

Saginaw Bay

Saginaw Bay was considered as a separate water body because of its severe algal enrichment problems, and the incorporation of phosphorus in the Saginaw Bay sediments as it affects the load to the main body of Lake Huron. The 1976 phosphorus load to Saginaw Bay was 1197 metric ton/yr. However, when this load was corrected for tributary flow, the base year load was found to be 870 metric ton/yr. The Task Group recommended a load of 440 metric ton/yr. This recommendation was made both because of the desired reduction in water supply taste and odor problems and because the level was realistically achievable. The phosphorus criterion of 15 μg/l was in the transition range between oligotrophic and eutrophic conditions.

Lake Erie

The target load for Lake Erie was based on minimization of the dissolved oxygen problem in the central basin. The phosphorus and chlorophyll-a values in the eastern and western basins would correspond to the loads necessary to achieve the desired central basin water quality. The 1972 agreement had as one of its objectives the restoration of year-round aerobic conditons in the hypolimnetic water of the central basin of Lake Erie. After considering all the benefits from a higher minimum level of dissolved oxygen, and the necessary reduction in phosphorus loads required to achieve this mimimun, the Task Group agreed to reaffirm the 1972 agreement objective. After analysis of model output [32] and new data available for Lake Erie, a load of 11,000 metric ton/yr was recommended. This loading was believed to be the maximum allowable load to maintain aerobic conditions in the hypolimnion during the summer. To maintain aerobic conditions in the central basin hypolimnion, an average hypolimnia dissolved oxygen of 4 mg/l will have to be maintained. The areal extent of anoxia is directly related to the mean minimum dissolved oxygen level in the hypolimnion [32]. Aerobic conditions will prevent the large flux of phosphorus from the sediment.

Lake Ontario

As a result of the 1972 investigations carried out during the International Field Year for the Great Lakes, it was concluded that the phytoplankton, zooplankton and benthic organisms had been severely degraded. To meet the phosphorus objective of 10 μg/l, which would enable restoration of a portion of the biota, a load of 7000 metric ton/yr was recommended.

Compatibility of Water Quality Objectives

During the development of the target loads, achievement of more than one desired water quality objective was not always possible. The attainment of year-round aerobic conditions in the central basin of Lake Erie would result in a total phosphorus concentration of 10 μg/l. Since the three Lake Erie subbasins are influenced by each other, the reduction to 10 μg/l total phosphorus in the central basin would result in concentrations of 20 and 8 μg/l total phosphorus in the western and eastern basins, respectively. These values differ from the nutrient objectives suggested by the Nutrient Work Group. However, the Task Group believed restoration of aerobic conditions to the central basin to be the prime concern.

The only other difference in the objectives suggested by the Nutrient Work Grop and those recommended by the Task Group was on Lake Superior. The Nutrient Work Group rounded the present concentration of 4.6 μg/l to 5 μg/l as an objective, whereas the Task Group, believing the concentration may decrease after 1 mg/l phosphorus is achieved at wastewater treatment plants $>$ 1 mgd, rounded the total phosphorus objective to 4 μg/l.

SOME PERSPECTIVES ON THE TARGET LOADS AND INPUT CONTROLS

In reviewing the target loads, including the objectives on which they were based and the manner in which they were developed, it is important to keep a proper perspective on how these target loads should be used. Also, the implications of different strategies that might be used to achieve the targets should be understood. Consequently, some "perspectives" on the target loads are discussed below.

Precision of the Target Loads

The target loads are based on the best available information and are believed to represent reasonable estimates of the loads necessary to reach the desired objectives. However, they should only be used as a measure of the approximate load reduction required. The target loads are designed to provide decision-makers with guidelines related to the degree or types of phosphorus removal that will be necessary.

The fact that the target loads should not be considered to be an exact quantity is illustrated by year-to-year variations in the load due to flow differences. While 1976 phosphorus loads were adjusted to a "base year" status in the development of the target loads, no mention was made of the expected year-to-year variability.

In order to provide insight into the year-to-year variability of land runoff inputs of total phosphorus to the Great Lakes, the normally expected upper and lower range of the diffuse tributary loads was estimated by Sonzogni et al. [33] and Chapra and Sonzogni [34], based on historical flow of data. The ranges estimated for each of the lakes, shown in Table IV, were derived by

Table IV. Estimated Range in U. S. Diffuse Tributary Loading
Due to "Normal" Year-to Year Variations

Lake	Range in Diffuse Phosphorus Tributary Load (metric tons/yr)
Superior	750-1200
Michigan	1400-2000
Huron	700-2000
Erie	3700-7000
Ontario	800-1400

assuming that the mean annual diffuse tributary load varies directly with the mean annual flow. In making these estimates (which should be considered only first approximations), the "normal" range in flow was assumed to be equal to the standard deviations of the mean annual tributary flow. The methodology used in making these estimates is discussed in further detail by Sonzogni et al. [33].

Table IV illustrates that the U. S. diffuse tributary loads to each of the lakes can vary significantly from one year to another, depending on meterological (and runoff) conditions.

During 1975 and 1976, U. S. Great Lakes tributary loads, with the exception of Lake Superior, were near or even exceeded the expected "normal" upper range. This was a result of exceptionally high U. S. Great Lake tributary flows during the 1975 and 1976 water years, particularly the Lake Huron and Ontario tributary flows. Consequently, from a management perspective, it is important to realize that the target loads are estimated for a normal year, and significant year-to-year variation in the load can be expected to occur.

Control Options

In deriving the target loads, the manner in which they are to be achieved was not specified. Although different types of point and diffuse source controls were briefly investigated, it was recognized that several alternatives exist for achieving the necessary reductions. However, the Pollution from Land Use Activities Reference Group (PLUARG) has investigated different possibilities for reaching the target loads.

An example of phosphorus control alternatives is presented for Lake Erie in Table V. The example is based on the "overview modeling" analysis done

Table V. U.S. Lake Erie Phosphorus Control Alternatives
(Example of Overview Modeling Results)

Remedial Measure[a]	Phosphorus Load Reduction (metric ton/yr)	Annual Cost Effectiveness ($/metric ton) removed from lake load
Municipal Point Sources		
1.0 mg/l phosphorus (or less if currently below 1 mg/l) to		
0.5 mg/l residual concentration	1200	10,000
0.5 mg/l to 0.3 mg/l	600	95,000
Rural Nonpoint Sources		
Level 1 (sound management)	350	Low
Level 2 (Level 1 + strip)		
Cropping, buffer strips, etc.)	550	65,000
Level 3 (Level 1 & 2 + high		
intensity land management)	750	175,000
Urban Nonpoint Sources		
Level 1 (street cleaning,		
reduction at source)	400	80,000
Level 2 (Level 1 + detention		
and sedimentation	1000	155,000
Water Conservation: Assume Wastewater Treatment Plant Effluent P Concentration Fixed at 1 mg/l		
10% reduction in flow	250	?
20% reduction in flow	500	?

[a]See Johnson et al. [35] for details of various remedial measures indicated.

for PLUARG. Note that Table V presents the load reduction (at the lake) and the cost-effectiveness of various phosphorus control options. For further details the reader is referred to Johnson et al. [35].

Table V shows that the most cost-effective option appears to be Level 1 rural nonpoint source control. Level 1 treatment consists of sound land management, which is in the land owners' self-interest, and should be applied for sound conservation, if not water pollution control purposes. It consists of measures such as general conservation plowing techniques, minimizing tillage, mulching, avoiding farming on slopes near streams, properly incorporating fertilizers and manure into the soil, and avoiding adding inorganic fertilizers in excess of that required for optimum crop growth. It can likely be implemented through training and education, utilizing existing programs. Consequently, the cost is small. At the same time, Level 1 treatment is not very effective at reducing phosphorus loads.

The next most attractive option appears to be limiting effluent phosphorus concentrations to 0.5 mg/l at municipal sewage treatment plants. It should be mentioned, however, that the cost-effectiveness of this option may be underestimated. Currently, a review of new cost information is underway for point-source phosphorus removal. Nevertheless, it is likely that municipal source control will remain a cost-effective option.

Municipal phosphorus effluent limitation down to 0.3 mg/l appears quite expensive. Similarly, other rural nonpoint options (Level 2 and Level 3) are relatively expensive per metric ton removed at the lake. It is questionable whether the cost of these options as shown in Table V could be justified solely for the sake of meeting a specified target load number.

It should be pointed out that preliminary information developed by the U.S. Army Corps of Engineers in their Lake Erie Wastewater Management Study is more optimistic about rural runoff control than is indicated by Table V. The study estimates that appropriate application of conservation tillage and no-till farming practices could result in a several thousand metric ton reduction in the total phosphorus load from rural diffuse sources. The study also indicates that the aforementioned farming practices may actually increase crop production, providing a rather cost-effective control option. The Lake Erie Wastewater Management Study results thus need to be carefully considered in the future.

The potential effect that water conservation would have on phosphorus loads was considered as part of the Great Lakes Basin Commission's Great Lakes Environmental Planning Study. If water conservation decreases the amount of wastewater that must be treated, and if it is assumed effluent phosphorus concentrations are fixed at 1 mg/l, then loads (load = flow x concentration) can presumably be reduced. Hypothetical flow reductions of 10 and 20% appear to result in significant phosphorus load reductions, indicating water conservation could be a viable option to consider. Unfortunately, the cost associated with water conservation has not been determined, but it does not appear that treatment plant costs for treatment of a more concentrated, but lower volume wastewater, would substantially increase. Obviously, factors other than water quality are important in evaluating the desirability of a water conservation program.

Importance of the 1 mg/l Phosphorus Limitation

Complete implementation of the 1-mg/l phosphorus effluent control over the next several years will result in a large and cost-effective load reduction, particularly for the lower lakes. For example, for Lake Erie, full implementation of the 1-mg/l phosphorus effluent requirement will achieve about 70% of the load reduction required to reach the target load. The effect of this action should be watched closely over the next several years. It should provide an indication of the reasonableness of the estimated target loads.

The 1-mg/l phosphorus effluent limitation will also be very important for the future, since without it, phosphorus loads would probably increase to an alarming level in the next several decades. This is illustrated by examining current and future municipal phosphorus loads under various control options using the overview modeling approach. Data on U.S. River Basin Group 2-3, which drains into eastern Lake Michigan, are used here as an example.

Municipal phosphorus reduction alternatives for River Basin Group 2-3 are presented in Table VI. Much of the data used to develop these projections

Table VI. Comparison of Municipal Phosphorus Reduction Alternatives for River Basin Group [2,3].

Alternative	Load to Lake (metric ton/yr)		
	Current	1990	2000
No Phosphorus removal (Assume average effluent concentration = 4 mg/l phosphorus)	920	1120	1280
"208" Projections	395	350	415
Phosphorus removal at all plants (to 1 mg/l)	230	280	320
Phosphorus removal at all plants (to 0.5 mg/l)	115	140	160

was derived from "208" areawide planning studies. Note from Table VI that loads are expected to increase in 1990 and 2000 as a result of projected urban population increases. However, as a result of phosphorus effluent limitations, the amount of future phosphorus discharges to Lake Michigan is minimized. Table VI also shows that, according to "208" projections, phosphorus discharges will fall below current levels by 1990, but by 2000 will rise to about current levels. Thus reductions in phosphorus loads by 1990, brought about by the installation of new phosphorus removal facilities (at the 1-mg/l phosphorus level), are offset by population increases occurring over the next 20 years.

Biological Availability of Phosphorus

The target loads are based on total phosphorus. The available phosphorus fraction can immediately stimulate biological activity. The release of biologically available phosphorus from the other fraction (total) is at a much slower rate, part of which never becomes available.

Based on studies of a limited number of rivers in the Great Lakes basin during runoff events [36,37], 40% or less of the suspended sediment phosphorus was estimated to be in a "biologically available" form. For many rivers, the "available" fraction of the phosphorus associated with the suspended solids is probably considerably less than 40%. Overall, probably no more than about 50 to 60% of the tributary total phosphorus (including soluble phosphorus) is likely to be "biologically available." The rate at which the immediately unavailable is converted to the available form is extremely important and must be addressed.

Considerable effort is currently being directed toward gathering data and modeling available phosphorus inputs. At this point, it seems that consideration of available phosphorus will not significantly change the degree of reduction of the total phosphorus loads required. However, it may indicate a need to control those sources that are high in "available phosphorus" relative to total phosphorus. At this time, point sources appear to have the highest "available phosphorus" to total phosphorus ratio. Consequently, control of point sources, particularly those that discharge directly to a lake, or in the downstream portion of a tributary watershed, may be particularly advantageous.

Most investigators have discounted shoreline erosion as a significant source of phosphorus, since the availability of shoreline erosion phosphorus is low. Most of this phosphorus is in the form of particulates, which rapidly settle out in nearshore zones and do not significantly affect the whole-lake phosphorus dynamics. However, when compared to available phosphorus loads from other sources, it becomes more deserving of attention.

Table VII compares the estimated available phosphorus shoreline erosion with other sources. Note that compared to the total phosphorus loads from all sources (except shoreline erosion), the "available phosphorus" shoreline erosion load is small. However, if for example, current loads were reduced to meet target loads, the relative importance of shoreline erosion load would be more significant. The estimated diffuse tributary "available phosphorus" load

Table VII. Importance of Shoreline Erosion Available Phosphorus
Relative to Other Phosphorus Inputs to Lake Erie
(Based on mid-1970 Data)

Source	Load (metric ton/yr)
Shoreline erosion	
Total P	11,000
Available P[a]	1400
Total P load from all sources except shoreline erosion	
Current load	18,500[b]
Target load	11,000
Diffuse tributary load	
Total P	9000[b]
Available P[c]	3600

[a]U.S. shoreline erosion available P assumed to be 5% of total; Canadian shoreline erosion during 1972 and 1973 as reported by Thomas and Haras [38].

[b]Mid-1970 load from Chapra and Sonzogni [34].

[c]Assuming 40% of diffuse tributary total P is available.

(Table VII). If the "available phosphorus" component of the atmospheric or point source load were estimated (or better known), the relative significance of the shoreline erosion phosphorus load to Lake Erie could be even better assessed.

Phosphorus Control and Toxic Substances

As a by-product of phosphorus load reductions, inputs of certain toxic substances are also likely to be decreased. Increased understanding of the eutrophication process should also improve our understanding of the environmental fate of toxic substances. More information is needed regarding the relationship between eutrophication corrective strategies and toxic pollution corrective strategies.

An interesting point with regard to toxic substance pollution of the Great Lakes is that sediment derived from nonpoint sources may help ameliorate the problem. Sediment can bind toxic pollutants and make them biologically unavailable or carry them to the sediment. Sediment input may also help dilute and "bury" toxic materials in lake sediment.

It is perhaps not coincidental that Lake Erie has the lowest PCB concentrations in fish [39], yet receives by far the largest tributary sediment load. The ratio of tributary sediment input to lake volume is given in Table VIII.

Table VIII. Ratio of Great Lakes Tributary Suspended Solids Load
to Lake Volume [39].

Lake	Tributary Suspended Solids Input/Lake Volume (metric ton/yr/km^3)	Ratio Normalized to Lake Superior
Superior	113	1.0
Michigan	144	1.3
Huron	298	2.6
Erie	13,523	119.7
Ontario	916	8.1

Note that Lake Erie receives over 100 times more tributary sediment input than Lake Superior on a volumetric basis. Although other factors are involved which apparently make Lake Erie less susceptible to PCB and toxic substances pollution relative to the other Great Lakes (e.g., productivity, sediment temperature, oxygen content, water residence time, particulate settling rate, the composition of the fishery), the sediment input is likely an important factor.

The possibility that sediment load is related to toxic substance levels leads one to speculate on the effect that reducing sediment loads, in the name of phosphorus control, may have on toxic contaminant levels. Perhaps a reduction of land runoff will improve the trophic status of Lake Erie, but in turn cause the lake to suffer more severe toxic pollution problems. Whether the sediment input could be reduced enough to make a difference is not clear, but the question deserves further attention.

Weighing the Costs of Phosphorus Control

As pointed out in the PLUARG final report [40], socially meaningful yardsticks are needed against which the costs of remedial programs should be weighed. Unfortunately, this is extremely difficult to do, especially in quantitative terms.

Eutrophication control will obviously benefit a number of uses of the Great Lakes. Some municipal water supplies will likely be improved and, in fact, the target load for Saginaw Bay was developed with the objective of reducing filter-clogging algae, taste and odor problems associated with water intakes in the bay. Recreational use is likely to be improved, and the fishery, including the economically valuable sport fishery, will likely also benefit. The benefit in terms of dollars has not yet been determined. Shore property values are also likely to be increased or maintained. Some estimates have been made on the effect of reducing *Cladophora* growths, which upon dying, can pile along shorelines causing very offensive conditions. Power and industrial uses of the lakes may also benefit, but the cost savings is not known.

There may also be by-products of Great Lakes eutrophication control which are often overlooked. Watershed remedial programs designed to protect the Great Lakes will likely result in improved river and inland lake water quality. Basic research into the eutrophication process will also increase our fundamental understanding of the lakes, which is needed from an overall management perspective. As already discussed, phosphorus control may have an important effect on toxic substances pollution. The relationship between eutrophication control and toxic substances abatement needs further study.

Finally, there are intangible yardsticks against which the costs of remedial measures need to be measured. Perhaps the most compelling yardstick is the fact that the Great Lakes are an international heritage—a great natural resource, consisting of 1/5 of the surface freshwater in the world. Unquestionably, the lakes should be maintained in a reasonable status. To what extent the public is willing to support remedial programs is yet to be satisfactorily determined.

ACKNOWLEDGMENTS

The development of the objectives reported in this paper was the joint project of the members of the Nutrient Objectives Workgroup. Thus, the authors gratefully acknowledge the members of this group, S. C. Chapra, P. Dillon, and F. H. Dobson.

The process by which Task Group III derived the proposed phosphorus target loads for the Great Lakes is detailed by Vallentyne and Thomas [41].

REFERENCES

1.   U.S. Department of State. "Great Lakes Water Quality Agreement with Annexes and Texts and Terms of Reference, Between the United States of America and Canada, signed by Ottawa, April 15, 1972," Washington, DC (1972).

2.   Schindler, D. W. "Evolution of Phosphorus Limitations in Lakes," *Science* 195:260-263 (1977).

3.   Wetzel, R. G. *Limnology,* Philadelphia: W. B. Saunders, (1975).

4.   Golterman, H. L. *Physiological Limnology,* (Amsterdam: Elsever Scientific Publishing Company, 1975).

5.   Beeton, A. M. "Eutrophication of the St. Lawrence Great Lakes," *Limnol. Oceanog.* 10:140-254 (1965).

6.   Beeton, A. M. "Changes in the Environment and Biota of the Great Lakes," in "Eutrophication: Causes, Consequences, Correctives, (Washington, DC: National Academy of Science (1969), pp. 150-187.

7.   Schelske, C. L. and E. S. Stoermer. "Phosphorus Silica and Eutrophication of Lake Michigan," in *Nutrients and Eutrophication*, Special Symposium, Vol. 1, G. E. Likens, Ed. (American Society of Limnology and Oceanography, 1972) pp. 157-171.

8.   Stadelmann, P. and A. Fraser. "Phosphorus and Nitrogen Cycle on a Transect in Lake Ontario during the International Field Year 1972-1973," Proc. 17th Conf. Great Lakes Res., Internat. Assoc. Great Lakes Res., (1974), pp. 92-108.

9.   Dobson, H. F. H., M. Gilbertson and P. G. Sly. "A Summary and Comparison of Nutrients and Related Water Quality in Lakes Erie Ontario, Huron, and Superior," *J. Fish. Res. Bd. Can.* 31:731-738 (1974).

10. Vollenweider, R. A., M. A. Munawar and P. Stadelmann "A Comparative Review of Phytoplankton and Primary Production in the Laurentian Great Lakes," *J. Fish. Res. Bd. Can.* 31:739-762 (1974).

11. Dobson, H. F. H. "A Trophic Scale for the Great Lakes," Canada Centre for Inland Waters, Burlington, Ontario. Unpublished (1976).

12. Schelske, C. L. and J. C. Roth.   "Limnological Survey of Lakes vity and Nutrients in Lake Michigan and Lake Superior," Proc. 13th Conf. Great Lakes Rés., Internat. Assoc. Great Lakes Res., (1970) pp. 93-105.

13. Schelske, C. L., and J. C. Roth. "Limnological Survey of Lakes Superior, Huron and Erie," Great Lakes Res. Div., Univ. Mich., Pub. No. 17, (1973).

13a. Great Lakes Water Quality Board. "Great Lakes Water Quality: Fourth Annual Report, Appendix B, Annual Report of Surveillance Subcommittee," International Joint Commission, Windsor, Ontario (1976).

14. Holland, R. E. and A. M. Beeton. "Significance to Eutrophication of Spatial Differences in Nutrients and Diatoms in Lake Michigan," *Limnol. Oceanog.* 17:88-96 (1972).

15. Rousar, B. C. and A. M. Beeton "Distribution of Phosphorus Silica Chlorophyll *a* and Conductivity in Lake Michigan and Green Bay," *Wisconsin Aca. Sci., Arts & Lett.* 61:117-140 (1973).

16. Holland, R. "Seasonal Fluctuations of Lake Michigan Diatoms," *Limnol. Oceanog.* 14:423-436 (1969).

17. Smith, V. E., K. W. Lee, J. C. Filkins, K. W. Hartwell, K. R. Rygwelski and J. M. Townsend. "Survey of Chemical Factors in Saginaw Bay (Lake Huron)," Ecological Research Series, U.S. Environmental Protection Agency, EPA-6003-77-125 (1977).

18. Postma, H. "Suspended Matter and Secchi Disc Visibility in Coastal Waters," *Neth. J. Sea Res.* 1:359-390 (1961).

19. Vollenweider, R. A. "The Scientific Basis of Lake and Stream Eutrophication, with Particular Reference to Phosphorus and Nitrogen as Eutrophication Factors," Tech. Rept. DAS/DSI/68.27, OECD, Paris, France (1968).

20. Dillon, P. J. "The Phosphorus Budget of Cameron Lake, Ontario: The Importance of Flushing Rate to the Degree of Eutrophy of Lakes," *Limnol. Oceanog.* 20:28-39 (1975).

21. Upper Lakes Reference Group. "The Waters of Lake Huron and Lake Superior. Vol. II (Parts A and B). Lake Huron, Georgian Bay, and the North Channel," International Joint Commission, Windsor, Ontario (1977).

22. Chapra, S. C. and A. Robertson. "Great Lakes Eutrophication: The Effect of Point Source Control of Total Phosphorus," *Science* 196: 1448-1450 (1977).

23. Thomas, N. "Physical-Chemical Requirements," in *Cladophora in the Great Lakes*," H. Shear and D. E. Konasewich, Eds., (Windsor, Ontario: International Joint Commission, 1975).

24. Chapra, S. C. "Total Phosphorus Model for the Great Lakes," *J. Env. Eng. Div.*, Amer. Soc. Civil Engr. 101:147-161 (1977);

25. Chapra, S. C. "Applying the Phosphorus Loading Concept to the Simulation and Management of Great Lakes Water Quality," prepared for Conference on Phosphorus Management Strategies for the Great Lakes, 11th Annual Cornell University Conference, Rochester, NY, April 17-20, 1979.

26. Stoermer, E. F., M. M. Bowman, J. C. Kington and A. L. Schaedel. "Phytoplankton Composition and Abundance in Lake Ontario During IFYGL, "Ecological Research Series, U.S. Environmental Protection Agency, EPA-600/3-75-004 (1975).

27. McNaught, D. C. and M. Buzzard. "Changes in Zooplankton Populations in Lake Ontario (1939-1972)," Proc. 16th Conf. Great Lakes Res., Internat. Assoc. Great Lakes Res., (1973) pp. 76-86.

28. McNaught, D. C., M. Buzzard and S. Levine. "Zooplankton Production in Lake Ontario as Influenced by Environmental Perturbations," U.S. Environmental Protection Agency, EPA-660/3-75-021 (1975).

29. Nalepa, T. F. and N. A. Thomas. "Distribution of Macrobenthic Species in Lake Ontario in Relation to Sources of Pollution and Sediment Parameters," *J. Great Lakes Res.* 2:150-163 (1976).

30. Chapra, S. C. and H. F. H. Dobson. "Quantification of the Lake Trophic Typologies of Naumann (Surface Quality) and Thienemann (Oxygen), with Special Reference to the Great Lakes," Great Lakes Env. Res. National Oceanic and Atmospheric Admin., Ann Arbor, MI (1979).

31. Bierman, V. J., Jr. "A Comparison of Models Developed for Phosphorus Management in the Great Lakes," presented at the Conference on Phosphorus Management Strategies for the Great Lakes, 11th Annual Cornell University Conference, Rochester, NY, April 17-20, 1979.

32. DiToro, D. M. et al. "Report on Lake Erie Mathematical Model," Ecological Research Series, U.S. Environmental Protection Agency, Grosse Ile, MI, (in preparation).

33. Sonzogni, W. C., T. J. Monteith, W. E. Skimin and S. C. Chapra. "Critical Assessment of U.S. Land-Derived Pollutant Loadings to the Great Lakes," Report to U.S. Environmental Protection Agency, prepared for U.S. Task D, Pollution from Land Use Activities Reference Group, International Joint Commission, Windsor, Ontario (1979).

34. Chapra, S. C. and W. C. Sonzogni. "Great Lakes Total Phosphorus Budget for the Mid-1970's," *J. Water Poll. Control Fed.* (in press).

35. Johnson, M. G., J. C. Comeau, T. M. Heidtke, W. C. Sonzogni and B. W. Stahlbaum, "Management Information Base and Overview Modeling," Report to the Pollution from Land Use Activities Reference Group, International Joint Commission, Windsor, Ontario (1978).

36. Armstrong, D. E., J. K. Perry and D. Flatness. "Availability of Pollutants Associated with Suspended or Settled River Sediments which Gain Access to the Great Lakes," Final Report on EPA Contract No. 68-01-4479, Water Chemistry Laboratory, University of Wisconsin, Madison (1979).

37. Logan, T. J., F. H. Verhoff and J. V. DePinto. "Biological Availability of Total Phosphorus," Technical Report Series, Lake Erie Wastewater Management Study, U. S. Army Corps of Engineers, Buffalo, NY (1974).

38. Thomas, R. L. and W. S. Haras. "Contribution of Sediment and Associated Elements to the Great Lakes from Erosion of the Canadian Shoreline," Report to the Pollution From Land Use Activities Reference Group, International Joint Commission, Windsor, Ontario (1979).

39. Konasewich, D. E., W. Traversy and H. Zar. "Great Lakes Water Quality Status Report on Organic Contaminants and Heavy Metals in the Lakes Erie, Michigan, Huron and Superior Basin," Implementation Committee, Great Lakes Water Quality Board, International Joint Commission, Windsor, Ontario, 1978.

40. Pollution from Land Use Activities Reference Group. "Environmental Management Strategy for the Great Lakes Ecosystem," Final Report to the International Joint Commission, Windsor, Ontario, 1978.

41. Vallentyne, J. R. and N. A. Thomas. "Fifth Year Review of Canada-United States Great Lakes Water Quality Agreement. Report of Task Group III, A Technical Group to Review Phosphorus Loadings," U.S. Department of State, Washington, DC (1978).

CHAPTER 5

ENVIRONMENTAL TRENDS IN LAKE MICHIGAN

D. C. Rockwell, C. V. Marion, M. F. Palmer,
D. S. DeVault and R. J. Bowden

Great Lakes National Program Office
U.S. Environmental Protection Agency
Chicago, Illinois

INTRODUCTION

Increased cultural demands on the Great Lakes have resulted in complex pollution problems which have become all too familiar. Efforts to protect and enhance the water quality of these unique ecosystems have resulted in the expenditure of billions of dollars at all levels of government and in the private sector in both the United States and Canada. Sewage treatment plants have been built or upgraded, cleaner burning fuels have been developed, auto emission standards have been established, the use and manufacture of many pollutants and hazardous materials have been curtailed, and massive programs for the treatment of industrial waste have been initiated by industry.

How effective has this effort been? Are the Lakes responding? To answer these questions the International Joint Commission has established a repeating nine-year cycle of intensive studies on each lake. This plan is based on the assumption that the open waters of the lakes are slowly changing in response to cultural and environmental impacts. During 1976 and 1977 the United States Environmental Protection Agency (EPA), in cooperation with the University of Michigan, undertook an intensive study of Lake Michigan. The results of this study and the changes which have been detected in the last 15 years are the subject of this chapter.

METHODS

The EPA Great Lakes National Program Office (GLNPO) conducted eight open-lake cruises in the southern basin of Lake Michigan in 1976 and four in 1977. Samples for chemical and phytoplankton analyses were collected in 8-liter opaque Niskin bottles. Sample depths were 2 m and 5 m, at the thermocline in the hypolimnion, and 1 m above the bottom substrate. Samples were collected at comparable depths during nonstratified periods. Station locations and cruise dates are indicated in Figure 1.

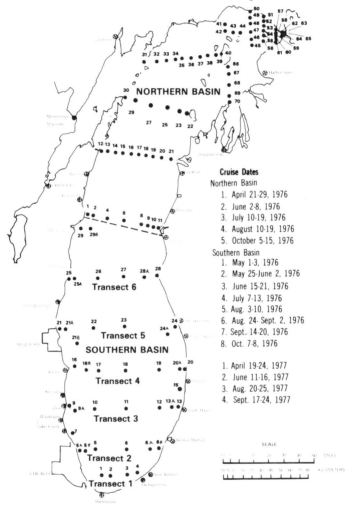

Figure 1. Lake Michigan survey cruise stations.

Temperature was determined by thermometer within one minute of sample collection. Vertical temperature profiles were made at all stations with an electronic bathythermograph. Profiles were made in both the downward and upward directions to account for hysteresis.

Secchi depth was estimated to the nearest 0.5 m with a standard 50-cm all-white Secchi disk.

Total phosphorus and total dissolved phosphorus samples were preserved with 2 ml/l concentrated H_2SO_4 and refrigerated at $4°C$. Total dissolved phosphorus samples were filtered within two hours of sample collection. Analyses were performed within 90 days of sample collection by the Automated Colorimetric Ascorbic Acid Reduction Method [1].

Total ammonia nitrogen was determined by the automated colorimetric phenate method within two hours of sample collection [1].

Dissolved reactive silicate analyses were made with Technicon Incorporated industrial method no. 105-71W/tentative, using a working concentration range of 0-5 mg/l as SiO_2. Dissolved reactive silicate analyses were made within two hours of sample collection.

Chloride concentrations were determined by the automated method utilizing a working concentration range of 0 to 20 mg/l as Cl [1]. Samples were analyzed within 90 days of sample collection.

Sulfate concentrations were determined by the automated methylthymol blue method using a working concentration range of 0 to 3 mg/l as SO_4 [2]. Sulfate samples were analyzed within 90 days of sample collection.

Chlorophyll-a samples were filtered and preserved with acetone within two hours of collection. Filtered and preserved samples were frozen and stored in the dark for no more than 30 days before completion of analysis. Chlorophyll-a samples are reported as corrected for pheophytin.

Phytoplankton samples were collected in 8-liter Niskin bottles lowered 1 m below the sampling depth, raised to the sampling depth and left opened for 10 seconds prior to sample collection. Samples depth were the same as those used for chemical analyses. A 960-ml subsample was withdrawn and immediately fixed with acid Lugol's solution. Identification and enumeration were performed using the inverted microscope method with 10-ml subsamples counted at 400X magnification [3]. Phytoplankton data are expressed in cells/ml. Filamentous blue-green algae were counted as 100 μ filament units.

For comparisons between EPA GLNPO data obtained in 1976 and 1977, the lake was arbitrarily divided into nearshore, offshore and open-lake areas. These were from 0-3 km, 3-8 km and >8 km from shore, respectively.

Phytoplankton data from the years 1971 [4], 1976 and 1977 were analyzed to determine possible changes in algae abundance (cells/ml) as well as changes in the relative abundance of the major taxa. Although the disadvantages of using cell counts as opposed to biomass are well known, the former unit was chosen. This was due both to the time required to make the necessary measurements for such a large data base and to provide for ready comparison with other southern basin studies, most of which have been expressed in cells/ml.

Microbiological samples were collected in sterilized, pre-evacuated Zobell bottles [2]. The following collection scheme was used:

1. For stations less than 11 m deep, a sample was taken at 2 m.
2. For stations from 12-20 m, samples were taken at 2 m below the surface and 1 m above the bottom.
3. For stations > 20 m deep showing a potential for thermocline development, samples were taken at 2 m and 20 m below the surface and 1 m above the bottom.

Fecal coliform concentrations were determined by the membrane filter technique. Colony counts were performed with a 10X stereo microscope. Counts were calculated in terms of fecal coliforms/100 ml of sample [2].

Determination of trophic indicator bacteria aerobic heterotrophs was performed with a modification of the filtration methodology detailed in *Standard Methods*. Under the modified procedure, membrane filters and total plate count agar (Standard Method agar) plates, presolidified in sterile disposable plastic petri dishes, type 50 x 15 mm, were used in place of agar pour plates. Plate incubation was carried out aerobically at 20°C for 48 hours. Colony counts were read with the aid of a 10X stereomicroscope. Counts were recorded as number of aerobic heterotrophic bacteria per ml of waste sample.

All samples were processed on the ship. Most samples were analyzed immediately following collection. When conditions such as unfavorable weather precluded immediate sample analysis, samples were refrigerated at 4°C until analysis could be performed.

Chemical and biological data were also generously provided by several other investigators, cited where appropriate. Statistical confidence levels were determined by the two-sample "t" test [5].

RESULTS

Chloride and Sulfate (conservative substances)

Both chloride and sulfate are useful in tracking man's impact on Lake Michigan. The natural background equilbrium concentration of chlorides may have been as low as 1.3 mg/l [6]. The concentration subsequently increased, reaching 3.0 mg/l by 1910. Offshore samples collected by the United States Department of the Interior averaged 6.5 mg/l in 1962-63.

Volume-weighted average results for 1976 and 1977 were 8.08 mg/l and 8.20 mg/l (Large Lakes Research Station-Grosse Ile).

During the last 15 years, the mean rate of chloride accumulation has been 0.10-0.13 mg/l/yr. The mean accumulation rate between 1860 to 1910 was estimated to be 0.035 mg/l/yr, and between 1910 to 1960 to be 0.07 mg/l/yr. Nearshore time-series data from three water-filtration plants are shown in Figure 2. The average, long-term, annual rates of increase over the different time periods shown are 0.10 ± 0.01 mg/l at Milwaukee's Linwood Filtration Plant, 0.12 ± 0.1 mg/l at Chicago's South Water Filtration Plant, and 0.15 ± 0.01 mg/l at Grand Rapids Lakeshore Filtration Plant. It is clear that the rate of accumulation is accelerating. The present lake chloride standards of Illinois (12 mg/l) will be exceeded early in the next century. If the rate is doubling every 50 years, chloride concentrations in Lake Michigan will reach Lake Erie levels during the first half of the 21st century.

Between 1976 and 1977, most of the observed chloride ion increase occurred over the winter, perhaps reflecting the impact of road-deicing compounds. In 1972-1973, salts used for road-deicing totaled 445,000 metric tons as chloride [7]. This source of chloride could account for approximately 40-45% of the total current load.

Figure 3 compares 1963 and 1976 chloride values from 11 different open-lake segments. The mean rate of accumulation over the 13-year period is higher at the northern and southern ends of the lake than in the middle of the lake (segments E and D). This is the abatement of brine discharges from the Frankfort-Manistee area, which caused higher concentrations during 1963. The nearshore zone at Frankfort-Manistee is the only area of the lake where 1976 average chloride concentrations were lower than the 1963 concentrations.

Rising levels of sulfate in the nearshore zones are of concern because sulfide (S^{2-}) inputs (principally from industrial sources) consume large amounts of oxygen in the water.

Sulfate time-series analyses at nearshore water intakes are plotted in Figure 4. The averaged annual rate of increase over the different time periods shown are 0.09 ± 0.02 mg/l at Milwaukee's Linwood Filtration Plant, 0.18 ± 0.01 mg/l Chicago's South Water Filtration Plant, and 0.31 ± 0.06 mg/l from Grand Rapids Lakeshore Filtration Plant. These increases are statistically significant at confidence levels greater than 95%. Closer examination of Milwaukee's data reveals that increases have occurred primarily since the early 1950s. The resulting average annual rate over this period at Milwaukee would be close to the rate observed at Chicago. Sulfate concentrations appear to be increasing in the nearshore waters at Grand Haven since 1973. A major source could be sodium sulfate, which is used primarily as a filler in home detergents. This component of major detergent formulations has doubled [8] in low-phosphate products.

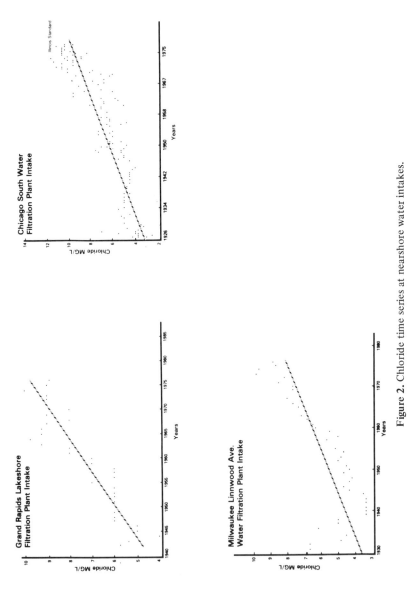

Figure 2. Chloride time series at nearshore water intakes.

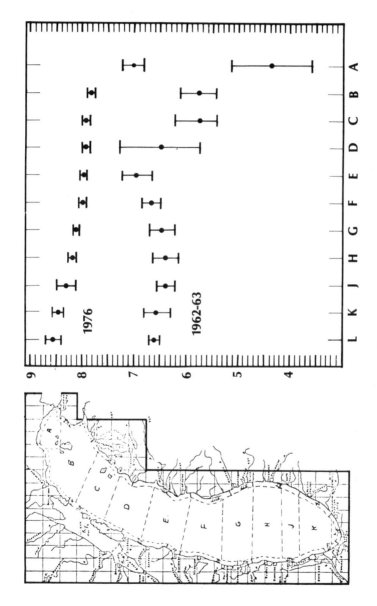

Figure 3. Chloride mg/l open lake areas mean ± three standard deviations.

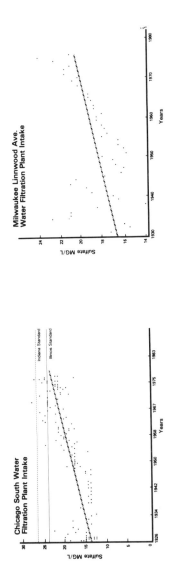

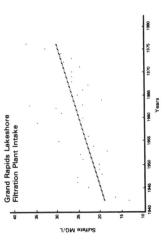

Figure 4. Sulfate time series at nearshore water intakes.

The concentration of sulfate in Lake Michigan increased more rapidly than any other ion between 1900 and 1960, based on data compiled and published by Beeton [9]. This data indicated an average rate of increase of 0.14 mg/l/yr (Figure 5). Monitoring of the Calumet area during 1965-1969

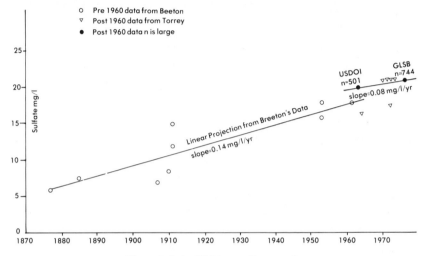

Figure 5. Lake Michigan sulfate trends.

showed a dramatic decrease in sulfate loads, due to changes in steel processing and effluent treatment [10]. The intensive surveys done by United States Department of Interior in 1962 63, and our 1976 survey, indicated that a linear projection of the open-lake rate increase is now estimated at 0.08 mg/l/yr over the last 15 years (Figure 5). This slowing in rate of sulfate accumulation in the open-lake after 1967 may be due to the abatement of sulfate discharged from the Indiana Harbor Canal.

The arithmetic mean value of sulfate in the open-water stations in the southern basin during 1976 was 21.10 ± 0.03 mg/l. The sulfate concentration in the southern basin is within 2.9 mg/l of the annual Illinois standard for Lake Michigan, which, at the current rate of increase of 0.08 mg/l yr, will be exceeded in 30 to 40 years.

Phosphorus

Phosphate phosphorus is one of the major nutrients required for plant nutrition and is utilized by plankton roughly in an atomic ratio of 1 to 15 to 106 with nitrogen and carbon, respectively. Since it is the most easily controlled of the major nutrients needed for plant growth, and usually in least supply, regulation of phosphorus is frequently the primary means for controlling eutrophication.

The impact of cultural pollution is most readily experienced in the nearshore zones of the lakes. Phosphate time-series records for the city of Chicago, begun in the early 1950s, show how rapidly this zone responds to nutrient enrichment and control. Figure 6 shows the accumulation of total phosphate between 1954 and 1979. In the 1940s, phosphate detergents were introduced into the market, and by the early 1950s, phosphate detergents dominated the detergent market. Concern over the rise of phosphates concentrations in the lake led to more frequent analyses in 1966. Models of Lake Michigan's 90% response time (for moving from one equilibrium condition toward a new equilibrium condition, after a change in the nutrient load) is estimated to be 16 years for phosphorus [11].

In 1972 the first partial bans on phosphate detergents were introduced in Indiana and the city of Chicago. Indiana's ban has more impact on the lake, because of the direct discharge of Indiana's northern counties into the lake, while most of Chicago's runoff and wastes are discharged away from the lake via inland waterways in Illinois. In 1973, Indiana's total ban on phosphate detergents took effect. Lake nearshore phosphate concentrations appear to have leveled off following 1973.

In 1975 and 1976, significant nutrient sources were diverted (Figure 6). Phosphate levels appear to have attained a maximum in recent times reaching values of 111 μg/l in April 1975, and 109 μg/l in November 1975. Since 1975, values have declined, with the lowest value (19 μg/l) reported in December 1978. The variance of this time series is large. An estimate was made (using Box-Jenkin methods), fitting a first-order autoegressive model on the first differences. The standard deviation of the residuals was 19 μg/l.

The distribution of average total phosphorus concentrations in 1976 and 1977 is shown in Figures 7 and 8. A reduction in the annual volume-weighted mean phosphorus concentration was seen in 92% of the stations in the southern basin, 58% of which were significant at the 95% confidence level. The changes in total phosphorus concentrations from inshore to offshore zones are consistent with important sources of total phosphorus to the lake. Important sources of the total phosphorus load to the lake in 1976 were Green Bay, the northern suburbs of Chicago, the Benton Harbor area, the Grand Haven-Muskegon area and the Ludington-Manistee area. Conspicuous by its absence was the Indiana Harbor area. In 1977 Milwaukee emerged as the apparent principal source of phosphorus to the southern basin.

The lowest total phosphorus concentrations in the open-lake transects occurred along the southern-most transect, with average values of 6.88 ± 0.22 μg/l in 1976. In 1977 this transect value was 5.20 ± 0.50 μg/l. Changes in the

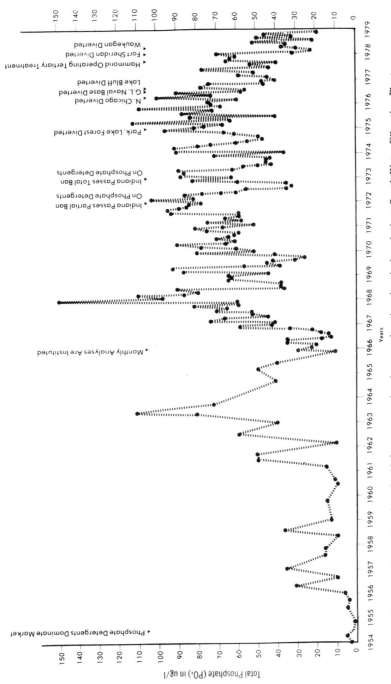

Figure 6. Total phosphate (ppb) from seasonal and comprehensive chemical analysis at South Water Filtration Plant, Chicago Water Purification Division.

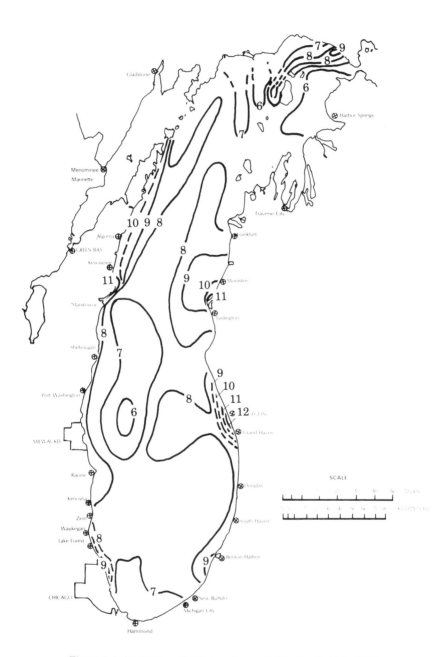

Figure 7. Lake Michigan total phosphorus distribution in μg/l–1976.

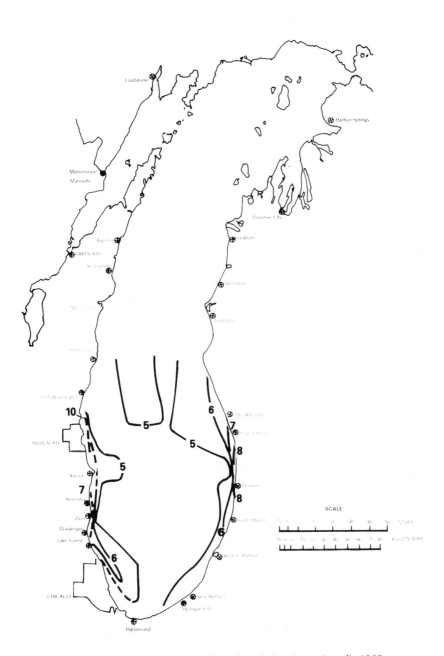

Figure 8. Lake Michigan distribution of total phosphorus in μg/1–1977.

expected pattern in the Calumet-Indiana Harbor area in 1976 and 1977 may be due to the detergent phosphate ban in Indiana. The 1976 ammonia values strongly suggest that reductions in phosphorus concentrations did not result from better wastewater treatment. The decrease in phosphorus concentrations observed in this area is significant, and is an indication that the policy of controlling phosphorus sources, can slow or even reverse eutrophication. Dissolved reactive phosphorus concentrations in the lake are usually less than 2 μg/l, which was the limit of detectability of the analysis.

Ammonia

In natural waters containing oxygen, ammonia is slowly oxidized to nitrite (NO_2) and nitrate (NO_3). This process allows one to estimate the "age" of pollution loadings. A high ratio of ammonia to nitrite indicates a recent source of pollution while a low ammonia to nitrate ratio indicates an earlier (i.e., older) input that has subsequently been oxidized. It is clear that ammonia is important in nearshore areas, where sources of pollution exist. Ammonia is also important in the open-lake areas which are remote from sources of pollution because, as a decomposition product, it can be an indicator of biological activity within the lake.

The annual station-average ammonia distribution for 1976 Lake Michigan data is shown in Figure 9. The impact of the Indiana Harbor Canal is evident in the distribution, with highest concentrations of ammonia near Indiana's eastern shoreline.

The 1977 annual station-average ammonia distribution is presented in Figure 10. Maximum annual ammonia concentrations were observed north of Milwaukee, between Port Washingron and Sheboygan, Wisconsin. Station-by-station comparison of annual volume-weighted mean concentrations showed reductions of total ammonia at 87% of the stations between 1976 and 1977. Fifteen percent were significant at the 95% confidence level. Two-thirds of these stations were in the two southernmost transects.

Turbidity

Measurements indicate that turbidity (Figure 11 and 12) decreases as one moves away from the nearshore zone toward the open waters. Greater water clarity is apparent in 1977, with the lowest values occurring near Milwaukee (Figure 11).

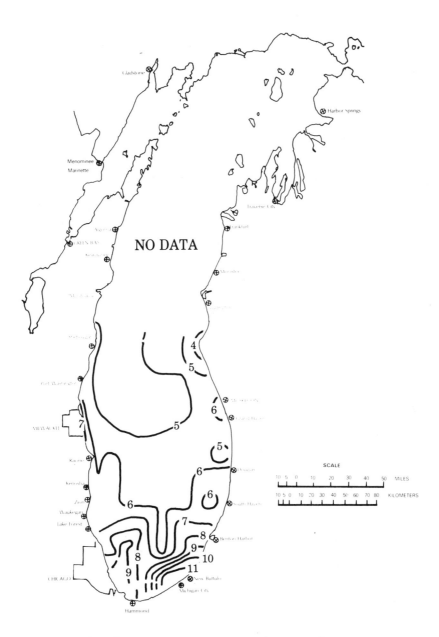

Figure 9. Lake Michigan ammonia distribution in $\mu g/l$–1976.

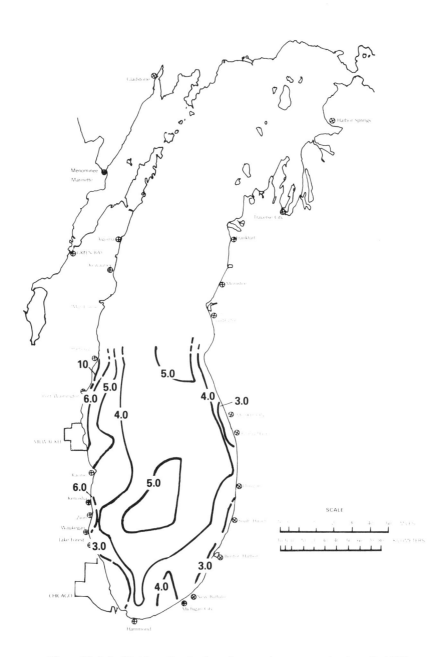

Figure 10. Lake Michigan distribution of ammonia concentration in μg/l–1977.

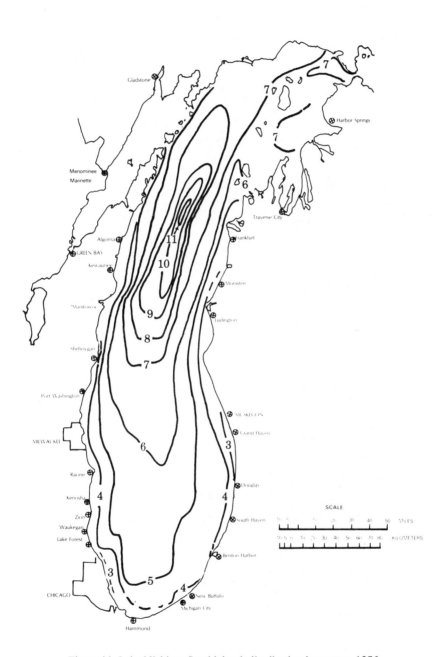

Figure 11. Lake Michigan Secchi depth distribution in meters–1976.

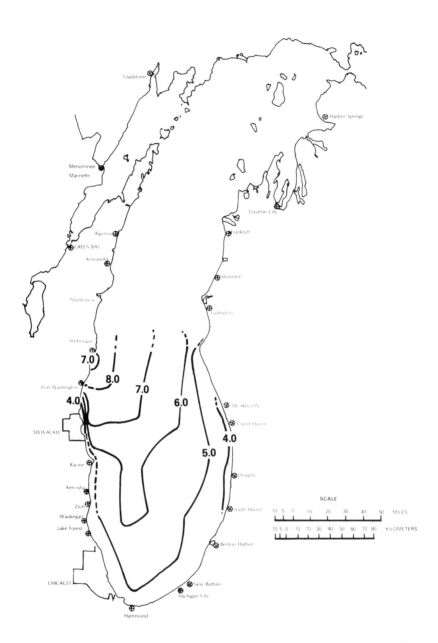

Figure 12. Lake Michigan distribution of transparency–Secchi depth in meters–1977.

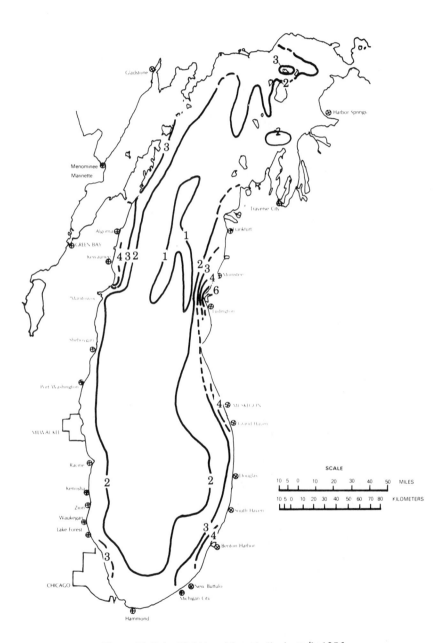

Figure 13. Lake Michigan chlorophyll *a* in μg/l–1976.

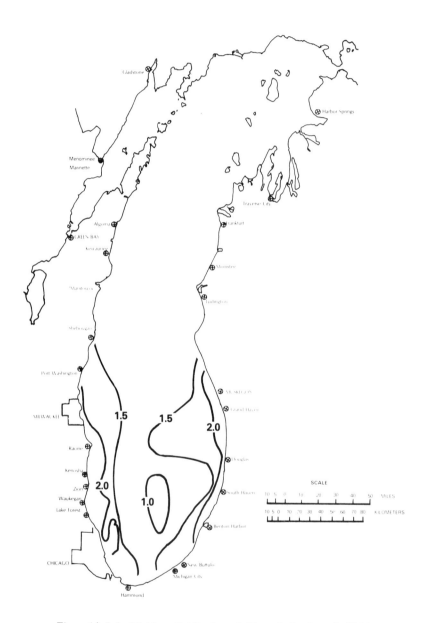

Figure 14. Lake Michigan distribution of chlorophyll *a* in μg/l–1977.

Chlorophyll-a

Algal production of chlorophyll-a concentrations in Lake Michigan are shown in Figure 13. The maximum values occur along the nearshore area south of the Green Bay outlet, towards Manitowoc Wisconsin, and north from Michigan City, Indiana, to Manistee, Michigan. In 1977, lower values were obtained at 85% of the stations, when compared with 1976 results (Figure 14).

In both years, chlorophyll-a concentrations along Indiana in the southern-most part of the basin do not attain the higher values found near other major tributaries.

A comparison of total phosphorus and chlorophyll-a over the 2-yr study period is shown in Figure 15. The separation between the curves exceeds three standard errors of the mean in the latter part of the season for phosphorus.

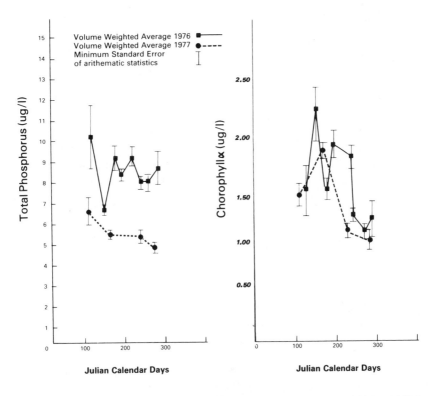

Figure 15. Total phosphorus and chlorophyll-a in Lake Michigan during 1976 and 1977.

Phytoplankton

Comparison of the data collected in 1971 by the Great Lakes Research Division, University of Michigan with our 1976 data from transect 2 (Table I) indicated that changes in the phytoplankton population occurred during that period. Diatoms decreased from an annual mean of 45.6% of the total population in 1971 to 17.0% in 1976. The decrease in the relative abundance of diatoms appears to be the result of increases in the total plankton population, as opposed to decreases in absolute diatom abundance. Most of this increase is the result of a marked increase in the abundance of the phytoflagellate population (see Table I for composition). This taxonomically heterogeneous group increased from approximately 12% of the total population in 1971 along transect 2 to over 50% in 1976. This group dominated the phytoplankton numerically in all areas of the lake in 1976 and 1977. While most of our data is taxonomically incomplete, Stoermer and Tuchmann [12] have reported large numbers of haptophytes in the southern-most area of the lake. Unfortunately, little is known of the freshwater ecology of this predominately marine taxa.

The changes between 1971 and 1976 may also result partially from differences in methodology between the two laboratories involved. However, the magnitude of the changes and the demonstrated ability of the Great Lakes Research Division to identify small flagellated forms [12,13] indicates that, while the data may not be exactly comparable, the differences are real.

In 1976 and 1977, (Tables II-IV) phytoplankton populations were generally highest in the nearshore area, and decreased toward the open lake. Diatoms usually reached their maximum relative and absolute abundance early in the sampling period, and decreased thereafter. Maximum diatom abundance was delayed until our second cruise in the offshore and open-lake region, possibly due to thermal inertia. The cyanophyta replaced the diatoms as the season progressed, with the highest relative and absolute abundance corresponding to the minimum levels of the diatoms. The phytoflagellates reached maximum abundance in June of both years.

Comparisons of the major algal groups (Tables II-IV) in comparable periods in 1976 and 1977 indicated that total (cells/ml) plankton, blue-green algae, and flagellates decreased significantly (95% confidence level) in all three areas of the lake in June 1977. Significant decreases in blue-green algae were also evident in August 1977 in the nearshore and open-lake. Diatom populations were above the 1976 levels in all lake areas during August and September 1977. Pennate diatoms were significantly (95% confidence level) higher in the nearshore, offshore and open-lake regions. The open-lake also exhibited significant (95% confidence level) increases in the centric diatoms.

Table I. Statistical Summary of Lake Michigan Phytoplankton; Transect 2, 1971[a] and 1976 (2-m Samples, Annual[b] Absolute and Relative Abundance)

Station	Year	Cells/ml	Green Algae		Blue-Green Algae		Diatoms		Flagellates	
			Average Absolute Abundance Cells (mg/l)	Average Relative Abundance (%)	Average Absolute Abundance Cells (mg/l)	Average Relative Abundance (%)	Average Absolute Abundance Cells (mg/l)	Average Relative Abundance (%)	Average Absolute Abundance Cells (mg/l)	Average Relative Abundance (%)
25[a]	1971	1116	193	17.3	224	20.1	627	56.2	72	6.4
5	1976	4295	177	4.1	767	17.9	547	12.7	2802	65.2
27[a]	1971	1200	105	8.7	332	27.7	642	53.5	68	5.7
5A	1976	5092	137	2.7	485	9.5	1037	20.4	3432	67.4
26[a]	1971	1050	150	14.3	177	16.8	665	63.3	59	5.6
5B	1976	5002	230	4.6	555	11.1	1232	24.6	2985	59.7
23[a]	1971	974	212	21.8	169	17.3	532	54.6	61	6.3
6	1976	3612	252	7.0	810	22.4	342	9.5	2250	62.3
22[a]	1971	1057	249	23.5	316	30.0	176	16.6	190	18.0
6A	1976	3095	222	7.2	705	22.8	335	10.8	1832	59.2
21[a]	1971	1007	224	22.2	204	20.3	297	29.5	281	27.9
6B	1976	5320	362	6.8	605	11.4	1267	23.8	3075	57.8

[a]Data from Stoermer, 1976 [4].
[b]Means for June, July, August, September.

Table II. Statistical Summary of Nearshore Lake Michigan Phytoplankton (2m and 5m averages)

	Diatoms		Flagellates		Blue-Greens		Greens		Total
	Average Cells/ml	Average Relative Abundance (%)	Average Cells/ml	Average Relative Abundance (%)	Average Cells/ml	Average Relative Abundance (%)	Average Cells/ml	Average Relative Abundance (%)	Average Cells/ml
May 25, 1976	1371	24.9	3232	58.7	661	12.0	283	5.1	4950
June 15, 1976	1137	15.7	5234	72.2	651	9.0	341	4.7	2363
July 7, 1976	1426	27.5	2990	57.7	354	6.8	409	7.9	5179
August 10, 1976	762	19.0	2080	51.8	943	23.5	234	5.8	4019
August 24, 1976	402	9.9	2113	52.2	1282	31.7	247	6.1	4044
September 14, 1976	460	12.9	2206	61.9	697	19.5	202	5.7	3565
April 19, 1977	1703	44.9	2002	52.8	76	2.0	259	6.8	4040
June 11, 1977	1599	38.9	2223	54.0	82	2.0	207	5.0	4111
August 20, 1977	995	25.4	1940	49.5	708	18.1	216	5.5	3359
September 17, 1977	860	20.3	2042	48.1	1181	27.8	180	4.2	4263

Table III. Statistical Summary of Offshore Lake Michigan Phytoplankton (2m and 5m averages)

	Diatoms		Flagellates		Blue-Greens		Greens		Total
	Average Cells/ml	Average Relative Abundance (%)	Average Cells/ml	Average Relative Abundance (%)	Average Cells/ml	Average Relative Abundance (%)	Average Cells/ml	Average Relative Abundance (%)	Average Cells/ml
May 25, 1976	1080	26.1	2467	59.7	477	11.5	229	5.5	4253
June 15, 1976	1224	20.57	3749	62.9	644	10.8	234	3.9	5851
July 7, 1976	625	21.4	1920	65.6	221	7.5	160	5.5	2926
August 10, 1976	366	9.1	2337	58.0	877	21.8	312	7.7	3892
August 24, 1976	195	4.5	2148	49.6	1751	40.4	224	5.2	4318
September 14, 1976	324	10.3	1876	59.6	773	24.5	176	5.6	3149
April 19, 1977	1074	41.5	1285	49.7	61	2.3	163	6.3	2583
June 11, 1977	982	25.7	2551	66.8	91	2.4	186	4.9	2810
August 20, 1977	661	18.9	1968	56.4	676	19.4	186	5.3	3491
September 17, 1977	680	20.3	1974	58.9	799	23.8	113	3.4	3566

Table IV. Statistical Summary of Openlake Lake Michigan Phytoplankton (2m and 5m averages)

	Diatoms		Flagellates		Blue-Greens		Greens		Total
	Average Cells/ml	Average Relative Abundance (%)	Average Cells/ml	Average Relative Abundance (%)	Average Cells/ml	Average Relative Abundance (%)	Average Cells/ml	Average Relative Abundance (%)	Average Cells/ml
May 25, 1976	755	21.6	2053	58.7	556	15.9	133	3.8	3500
June 15, 1976	1000	18.2	3680	67.0	537	9.8	100	1.8	5493
July 7, 1976	708	21.1	1990	59.2	387	11.5	274	8.2	3359
August 10, 1976	184	5.0	1535	41.8	1717	46.8	239	6.5	3671
August 24, 1976	105	2.7	1808	47.4	1752	4.6	151	4.0	3816
September 14, 1976	100	3.5	1869	66.0	711	25.1	151	5.3	2831
April 19, 1977	871	41.6	1008	48.2	63	3.0	102	4.9	2092
June 11, 1977	1021	32.9	1763	56.8	93	3.0	199	6.4	3106
August 20, 1977	381	11.8	1595	49.2	1186	36.6	131	4.0	3241
September 17, 1977	347	11.8	1658	56.4	840	28.6	82	2.8	2938

Analyses of the species observed in 1977 (Tables V-VIII) showed that several mesotrophic-eutrophic species are abundant in all areas of the lake. These include *Diatoma tenue var. elongatum, Tabellaria fenstrate, and Fragilaria intermedia* [14,15]. *Fragilaria crotonensis* was the dominant diatom throughout the 1977 study period.

Heterotrophic Bacteria

Heterotrophic bacteria, because of their extreme sensitivity to fluctuating nutrient levels, are excellent indicators of organic pollution and eutrophication [16].

These capabilities were used in conjunction with a trophic status evaluation system which included total phosphorus, chlorophyll and Secchi depth. The systems which appear to be most applicable to the Great Lakes are that of the Upper Lakes Reference Group [17] developed by Dobson [18] and the system derived by Rast and Lee [19]. Weighing equally each of these chemical systems, and including aerobic heterotrophs as a fourth component, trophic diagrams were developed for each study area. Table IX provides a summary of the transition values between various trophic states developed by each observer for total phosphorus, chlorophyll-*a*, and Secchi depth. Table IX also contains trophic transition values for aerobic heterotrophs developed by GLNPO, Surveillance and Research Staff. Using the above system, most of the open-water area of Lake Michigan was characterized as oligotrophic during both 1976 and 1977 (Figures 16 and 17). A mesotrophic zone extended around the entire shoreline during the 1976 study.

In 1977, however, the zone that was monitored at the open-lake stations which were nearest to the shore, and which had exhibited a mesotrophic status in 1976, now indicated an oligotrophic status. The open-lake stations nearest to the shore, and which still showed a mesotrophic status, were observed from Port Washington, Wisconsin to below Milwaukee, and from Kenosha to Chicago, Illinois, on the west shore, and from White Hall to New Buffalo, Michigan, on the east shore of the lake. Dotted lines on Figure 17 indicated a probable mesotrophic zone not determined during the 1977 lake study.

Heterotrophic bacterial densities in the oligotrophic zone were found to be consistent with the trophic status of the study areas. Heterotrophic geometric mean values for the open-lake, mainly oligotrophic, zone in 1976 and 1977, respectively, were 9/ml (2-33) (one standard deviation) and 8/ml (2-35). In 1976, the nearshore aerobic heterotrophs averaged 25/ml (6-100). In 1977, the heterotrophic bacterial density decreased to 19/ml (5-79).

In the mesotrophic zone, the geometric mean value for aerobic heterotrophs was 21/ml (5-87). In 1977 the geometric mean heterotrophic bacterial count showed an apparent increase to 30/ml (8-110). However, this

Table V. Summary of 1977 Lake Michigan Assemblage and Tallies of Phytoplankton (cells/ml)

Genera spp.	Nearshore Mean	Offshore Mean	Open Lake Mean	April Mean	June Mean	August Mean	September Mean
Centric Diatoms							
Cyclotella sp.	125.6	75.3	72.0	103.0	80.0	52.1	138.5
Melosira sp.	283.9	208.8	275.7	424.7	215.7	126.2	120.7
Stephanodiscus sp.	22.5	23.8	32.2	30.0	32.7	0	0
Pennate Diatoms							
Synedra acus	89.4	58.7	44.9	67.5	70.3	49.6	59.1
Synedra ulna	82.9	67.0	39.0	67.4	77.4	33.8	30.0
Rhizosolenia eriensis	67.5	50.6	54.3	46.7	61.2	39.1	80.0
Rhizosolenia longiseta	86.3	61.9	85.9	51.2	112.8	30.0	56.3
Asterionella formosa	123.9	66.3	95.9	79.3	88.6	107.1	99.2
Nitzschia sp.	84.3	40.0	33.5	64.4	63.8	40.0	43.6
Nitzschia palea	20.0	58.3	30.0	23.4	30.0	30.0	45.0
Nitzschia acicularis	47.5	30.0	28.0	30.0	30.0	45.0	30.0
Fragilaria crotonensis	687.2	509.6	380.7	740.0	494.8	464.6	376.5
Diatoma Tenue V. elongatum	100.0	54.6	53.0	52.9	70.5	114.6	53.8
Fragilaria intermedia	240.0	370.0	160.0	196.0	348.0	245.0	0
Tabellaria fenestrata	147.0	154.1	110.0	133.5	157.3	123.3	113.6
Tabellaria flocculosa	170.0	60.0	150.0	0	60.0	0	250.0
Cocconeis sp.	30.0	0	20.0	23.3	0	0	0
Navicula sp.	0	0	20.0	20.0	0	0	0
Synedra sp.	90.0	30.0	26.7	33.3	0	0	0
Nitzschia sigmoidea	60.0	0	35.0	40.0	45.0	0	0
Nitzschia linearis	40.0	60.0	30.0	0	40.0	30.0	0
Cymatopleura solea	30.0	0	30.0	0	30.0	0	0
Amphiprora sp.	30.0	0	0	0	30.0	0	0

Table VI. Summary of 1977 Lake Michigan Assemblage and Tallies of Phytoplankton (number of taxon occurrences)

Genera spp.	Nearshore Tally	Offshore Tally	Open Lake Tally	Tally Total	April	June	August	September
Centric Diatoms								
Cyclotella sp.	36	34	49	119	27	28	38	26
Melosira sp.	28	34	37	99	34	30	21	14
Stephanodiscus sp.	12	8	9	29	18	11	0	0
Pennate Diatoms								
Synedra acus	33	38	41	112	32	33	25	22
Synedra ulna	21	23	20	64	23	27	8	6
Rhizosolenia eriensis	16	16	14	46	9	17	10	10
Rhizosolenia longiseta	24	21	17	62	17	29	8	8
Asterionella formosa	33	40	39	112	29	28	31	24
Nitzschia sp.	28	22	20	70	23	21	12	14
Nitzschia palea	1	6	1	8	3	1	2	2
Nitzschia acicularis	4	4	5	13	3	3	4	3
Fragilaria crotonensis	32	26	29	87	22	21	24	20
Diatoma Tenue V. elongatum	24	22	20	66	17	20	13	16
Fragilaria intermedia	3	5	4	12	5	5	2	0
Fragilaria fenestrata	23	17	19	59	17	22	9	11
Tabellaria flocculosa	2	1	1	4	1	1	2	2
Cocconeis sp.	1	0	2	3	3	0	0	0
Navicula sp.	0	0	1	1	1	0	0	0
Synedra sp.	1	2	3	6	6	0	0	0
Nitzschia sigmoidea	3	0	2	5	3	2	0	0
Nitzschia linearis	6	1	4	11	0	9	2	0
Cymatopleura solea	1	0	1	2	0	2	2	0
Amphiprora sp.	1	0	0	1	0	1	0	0

Table VII. Summary of 1977 Lake Michigan Phytoplankton Geographic and Seasonal Taxon Means (cells/ml)

Genera spp.	Nearshore Mean	Offshore Mean	Open Lake Mean	April Mean	June Mean	August Mean	September Mean
Cyanophyton							
Anacystis sp.	761.7	418.7	517.1	33.8	30.0	426.4	741.1
Aphanothece sp.	175.8	151.2	203.7	0	0	230.8	103.7
Chroococcus sp.	57.0	90.2	70.5	20.0	0	69.1	80.7
Anabaena sp.	62.9	30.0	25.0	0	30.0	35.0	55.0
Oscillatoria limnetica	68.3	40.0	58.4	57.5	56.1	45.0	65.6
Gomphosphaeria lacustris	55.8	56.9	67.2	0	0	63.5	63.8
Coelosphaerium kuetzingianum	43.3	36.4	28.6	35.0	30.0	40.0	38.2
Schizothrix calcicola	40.0	42.7	43.8	35.5	55.0	38.6	30.0
Microcoleus lynbyaceus	42.0	68.9	35.0	33.3	40.0	30.0	84.3
Schizothrix sp.	45.0	30.0	60.0	30.0	60.0	0	60.0
Chlorophytes							
Quadrigula lacustris	47.8	51.5	57.3	36.0	58.7	55.8	57.7
Scene desmus quadricauda	75.6	37.5	25.0	27.5	54.0	75.7	30.0
Scenedesmus sp.	41.1	32.2	35.8	43.1	35.9	32.9	40.0
Scenedesmus dimorphus	30.0	30.0	0	0	30.0	30.0	0
Scenedesmus abundens	0	30.0	0	0	0	30.0	0
Scenedesmus bijuga	30.0	0	0	30.0	0	30.0	0
Ankistrodesmus falcatus	131.9	99.7	90.8	133.3	133.3	51.3	56.1
Oocystis sp.	87.1	72.7	47.4	57.1	72.0	85.7	49.5
Closteriopsis sp.	23.3	30.0	30.0	20.0	30.0	0	0
Wostella sp.	0	90.0	30.0	0	60.0	0	0
Kirchneriella lunaris	0	30.0	0	0	0	30.0	0
Cosmarium sp.	30.0	30.0	37.5	0	30.0	30.0	45.0
Tetraedron sp.	30.0	30.0	30.0	0	30.0	0	30.0
Golenkinia radiata	0	30.0	0	0	0	30.0	0

Selenastrum sp.	0	0	0	0	0	90.0	0
Pediastrum sp.	30.0	90.0	0	0	30.0	30.0	0
Dictyosphaerium sp.	30.0	30.0	0	0	30.0	30.0	30.0
Actinastrum sp.	30.0	0	0	0	30.0	30.0	0
Gloeocystis sp.	40.0	0	80.0	60.0	0	0	0
Coelastrum sp.	30.0	0	30.0	30.0	0	30.0	30.0
Elaketothrix gelatinosa	30.0	30.0	30.0	30.0	0	0	30.0
Lagerheimia sp.	30.0	30.0	30.0	0	30.0	30.0	30.0
Micractinium sp.	40.0	30.0	45.0	0	30.0	45.0	30.0
Crucigenia sp.	90.0	87.5	32.0	26.7	30.0	75.7	67.5
Staurastrum sp.	0	30.0	0	0	30.0	0	0
Flagellates							
Dinobryon sp.	219.1	361.1	205.5	412.1	403.1	135.5	162.6
Cryptomonas ovata	105.6	99.2	79.6	79.7	116.0	98.2	82.8
Cryptomonas erosa	50.0	42.7	58.6	58.0	56.3	30.0	30.0
Chlamydamonas sp.	60.0	30.0	20.0	48.3	0	0	0
Chrysococcus sp.	0	0	20.0	20.0	0	0	0
Trachelomonas sp.	0	30.0	30.0	30.0	0	0	0
Phacus sp.	0	40.0	0	40.0	0	0	0
Peridinium sp.	25.0	0	26.7	20.0	30.0	0	0
Mallomonas sp.	39.0	30.0	42.0	20.0	30.0	38.6	32.3
Ceratium hirundinella	30.0	30.0	30.0	0	0	28.2	39.7
Gonium sp.	30.0	0	0	0	0	30.0	0
Euglena sp.	40.0	25.0	0	43.9	30.0	0	0
Misc. flagellates	1,094.1	875.1	740.6	714.6	936.8	765.7	1,079.4
Cryptomonas sp.	502.1	431.5	474.1	365.2	443.3	485.1	563.5

Table VIII. Summary of 1977 Lake Michigan Assemblange and Tallies of Phytoplankton (number of taxon occurrences)

Genera spp.	Nearshore Tally	Offshore Tally	Open Lake Tally	Tally Total	April	June	August	September
Cyanophyton								
Anacystis sp.	18	30	36	84	8	5	34	37
Aphanothece sp.	12	22	27	61	0	0	34	27
Chroococcus sp.	10	12	19	41	1	0	22	18
Anabaena sp.	7	6	6	19	0	5	6	8
Oscillatoria limnetica	29	32	32	93	24	33	16	18
Gomphosphaeria lacustris	13	20	25	58	0	0	26	32
Coelosphaerium kuetzingianum	9	14	7	30	5	5	9	11
Schizothrix calcicola	7	11	8	26	11	16	7	2
Microcoleus lynbyaceus	15	9	2	26	4	9	6	7
Schizothrix sp.	4	4	5	13	2	8	0	3
Chlorophytes								
Quadrigula lacustris	23	27	26	76	15	23	24	13
Scene desmus quadricauda	9	8	2	19	4	5	7	5
Scenedesmus sp.	19	23	13	55	13	20	15	6
Scenedesmus dimorphus	2	1	0	3	0	2	1	0
Scenedesmus abundens	0	1	0	1	0	0	1	0
Scenedesmus bijuga	2	0	0	2	1	0	1	0
Ankistrodesmus falcatus	32	31	36	99	33	33	15	18
Oocystis sp.	14	22	19	55	7	5	23	20
Closteriopsis sp.	3	1	4	8	2	6	0	0
Wostella sp.	0	1	1	2	0	2	0	0
Kirchneriella lunaris	0	1	0	1	0	0	1	0
Cosmarium sp.	1	1	4	6	0	2	2	2
Tetraedron sp.	3	1	1	5	0	4	0	1
Golenkinia radiata	0	1	0	1	0	0	1	0

Species	1	2	3	4	5	6	7	8
Selenastrum sp.	0	1	0	1	0	0	1	0
Pediastrum sp.	4	0	0	4	0	2	2	0
Dictyosphaerium sp.	6	3	0	9	0	4	2	3
Actinastrum sp.	2	0	1	2	2	1	1	0
Gloeocystis sp.	1	0	2	2	1	0	0	0
Coelastrum sp.	1	1	1	3	2	0	1	1
Elaketothrix gelatinosa	1	1	3	3	0	0	0	1
Lagerheimia sp.	4	1	2	8	0	1	3	4
Micractinium sp.	3	4	10	6	3	1	4	1
Crucigenia sp.	6	1	0	20	0	2	7	8
Staurastrum sp.	0	1	0	1	0	1	0	0
Flagellates								
Dinobryon sp.	33	38	33	104	19	29	33	23
Cryptomonas ovata	34	39	48	121	30	25	34	32
Cryptomonas erosa	12	11	22	45	20	16	4	5
Chlamydamonas sp.	4	1	1	6	6	0	0	0
Chrysococcus sp.	0	0	2	2	2	0	0	0
Trachelomonas sp.	0	1	1	2	2	0	0	0
Phacus sp.	0	1	0	1	1	0	0	0
Peridinium sp.	2	0	3	5	2	3	0	0
Mallomonas sp.	10	10	10	30	1	2	14	13
Ceratium hirundinella	5	7	11	23	0	0	17	6
Gonium sp.	1	0	0	1	0	0	1	0
Euglena sp.	4	2	0	6	5	1	0	0
Misc. flagellates	39	45	54	138	33	34	35	36
Cryptomonas sp.	38	46	56	140	33	33	37	37

Table IX. Trophic Status Transition Values for Lake Michigan

| | Trophic Status Transition Values | |
| | Oligotrophic/ | Mesotrophic/ |
Source	Mesotrophic	Eutrophic
Dobson [18]		
Summer Total P (µg/l)	8	19
Chlorophyll *a* (µg/l)	2	5
Secchi Depth (m)	6	3
Rast and Lee [19]		
Annual Total Phosphorus (µg/l)	10	20
Summer Mean Epilimnion Chlorophyll *a* (µg/l)	2	6
Secchi Depth (m)	4.6	2.7
Upper Lakes Reference Group, 1976–Volume I:		
The Waters of Lake Huron and Lake Superior [17][a]		
Total Phosphorus (µg/l)	6.5	14.1
Chlorophyll *a* (µg/l)	2.4	7.8
Secchi Depth (m)	8.6	2.9
Aerobic Heterotrophs–nearshore		
($<$ 50 ft in depth or $<$ 2 miles from shore)	120	2000
Aerobic Heterotrophs–offshore		
($>$ 50 ft in depth or $>$ 2 miles from shore)	20	200

[a]Estimates of the midrange of each parameter made from Figure 7.1 of Reference 17.

increase was attributed to a narrower, more nutrient-enriched, bacterially productive, mesotrophic zone which was much closer to the shoreline than that observed in 1976. Bacteria are generally more numerous near the shoreline [20]. Transitory heterotrophic bacteria of a health-oriented significance (ie., fecal coliforms) showed a geometric mean value of 1/100 ml in both the oligotrophic and mesotrophic zones in 1976. At most open-lake stations, however, no detectable levels of fecal coliforms were observed. Coliform data for 1977 were not available for this study.

Lack of past data for comparison restricts our ability to determine any long-term trends in the bacterial water quality of Lake Michigan. However, the heterotrophic bacterial data compiled during this study correlates significantly with the other trophic status observations, which indicate a decrease in the organic loadings to the lake from 1976 to 1977.

Dissolved Reactive Silica

The only known significant mechanism of silica removal is incorporation into diatoms. Figure 18 compares the annual silica cycle to the surface waters at Lake Michigan for the years 1954, 1965 and 1976. There has been a significant decrease in silica in the surface waters since 1954. A comparison

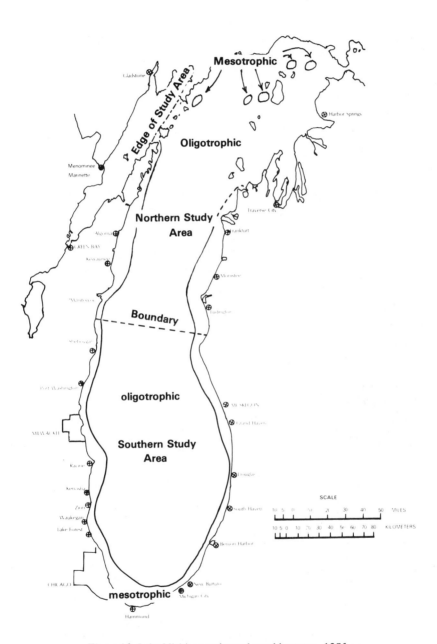

Figure 16. Lake Michigan estimated trophic status—1976.

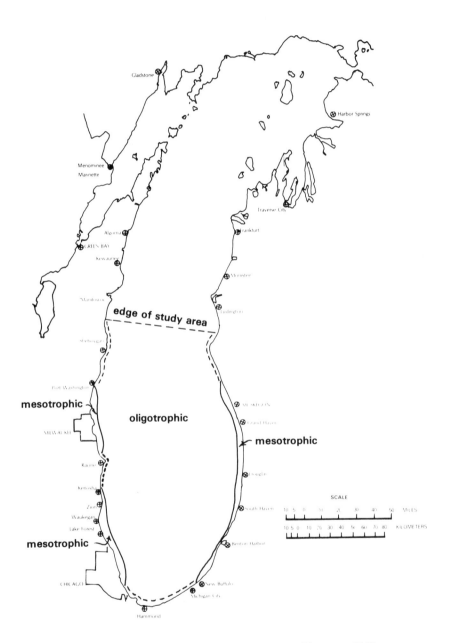

Figure 17. Southern Lake Michigan estimated trophic status–1977.

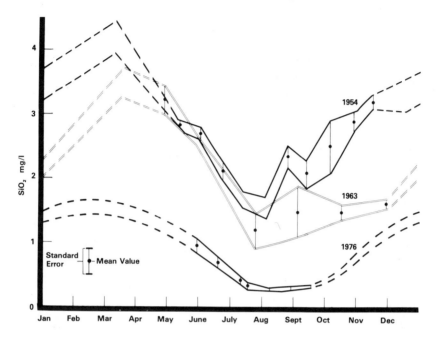

Figure 18. Dissolved reactive silica in Lake Michigan surface water—1954, 1963 and 1976.

of 1976 and 1977 annual dissolved reactive silica values on a station-by-station basis revealed an increase in annual silica levels at the nearshore stations, principally along the Illinois-Indiana shoreline. A significant increase in silica concentrations at the 95% confidence level was shown at 15% of the southern basin stations. These stations are within the contour illustrated in Figure 19. This represents an area of 2700 km² or about 9.3% of the southern basin.

DISCUSSION

The long term deterioration of water quality in the open lake is indicated by increasing conservative ion concentrations, increasing phytoplankton populations, and decreasing dissolved reactive silica concentrations. While the open-lake waters were still evaluated as oligotrophic [17-19] in 1976 and 1977, several mesotrophic-eutrophic phytoplankton species were abundant. Blue-green algae occurred in all areas of the lake, comprising as much as 46% of the total population in some areas. Historically, the blue-greens have comprised an insignificantly small portion of the Lake Michigan phytoplankton population [15,21].

Schelske and Stoermer [22] have suggested that continued nutrient

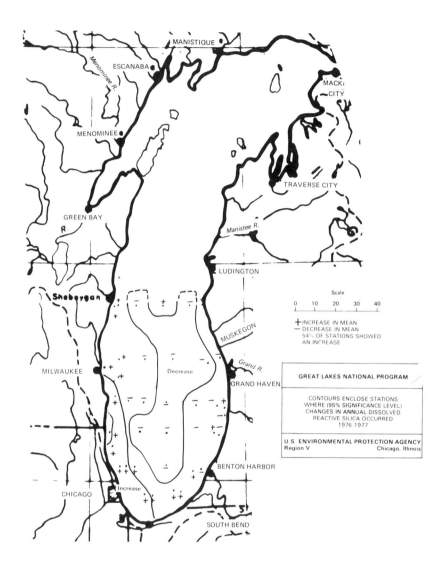

Figure 19. Volume-weighted changes in mean concentrations from 1976 to 1977.

enrichment and subsequent increased silica utilization by the diatom population could result in silica being depleted to such an extent that algal forms not requiring silica (i.e., green and blue-green algae) would increase in abundance. Our data suggests that through 1977, silica depletion was becoming more severe in the open-lake waters and blue-green algae have become more abundant.

Changes occurring over one year do not indicate a trend. However, the open-lake data were encouraging. Trophic conditions were substantially better in 1977 than in 1976 throughout much of the southern basin. Decreases in phosphorus, ammonia, chlorophyll-a, phytoplankton cell counts and increased silica concentration were observed in 1977. These improvements occurred throughout the southern basin. However, they were more pronounced in the nearshore zone of the southern and southwestern portions of the basin. This portion of the lake experienced significant (95% confidence level) increases in dissolved reactive silica concentration, as well as increased diatom population, in 1977.

Althouth our data did not cover the spring diatom pulse, other investigators [23,24] have reported marked decreases in both cell numbers and biomass in the Chicago nearshore area. Lower diatom populations in the early spring may partially explain the increased silica concentrations observed in 1977.

Data from the Chicago Water Filtration Plant [25] indicated that improvements in water quality in the nearshore zone may have begun prior to our 1976 study. Figure 20 illustrates the decreases in the amounts of chlorine and carbon required to produce potable water that have occurred since the mid-1970s. The phosphorus times-series (Figure 6) from the plant indicates a leveling-off of the increasing concentration in 1973, with definite decreases beginning in 1975. Phytoplankton populations also began decreasing in 1973, particularly among the eutrophic diatoms such as *Stephanodiscus binderanus* [23,24] .

Several factors appear to have combined to produce the improvements in water quality, particularly in the nearshore zone. In 1972, Indiana passed phosphate detergent ban legislation. This reduced detergent phosphorus content in two stages to 0.5% (by weight) in 1973. The timing of this ban coincides with the leveling-off of phosphorus concentrations and decreases in phytoplankton biomass at Chicago. The 1976 ammonia data (Figure 9) suggests that the decreases in the phosphorus imputs from Indiana did not result from better treatment of wastewater.

Another factor in the improved water quality was the diversion of eight municipal wastewater treatment plants in Lake Country, Illinois, in 1975 and 1976 (Figure 6). The diversion of these plants can be expected to provide long-term relief as long as treatment at other facilities does not deteriorate or become negated by increased population.

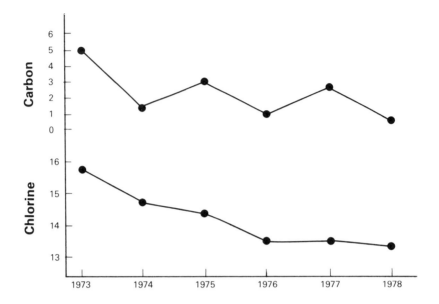

Figure 20. Chicago Water Purification Division South Water Filtration Plant–dosage, pounds/million gallons/yr.

Unfortunately, the largest contribution to the open-lake phosphorus reduction could have been the severe winter of 1976-1977, which may have temporarily augmented the effects of phosphorus bans and treatment plant diversions. The heavy ice cover may have assisted in nutrient removal through plankton die-off and subsequent sedimentation. Ice cover would have prevented resuspension of sediments and the recycling of some of the phosphorus attached to the sediments.

Although nutrient removal has contributed to the slowing or partial reversal of cultural eutrophication in Lake Michigan, potential problems are appearing. Present concentrations of chlorides and sulfates are well below the drinking water standard of 250 mg/l. However, there remains the unpleasant possibility that increased concentrations of conservative ions may lead to fundamental, possibly irreversible, changes in the lake's natural biological components. While the evidence is not yet complete, certain diatoms, blue-green algae, and flagellate species *may* presently be increasing as a result of increasing conservative ion concentrations. The extent, severity and desirability of these changes, should they occur, are not known.

ACKNOWLEDGMENTS

The authors would like to express their gratitude to Dr. Eugene Stoermer and Dr. William Danforth for furnishing both expert advice and data, to

Mr. Terry Moan who performed much of the phytoplankton data, and to Ms. Carole Bell for typing the manuscript.

REFERENCES

1. "Methods for Chemical Analysis of Water and Wastes," U.S. Environmental Protection Agency, Cincinnati, OH (1974).
2. *Standard Methods for the Examination of Water and Wastewater,* Fourteenth Edition, (Washington, DC: American Public Health Association, 1975).
3. "Biological Field and Laboratory Methods," U.S. Environmental Protection Agency, Cincinnati, OH (1973).
4. Stoermer, E. F., University of Michigan, Ann Arbor MI. Personal Communication (1978).
5. Zar, J. H. *Biostatistical Analysis* (Englewood Cliffs, NJ: Prentice-Hall Inc., 1974).
6. Ackermann, W. C., R. H. Harmeson and R. A. Sinclair. "Some Long-Term Trends in Water Quality of Lakes and Rivers", *Am. Geophys. Union* 15:516-522 (1970).
7. Doneth, J. et al. "Material Usage in the U.S. Great Lakes Basin," Report Prepared for Task B, Pollution from Land Use Activities Reference Group, International Joint Commission, (Windsor, Ontario 1975)
8. Fuchs, R. J. "Trends in the Use of Inorganic Compounds in Home Laundry Detergents in the United States," *Chem. Times Trends* pp. 36-41 (1978).
9. Beeton, A. M. "Eutrophication of the St. Lawrence Great Lakes," *Limnol. Oceang.* 10:240-254 (1965).
10. "Conference on Pollution of Lower Lake Michigan, Calumet River, Grand Calument River, Little Calumet River, and Wolf Lake, Illinois and Indiana. Water Quality in the Calumet Area". Technical Committee on Water Quality (1970).
11. Chapra, S. C. "Total Phosphorus Model for Great Lakes," *J. Environ. Eng. Div., Am. Soc. Civil Eng.* 103:147-161 (1977).
12. Stoermer, E. F. and M. L. Tuchman. "Phytoplankton Assemblage of the Nearshore Zone of Southern Lake Michigan," U.S. Environmental Protection Agency (in press).
13. Stoermer, E. F., M. M. Bowman, J. C. Kingston and A. L. Schaedel. "Phytoplankton Composition and Abundance in Lake Ontario During IFYGL," U.S. Environmental Protection Agency, Corvallis, Oregon (1975).
14. Stoermer, E. F. and J. J. Yang. "Distribution and Relative Abundance of Dominant Plankton Diatoms in Lake Michigan," Publication #16, Great Lakes Research Division, University of Michigan, Ann Arbor, Michigan (1970)

15. Tarapchak, S. J. and E. F. Stoermer. "Environmental Status of the Lake Michigan Region. 4. Phytoplankton of Lake Michigan," Argonne National Laboratory, Argonne, Illinois (1976).
16. Rao, S. S., and C. G. M. Berkel. "Bacterial Index Ratio: A Novel Water Quality Assessment Technique," *J. Great Lakes Res.* 4:106-109 (1978).
17. Upper Lakes Reference Group. "The Waters of Lake Huron and Lake Superior, Summary and Recommendations," International Joint Commission, Windsor, Ontario (1976).
18. Dobson, H. F. H. "Eutrophication Status of the Great Lakes," Canada Centre for Inland Waters, Burlington, Ontario. Unpublished.
19. Rast, W., and G. F. Lee. "Summary Analysis of the North American OECD Eutrophication Project: Nutrient Loading-Lake Response Relationships and Trophic Status Indices," U.S. Environmental Protection Agency, Corvallis, OR, Ecological Research Series Report EPA 600/3-78-008 (1978).
20. Taylor, C. B. "Bacteria of Fresh Water. I. Distribution of Bacteria in English Lakes," *J. Hyg.* 40:616-640 (1940).
21. Damann, K. E. "Plankton Studies of Lake Michigan. II. Thirty-three Years of Continuous Plankton and Coliform Bacteria Data Collected From Lake Michigan at Chicago, Illinois," *Trans. Micro. Soc.* 79(4): 397-404 (1960).
22. Scheleski, C. L., and E. F. Stoermer. "Eutrophication, Silica Depletion, and Predicted Changes in Algae Quality of Lake Michigan," *Science* 173:423-424 (1971).
23. Makarewicz, J. C. State University of New York, Brockport, New York. Personal Communication (1979).
24. Danforth, W. F., and W. Ginsburg. "Recent Changes in the Phytoplankton of Lake Michigan near Chicago," (in press).
25. Costello, J., Purification Division, South Water Filtration Plant, Chicago, Illinois. Personal communication (1979).

SECTION II

MODELING OF PHOSPHORUS DYNAMICS
IN THE GREAT LAKES

CHAPTER 6

APPLICATION OF THE PHOSPHORUS LOADING
CONCEPT TO THE GREAT LAKES

S. C. Chapra

Great Lakes Environmental Research Laboratory
National Oceanic and Atmospheric Administration
Ann Arbor, Michigan

INTRODUCTION

The phosphorus loading concept provides various models that predict lake trophic state (see Reckhow [1] for a review). Although some of these models have a theoretical basis [2,3], they all are essentially statistical in that they represent generalizations drawn from observations, and thus, variability can be associated with their predictions [4]. However, since the models are typically derived from data for many lakes (cross-sectional samples), rather than from many measurements for a single lake (time-series or longitudinal samples), the total prediction uncertainty can be treated as two independent components (Figure 1). The first component, called "perceptual error," reflects man's ability to observe the true trophic state of a particular lake and is primarily a composite of measurement error (both of the analytical technique and of the sampling design) and year-to-year variability (primarily because of meterological differences). The second component, termed "lake uniqueness," is related to the fact that lakes are not chemostats. In other words, even if two lakes have very similar characteristics in the context of the phosphorus loading concept (i.e., equal loading, residence time, depth, etc.), they will differ because of biological, chemical and physical factors not considered in the simple models (see Shapiro [5] for a stimulating exposition of this thesis).

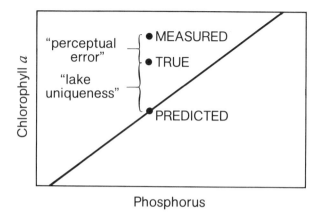

Figure 1. A P/chlorophyll plot showing two components of prediction uncertainty. Note that the uncertainty due to perceptual error will be reduced by additional measurements, but uncertainty due to natural differences between lakes will be maintained at a constant level.

The significance of distinguishing between these components is that, even though an infinite number of measurements would reduce the perceptual error (of the mean) for a particular lake to zero, that single lake would continue to diverge from the ideal lake represented by the model prediction. Although an analysis of variance to separate these effects has yet to be performed, evidence from individual lakes [5] suggests that divergence from statistical relationships due to lake uniqueness can be significant.

For the above reasons, it is assumed that direct application of the phosphorus loading concept to the Great Lakes would include uncertainty because of their uniqueness. Since management decisions for a system of such magnitude can result in expenditures of billions of dollars, it was considered prudent to minimize this uncertainty by using many of Vollenweider's concepts to derive a model exclusively from Great Lakes data. In other words, the present exercise describes a "Great Lakes-specific" phosphorus loading model.

This work was performed in support of the Fifth Year Review of the Canada-United States Great Lakes Water Quality Agreement. As such, it represents the state-of-the-art and the state-of-the-data in late 1977. In addition, because the Fifth Year Review emphasized problem areas, the present exercise is limited to Lakes Erie and Ontario, which, aside from such localized segments as Saginaw Bay in Lake Huron and Green Bay in Lake Michigan, have been more severely eutrophied than the other Great Lakes. Applications to the entire system [6] and to localized areas [8] are presented elsewhere.

MODEL DESCRIPTION

Whereas previous attempts to apply the phosphorus loading concept to the Great Lakes [8,9] have dealt exclusively with the dynamics of total phosphorus, the model has been extended to predict several variables that more directly represent the deleterious effects of eutrophication. The approach, as schematized in Figure 2, consists of a number of submodels and

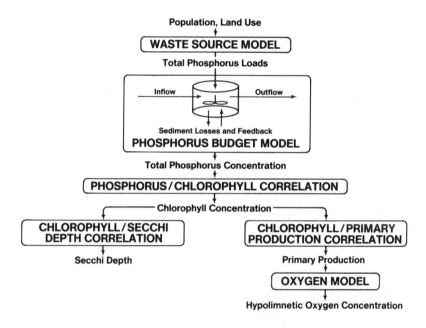

Figure 2. Schematic of the P loading model for the Great Lakes. The approach consists of a number of submodels and correlations that form a hypothesized causal chain starting with human development of the drainage basin and ending with such symptoms of eutrophication as Secchi depth and hypolimnetic dissolved oxygen.

correlations that form a hypothetical causal chain starting with human development of the drainage basin and ending with such symptoms of eutrophication as reduced Secchi depth and depletion of hypolimnetic dissolved oxygen. Each step in the chain is discussed below.

Waste Source Model

The waste source model is used to calculate phosphorus loadings on the basis of variables reflecting the cultural and natural characteristics of each lake's drainage basin. In the present exercise this step was bypassed because direct measurements of loading for the period of the simulation are available (Table I).

Table I. Parameters for Total Phosphorus Budget Model of the Lower Great Lakes

Parameter	Symbol	Units	Lake Erie Western	Central	Eastern	Lake Ontario
Mean Depth	z	m	7.6	17.8	27.0	86.0
Surface Area	A_s	km^2	3680	15,300	6150	19,000
Volume	V	km^3	28	274	166	1634
Outflow (1970-76)	Q	km^3/yr	198	207	212	252
(Long Term)			171	178	182	212
Bulk Diffusion Coefficient	E'	km^3/yr	140	490		
P Concentration (1970-76)	p	$\mu g/l$	39.3	19.4	17.2	21.0
Loading (1970-76)	W	t/yr	16,500[a]	4900	2100	10,500[b]
(Base)			14,499[a]	4007	1463	8613[b]
Feedback	F	t/yr	800	5200	1400	1400
Apparent Settling Velocity						
(No Feedback)	v	m/yr	60.3	29.0	23.3	22.2
(Feedback)	v'	m/yr	65.8	46.4	36.6	25.7

[a]Includes the Lake Huron outflow.
[b]Does not include Lake Erie contribution.

Phosphorus Budget Model

The budget model is designed to calculate in-lake phosphorus concentration as a function of loading on an annual time frame; it ignores within-year variability. Mathematically, the model consists of a set of coupled differential equations that result when mass balances are written around each of the major sections of the Great Lakes idealized as completely mixed systems (see Figure 3 for the segmentation scheme for the lower Great Lakes). Each equation accounts for the major sources and sinks of phosphorus, such as loading, exchange between lakes and sediment-water interactions. In an earlier version [8], sediment losses were parameterized with an apparent settling velocity of 16 m/yr, which was derived from budget data collected in lakes in southern Ontario [3]. This value worked adequately when compared with Great Lakes data available several years ago; however, recent information permits a more refined estimate, using a mass balance to calculate an apparent settling velocity for each segment. A general phosphorus budget for a completely mixed segment can be written as follows:

$$V\frac{dp}{dt} = W + Q_b p_b - Qp + E'(p_b - p) - vA_s p + F \qquad (1)$$

Figure 3. Segmentation scheme for the lower Great Lakes.

where
V = segment volume
p = mean annual total phosphorus concentration of the segment
t = time
W = rate of mass loading of total phosphorus to the segment
Q_b = advective water flow from an adjacent segment
P_b = total phosphorus concentration of the adjacent segment
Q = advective water flow leaving the segment
E' = a bulk diffusion coefficient
v = apparent settling velocity
A_s = segment surface area
F = the rate of feedback of phosphorus from sediments of the lake

In cases where the segment is adjacent to more than one other segment (e.g., the central basin of Lake Erie has interfaces with both the western and eastern basins), additional diffusion terms can be included.

In the present analysis, two cases for sediment-water interactions are used (Figure 4), one calculated with measured feedback rates v' and the other with no sediment feedback v. The two cases are included because there are indications [10,11] that sediment feedback rates gradually diminish during lake recovery. Since the actual recovery would probably fall between the feedback and the no-feedback cases, both are used to give a range, rather than a single prediction, by solving Equation 1 for both cases. Equation 1 is written for each segment of the lower Great Lakes. Time variable solutions are determined with numerical methods, and steady-state solutions are

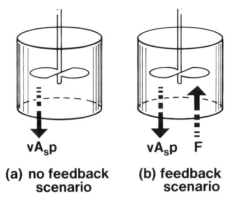

**(a) no feedback
scenario**

**(b) feedback
scenario**

SEDIMENT WATER INTERACTIONS

Figure 4. Two cases, (a) no feedback and (b) feedback, for the characterization of sediment-water interactions of total phosphorus.

determined with Thomann's approach [6,12]. In either case, the purpose of the phosphorus budget model is to transform loadings into in-lake concentrations of total phosphorus. These concentrations can, in turn, be related to other trophic variables by the following correlations.

Surface Water Quality Correlations

The correlations between the surface water quality variables (i.e., total phosphorus, chlorophyll-a, Secchi depth and primary production) are derived elsewhere [13]. The relationships (Figure 5) are:

$$\text{Chl-a} = 0.259 \ p \tag{2}$$

$$S = \frac{16.2}{1 + 0.783 \ \text{Chl-a}} \tag{3}$$

$$\text{Pr} = 420 \ (1 - e^{-0.148 \ \text{Chl-a}}), \tag{4}$$

where Chl-a = mean summer, surface (1-m depth), total (i.e., uncorrected for phaeopigments) chlorophyll concentration in offshore water

S = mean summer Secchi depth in offshore water

Pr = mean annual surface primary production

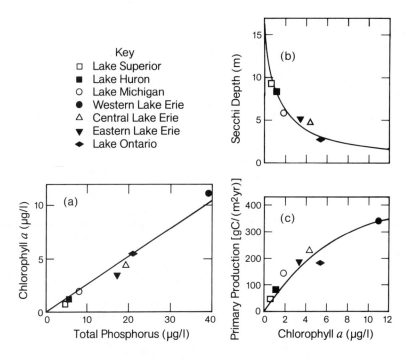

Figure 5. Correlations between surface water quality variables.

Hypolimnetic Oxygen Model

Oxygen Depletion

During the period of strong summer stratification, a lake can be idealized as two well-mixed layers of constant thickness separated by a region of minimal mixing—the thermocline. The total (i.e., uncorrected for turbulent exchange across the thermocline) volumetric oxygen depletion rate $D_{v,T}$ for a time period of length t can be calculated for the hypolimnion as

$$D_{v,T} = \frac{C_{h,o} - C_{h,t}}{t} \tag{5}$$

where $C_{h,o}$ and $C_{h,t}$ are the dissolved oxygen concentrations of the hypolimnion at the beginning and end of the time period, respectively. To separate the effects of organic oxidation and mixing requires an estimate of the turbulent exchange between the layers. This can be determined from a heat balance for the hypolimnion as follows:

$$V_h \frac{dT_h}{dt} = E'' (T_e - T_h) \tag{6}$$

where V_h = the volume of the hypolimnion
T_h = the temperature of the hypolimnion
t = time
T_e = the temperature of the epilimnion
E'' = vertical bulk diffuse coefficient

If the epilimnion temperature is assumed to be constant during the period of the calculation (i.e., $T_e = \bar{T}_e$), Equation 6 can be solved as follows:

$$T_{h,t} = T_{h,o}\, e^{\left(\dfrac{-E''}{V_h}\,t\right)} + \bar{T}_e \left(1 - e^{\left(\dfrac{-E''}{V_h}\,t\right)}\right) \tag{7}$$

where $T_{h,o}$ and $T_{h,t}$ are the hypolimnion temperatures at the beginning and end of the time period, respectively, Equation 7 can be rearranged to yield:

$$E'' = \frac{V_h}{t}\ln\frac{T_{h,o} - \bar{T}_e}{T_{h,t} - \bar{T}_e} \tag{8}$$

A mass balance for hypolimnetic dissolved oxygen can then be written as:

$$V_h\frac{dc_h}{dt} = E''\,(C_e - C_h) - V_h D_v \tag{9}$$

where C_h and C_e are the oxygen concentrations of the hypolimnion and epilimnion, respectively, and D_v is the corrected volumetric rate of oxygen depletion (i.e., due to oxidation alone), which can be related to the areal depletion rate D_a by:

$$D_a = z_h D_v \beta \tag{10}$$

where z_h is the mean thickness of the hypolimnion and β (0.1) is a correction factor. If the epilimnion concentration is assumed to be a constant $\bar{C}_e$ for the calculation period, Equation 10 can be solved as:

$$C_{h,t} = C_{h,o}\, e^{-\dfrac{E''}{V_h}\,t} + \left(\bar{C}_e - \frac{V_h}{E''}\,D_v\right)\left(1 - e^{-\dfrac{E''}{V_h}\,t}\right) \tag{11}$$

where $C_{h,o}$ and $C_{h,t}$ are the hypolimnion concentrations at the beginning and end of the time period, respectively. Equation 11 can then be rearranged to calculate the corrected volumetric depletion rate,

$$D_v = -\frac{E''}{V_h} \frac{C_{h,t} - C_{h,o} \; e^{\left(-\frac{E''}{V_h} t\right)} - \bar{C}_e \left(1 - e^{-\frac{E''}{V_h} t}\right)}{1 - e^{\left(-\frac{E''}{V_h} t\right)}} \qquad (12)$$

Relationship of Depletion to Surface Production

The relationship of oxygen deficits to lake productivity dates to the investigations of Birge and Juday [14] and Thienemann [15]. Although the mechanisms governing the relationship are complex, it is generally accepted that the areal deficit rate of an autotrophic lake of moderate depth increases with increasing productivity, since the deficit is primarily the result of heterotrophic assimilation of organic matter that has fallen from the trophogenic zone [16]. As a first approximation, it is assumed that the relationship is linear, that is:

$$D_a = \theta \, \alpha \, Pr \qquad (13)$$

where D_a = the areal oxygen depletion rate
θ = the efficiency of the hypolimnetic oxidation of organic matter derived from the surface
α = a correction factor (equal to 0.029)

This correction factor keeps the units consistent and accounts for the conversion of carbonaceous material to oxygen demand if the oxidation is idealized [17] as:

$$C_{106} H_{263} O_{110} N_{16} P_1 + 138 \, O_2 \qquad (14)$$

Equations 10, 11 and 13 can then be combined to determine the minimum concentration of hypolimnetic oxygen (i.e., just before overturn) as:

$$DO_{min} = C_{h,o} \; e^{\left(-\frac{E''}{V_h} t_s\right)} + \left(\bar{C}_e - \frac{V_h \, \theta \alpha Pr}{E'' z_h \beta}\right) \left(1 - e^{-\frac{E''}{V_h} t_s}\right) \qquad (15)$$

where t_s is the period of thermal stratification.

Using data (Table II and Figure 6) and the above equation yields the following relationship for central Lake Erie:

$$DO_{min} = 10.22 - 0.054 \, Pr \qquad (16)$$

Table II. Data from 1967 Through 1972 and Calculations of Oxygen Depletion for
the Central Basin of Lake Erie During the Stratified Period

t (June 1 to September 2)	3.07 months
$C_{h,o}$	10.72 mg/l
$C_{h,t}$	0.60 mg/l
$D_{v,t}$ (Equation 5)	3.30 mg/l-mo
V_h	40 km^3
$T_{h,o}$	7.77°C
$T_{h,t}$	12.50°C
$\bar{T}_e$	19.94°C
E'' (Equation 8)	6.42 km^3/mo
$\bar{C}_e$	9.66 mg/l
D_v (Equation 12)	4.01 mg/l-mo
z_h	4 m
D_a (Equation 10)	1.60 mg/cm^2-mo
t_s (June 1 to September 30)	4 mo
Pr (Equation 4)	220 g C/m^2-yr
θ	0.25
DO_{min}	10.22 − 0.054 Pr

SIMULATIONS

Rather than reporting numerous simulation results, select examples are
included to demonstrate the model's use in a management context. In all
cases, since the exercise is intended to determine the average response to
future control measures, long-term average water flows are used. This is
particularly important since the period for which the model was calibrated
(early 1970s) was inordinately wet. In a similar fashion, base loads that
reflect average conditions of land runoff were used (Table I).

Results for Lake Erie

Figure 7 presents total phosphorus predictions (for both the feedback and
the no feedback cases) as a function of loading, along with additional scales
showing the response of the other trophic state variables. Note that because
of the present waste management strategy, the Lake Erie load is given for
the whole lake, rather than by basin.

Figure 7 can be used in two ways. First, it shows the loading necessary
to meet a prescribed water quality objective. For example, if it were desirable
to raise critical oxygen levels in the central basin to 0 mg/l, loading would

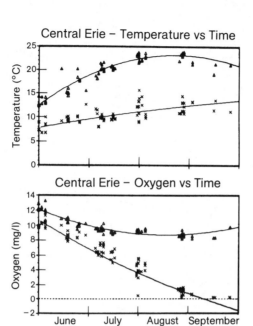

Figure 6. Temperature and oxygen concentration for central Lake Erie from 1967 through 1972 for the summer stratified period (June through September). Triangles designate surface data (1-m depth) and x's designate bottom data ($\geq$ 22-m depth).

have to be reduced to approximately 16,000-18,000 metric ton/yr. Alternatively, suppose that economic considerations dictated a loading of 13,000 metric ton/yr. The resulting central basin oxygen concentration would range from 0.9 to 2.5 mg/l.

Results for Lake Ontario

The loading design for Lake Ontario is complicated by the fact that it is so significantly affected by Lake Erie. One way to present model results for such a coupled system is as in Figure 8, which gives the total phosphorus concentrations in Lake Ontario resulting from combinations of loading levels for the two lakes. Thus, the plot shows that if it were decided that the Lake Erie and Ontario loads were to be reduced to 13,000 and 4000 metric ton/yr, respectively, a concentration of approximately 10 μg/l would result.

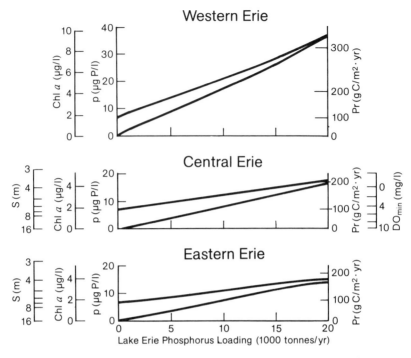

Figure 7. Calculated response of the three basins of Lake Erie to loading.

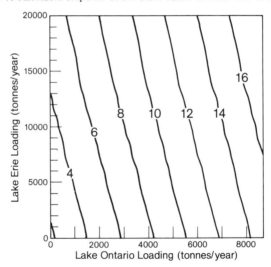

Figure 8. Calculated in-lake concentration of total phosphorus in Lake Ontario (the contours) as a function of loading for Lakes Erie and Ontario (average of feedback and no feedback cases).

Results for the Lower Lakes as a System

A more general presentation of model results than the foregoing are the matrices in Table III. The top section (Table III) is a spatial system response

Table III. System Response Matrices for the Lower Great Lakes for the
No-Feedback Case (Loading Change = 1000 Metric Tons/Yr)

	Lake Response[a] (μg P/l)			
	Western Erie	Central Erie	Eastern Erie	Lake Ontario
Western Erie	2.41	0.35	0.17	
Central Erie	0.78	1.15	0.55	
Eastern Erie	0.57	0.84	1.84	
Lake Ontario	0.16	0.24	0.53	1.58

	90% Response Time[b]			
	Western Erie	Central Erie	Eastern Erie	Lake Ontario
Western Erie	0.2	1.0	1.3	
Central Erie	1.0	0.9	1.3	
Eastern Erie	1.3	1.3	0.9	
Lake Ontario	6.7	6.6	6.3	5.9

[a]Spatial system response matrix. Each element represents the change in each segment's concentration due to a unit loading change in any other segment.
[b]Temporal system response matrix. Each element represents the time required for each segment to reach 90% of a new steady state following a step change in loading to any other segment.

matrix that gives the steady-state response of each segment to a 1000 metric ton/yr load change in any other segment [12]. Thus, a 1000 metric ton/yr loading change in western Erie would result in concentration changes of 2.41, 0.78, 0.57 and 0.16 μg/l in the western, central and eastern basins of Lake Erie and in Lake Ontario, respectively.

Table III also shows a temporal system-response matrix which gives the time required for each segment to reach 90% of a new steady-state condition following a step change in the load to any other segment. Thus, it would take Lake Ontario 6.7 years to achieve 90% of an improvement of its water quality following a loading reduction to western Lake Erie.

The preceding examples were intended to show how the model can be used in one part of the decision-making process; namely, the prediction of lake response to changes in loads. In all such applications, it is imperative that the model's limitations be made clear to managers.

DISCUSSION

In recent years, controversy has developed around the use of mathematical models for lake water quality management. In the extremes, the argument pits those who oversell models as infallible predictors of lake water quality against those who throw up their hands and contend that limnological systems are much too complex to be characterized quantitatively. As is usually the case, the truth lies somewhere between the extremes.

For example, models should never be represented as providing the complete answer to a water quality question. In fact, models typically address but one facet of the problem (i.e., the prediction of the response of a water body to external changes). There are a whole host of social, economic and political considerations that also must be brought to bear on such decisions. In addition, even in the narrower context of predicting a lake's response, models are, by definition, idealizations and as such they represent an approximation of reality. In this sense, model predictions should always be supplemented and interpreted by experts with a strong scientific understanding of the subtleties that might be beyond the scope of the idealized model.

At this point, an antimodeler might suggest that models be excluded altogether and that managers rely exclusively on the judgment of the expert. Although this is an extreme viewpoint, it has some merit in the sense that there may be gifted limnologists whose conceptual understanding of lake ecosystems are more advanced than the state-of-the-art computer models. However, there are two reasons why models are needed in spite of this.

First, no matter how astute, no expert can muse or speculate in more than one or two significant figures. In other words, when it comes down to deciding how many metric tons of phosphorus to remove from the Detroit treatment plant, everyone must take up pencil and pad (or digital computer) and make a calculation; and that, in the present context, is all a model really is—a calculation.

The second way in which models are necessary is by serving as extensions of intuition by using mathematics to organize the voluminous information of a complex problem into manageable terms. That is, there are some systems that are so complicated that human comprehension or intuition, however subtle, tends to break down. For example, the effect of the Detroit waste treatment plan on the water quality of Lake Ontario is not intuitively obvious.

The present paper has described a modeling approach that I hope will contribute to the rational management of Great Lakes eutrophication. In this context, one of its major strengths is its simplicity. All too often, models are so complex that their credibility becomes, to a certain degree, a matter of faith. Consequently, arguments over their use, as described at the onset of this discussion, often take an extreme form. It is to be hoped that the present model is straightforward enough that it will be neither oversold nor underestimated. To further insure a balanced appraisal of the results, the remainder of the discussion is devoted to major shortcomings of the approach.

Data Quality

Although all models are based on observations, the highly empirical nature of the present approach makes it especially dependent on the quality of the data from which it was developed. This is compounded by the fact that, because of their immensity, the Great Lakes are difficult to measure. In particular, lake phosphorus loads are extremely hard to estimate. For this reason, it must be reiterated that the model is a reflection of the state-of-the-data that was available in 1977. Better observations in the future could modify and improve the present results.

Uncertainty

Although the present approach is statistical, it is deterministic in that a single prediction results from any set of independent variables. Recently, it has been suggested that probabilistic information accompany these predictions [4]. Although additional research is needed to definitively assess such uncertainty, an analysis of Lake Ontario data from 1969 through 1978 has yielded an estimate of the standard error of from 1-2 μg P/l for predictions using the present approach [18]. If this is taken to be a valid estimate, it means that if a manager wants to be 90% certain of attaining a total phosphorus water quality goal in Lake Ontario, he will have to design his loading reduction program for an inlake concentration that is approximately 1-2 μg/l lower than the goal. Even though the above estimate is somewhat tentative, it is valid to draw the general conclusion that the consideration of uncertainty means that more stringent loading reductions will be needed to meet water quality objectives with greater than 50% certainty.

Availability

The present model assumes that water quality is directly proportional to the lake's total phosphorus level. Recently, Schaffner and Oglesby [19,20] have presented a modeling approach which considers that all of the phosphorus may not be available for algae assimilation. Although this approach is unverified (in the sense that it has not been tested on an independent data set) and possibly specific (in that it was developed for 13 lakes in New York State), its general implications for Great Lakes management should be explored.

The major difference between the Schaffner-Oglesby approach and total phosphorus models (Table IV) is in the calculation of phosphorus loads. In

Table IV. Comparison of Loading Factors for Total Phosphorus Models vs "Biologically Available Phosphorus" Using the Approach of Oglesby-Schaffner [19] $(mg P/m^2/yr)$

	Total Phosphorus Models	Oglesby-Schaffner Approach
Point Sources	(same)	
Atmospheric Sources	(same)	
Urban Land Runoff	(same)	
Forest Land Runoff	11.7	8.3
Active Agriculture	46	13.3
Forest and Pasture (or inactive agriculture)	23.3	

particular, a smaller portion of agricultural land runoff is considered to be available. This generally increases the desirability of point source control, which would make up a more substantial fraction of the part of the loading determining lake quality. Further, since less loading would affect lake quality, the Schaffner-Oglesby approach makes prospects for lake rehabilitation more optimistic, in that less treatment would be necessary to yield improvements.

In relation to the Great Lakes, it can be shown [18] that such conclusions are not very important for Lake Ontario, where land runoff sources do not represent a significant fraction of the loading. However, for Lake Erie and for nearshore areas near river mouths, the effects could be sizable. Additional research into the differentiation of total and available phosphorus and the development of dynamic available phosphorus models are needed to assess these effects.

Space and Time Scales

The use of an annual time frame and complete mixing obscures detail on the finer time and space scales that might be of interest to decision-makers. For example, peak phytoplankton values and nearshore concentrations represent extreme stresses of eutrophication that could be more meaningful to the public that could summer and whole-lake averages. Answers to these specific questions require a different modeling strategy.

Mortimer [21] has offered the aphorism that, when building lake models, "two lakes are better than one." In other words, to be truly valid, a model should be capable of broad cross-sectional application. To this, I would add that "2 years are better than one" in that a model must not only be calibrated for a single year of data, but should be tested or verified against additional years. Finally, both in a management and a research context, "two models are better than one." Several models, addressing different scales of the same phenomenon are valuable in the sense that their consistency lends support to their validity, and their differences are useful in learning more about their effectiveness in simulating the system. The present model is but one of several that have been developed for Great Lakes water quality simulation. It is to be hoped that the coordinated use and development of these models, as described in this volume, will further strengthen rational water quality decision-making for the Great Lakes.

ACKNOWLEDGMENTS

This paper is GLERL Contribution No. 182, Great Lakes Environmental Research Laboratory, National Oceanic and Atmospheric Administration, Ann Arbor, Michigan.

REFERENCES

1. Reckhow, K. H. "Quantitative Techniques for the Assessment of Lake Quality," Michigan Department of Natural Resources, Lansing, MI (1978).
2. Vollenweider, R. A. "Input-Output Models (with Special Reference to the Phosphorus Loading Concept in Limnology)," *Schweiz. Z. Hydrol.* 37(1):53-84 (1975).
3. Chapra, S. C. "Comment on 'An Empirical Method of Estimating the Retention of Phosphorus in Lakes' by W. B. Kirchner and P. J. Dillon," *Water Res.* 11(6):1033-1034 (1975).
4. Chapra, S. C., and K. H. Reckhow. "Expressing the Phosphorus Loading Concept in Probabilistic Terms," *J. Fish. Res. Bd. Can.* 36(2): 225-229 (1979).
5. Shapiro, J. "The Need for More Biology in Lake Restoration," Contribution No. 183, Limnological Research Center, University of Minnesota, Minneapolis, MN (1978).
6. Chapra, S. C., and W. C. Sonzogni. "Great Lakes Total Phosphorus Budget for the Mid-1970's," *J. Water Poll. Control Fed.* (in press).
7. Chapra, S. C. "Applying Phosphorus Loading Models to Embayments," *Limnol. Oceanog.* 24(1):168-171 (1969).
8. Chapra, S. C. "Total Phosphorus Model for the Great Lakes," *J. Eng. Div., Am. Soc. Civil Eng.* 103(EE2):147-161 (1977).

9. Chapra, S. C., and A. Robertson. "Great Lakes Eutrophication: The Effect of Point Source Control of Total Phosphorus," *Science* 196(4297):1448-1450 (1977);

10. Ahlgren, I. In: *Interactions Between Sediments and Freshwater,* H. L. Golterman, Ed. (Wagenigen, Netherlands: PUDOC, 1977), p. 372.

11. Lorenzen, M., D. J. Smith and L. V. Kimmel. In: *Modeling Biochemical Processes in Aquatic Ecosystems*, R. P. Canale, Ed. (Ann Arbor, MI: Ann Arbor Science Publishers, Inc., 1976), p. 75.

12. Thomann, R. V. "Mathematical Model for Dissolved Oxygen," *J. San. Eng. Div., Am. Soc. Civil Eng.* 89(SA1):1-30 (1963).

13. Chapra, S. C., and H. F. H. Dobson, "Quantification of the Lake Trophic Typologies of Naumann (Surface Quality) and Thienemann (Oxygen) with Special Reference to the Great Lakes," Great Lakes Environmental Research Laboratory, National Oceanic and Atmospheric Administration, Ann Arbor, Michigan (1979).

14. Birge, E. A., and C. Juday. "The Inland Lakes of Wisconsin. The Dissolved Gases of the Water and Their Biological Significance," *Bull. Wis. Geol. Nat. Hist. Survey* 22:1-259 (1911).

15. Thienemann, A. "Der Sauerstoff im eutrophen und oligotrophen See. Ein Beitrag zur Seetypenlehre." *Binnenzewässer* 4:1-175 (1978).

16. Hutchinson, G. E. *A Treatise on Limnology. I. Geography, Physics and Chemistry* (New York: John Wiley and Sons, Inc., 1957), p. 600 ff.

17. Stumm, W., and J. J. Morgan. *Aquatic Chemistry* (New York: John Wiley and Sons, Inc., 1972), p. 429.

18. Chapra, S. C. "Simulation of Recent and Projected Total Phosphorus Trends in Lake Ontario," Great Lakes Environmental Research Laboratory, National Oceanic and Atmospheric Administration, Ann Arbor, Michigan (1979).

19. Schaffner, W. R., and R. T. Oglesby. "Phosphorus Loadings to Lakes and Some of Their Responses. Part I. A New Calculation of Phosphorus Loading and its Application to 13 New York Lakes," *Limnol. Oceanog.* 23(1):120-134 (1978).

20. Oglesby, R. T., and W. R. Schaffner. "Phosphorus Loadings to Lakes and Some of Their Responses. Part 2. Regression Models of Summer Phytoplankton Standing Crops, Winter Total P and Transparency of New York Lakes," *Limnol. Oceanog.* 23(1):135-145 (1978).

21. Mortimer, C. H. In: *American-Soviet System on the Use of Mathematical Models to Optimize Water Quality Management*, U.S. Environmental Protection Agency, EPA-600/9-78/024, U.S. Government Printing Office, p. 316 (1978).

CHAPTER 7

DYNAMIC PHYTOPLANKTON-PHOSPHORUS MODEL OF LAKE ONTARIO: TEN-YEAR VERIFICATION AND SIMULATIONS

R. V. Thomann and J. S. Segna*

Environmental Engineering and Science
Manhattan College
Bronx, New York

INTRODUCTION

Only five years have passed since reports were presented on the initial work of constructing a dynamic mathematical model of the phytoplankton of Lake Ontario. Yet, since that time, there has been a considerable increase in the level of understanding of phytoplankton-nutrient interactions and in the availability of a long-term data base on lake water quality. Furthermore, there has been a continual interplay between the results of a variety of quantitative analyses, the prediction of lake status and the implementation of nutrient control programs. One result of this interaction between the decision-making community of the Great Lakes and the community of limnological analysts has been the new Great Lakes Water Quality Agreement of 1978 [1], which reaffirms and extends the earlier agreement of 1972. Some of the new target loads for phosphorus inputs to the lakes resulted from a detailed review and comparison of several earlier models and analyses [2].

This chapter is divided into three components: (1) a brief review of the

---

*Present address: Tetra Tech Inc., Melville, New York 11746.

past dynamic modeling effort; (2) a modeling analysis of 10 years of data on Lake Ontario, with detailed analysis of the verification status of the earlier model and an updated model; and (3) preliminary projections on a short-term and long-term basis using the updated model.

THE RECENT PAST

The emphasis in this chapter is on the dynamic model of phytoplankton and nutrients of Lake Ontario as developed at Manhattan College; not to the exclusion of other work, but rather to provide a defined focus for the analysis. The paper is not intended therefore to be a review of all work related to dynamic phytoplankton models of Lake Ontario. For example, considerable effort has been expended by the Great Lakes Environmental Research Laboratory of the National Oceanic Atmospheric Association and the Canada Centre for Inland Waters, both of which have furthered the understanding of phytoplankton seasonal dynamics in Lake Ontario. Neither effort, however, included the use of the models for year-to-year and long-term projections. A review of this other work and its contribution to increasing scientific understanding is given by Thomann et al. [3].

The first Manhattan Model (Lake 1) was a two-layered, horizontally well-mixed model of Lake Ontario, containing a phytoplankton-nutrient kinetic structure consisting of eight state variables [4]. The eight state variables were:

1. phytoplankton chlorophyll-a;
2. herbivorous zooplankton carbon;
3. carnivorous zooplankton carbon;
4. "slowly available" phosphorus (dissolved and particulate organic and particulate inorganic);
5. available phosphorus;
6. "slowly available" nitrogen;
7. ammonia nitrogen; and
8. nitrate nitrogen.

The model was calibrated to an average of four years of data (1967-1970) and the seasonal behavior of the phytoplankton system was described. The results indicated that spring growth and peak chlorophyll concentrations are related primarily to increasing light and temperature, and that phosphorus limitation essentially controls the spring peak. Complex interactions between zooplankton grazing and subsequent recycling of nutrients due to excretion and plankton respiration result in a late summer increase in phytoplankton biomass. The principal kinetic considerations that were later shown to be critical in making long-term projections were:

1. net loss rates of sinking phytoplankton and particulate phosphorus forms;
2. the relative availability of the "slowly available" phosphorus compounds; and
3. the form of the nutrient limitation term (i.e., multiplication of nutrient limitation terms versus choosing the minima of the nutrient limitation terms).

The model was then used to prepare a series of long-range projections on the efforts of nutrient reductions on phytoplankton levels [5-7]. Because of the uncertainty in the numerical values of some of the key parameters, the uncertainty of the kinetic structure and the fact that the lake had not been observed for a sufficient period at markedly different input load conditions, three kinetic scenarios were used: pessimistic, reasonable and optimistic. Subsequently, Bierman [8] showed that what was considered "reasonable" in the 1976 work was quite the contrary. Rather, the "optimistic" kinetic set appeared to be in better agreement with the subsequent lake behavior. This is further confirmed by the results presented below. The optimistic kinetic condition was constructed to keep the lake in equilibrium with a load of 12,400 metric ton/yr. The loss rate for both the slowly available phosphorus and the readily available phosphorus was set at 0.001/day and phytoplankton settling was set at 0.1 m/day. It was estimated from the Lake 1 model under optimistic kinetics, that the range in total phosphorus in the epilimnion would be from about 17 μg P/l at 12,400 metric ton/yr to 12 μg P/l at the 1972 Water Quality Agreement target load of 9000 metric ton/yr. Annual average phytoplankton would range from 4 μg/l chlorophyll-a/l to 3 μg chlorophyll-a/l for the same loads. Estimates were also made for a lower bound condition, termed the Pastoral Load. This load was estimated at 3400 metric ton/yr and lake response for phosphorus and phytoplankton was calculated at 5 μg P/l and 0.8 μg chlorophyll-a/l, respectively. All of this earlier work was based on the comparison of the Lake 1 model to the 1967-1970 data average. To extend the analysis further and to increase model credibility, a verification analysis was performed on 10 years of data and using an updated kinetic structure.

DATA ANALYSIS, 1967-1976, AND MODEL CALIBRATION

Data Base

A detailed review of the 1967-1976 data base is given by Segna [9]. The bulk of the lake data used for the 10-yr analysis was obtained by the Canada Centre for Inland Waters (CCIW). Although cruise tracks change from year to year, both in size (number of stations) and location, the CCIW cruises provide excellent spatial coverage for most of the surface of the lake, and for certain parameters, useful data were also available with depth. For this work, only main lake stations with sounding depths greater than 50 m are considered. The data are compiled by depth intervals corresponding to the model segmentation of 0-17 m (Segment 1) and greater than 17 m (Segment 2), and are then reduced to monthly averages with means and standard deviations for each parameter. The monthly averages were usually a single cruise average.

The variables used in the calibration and verification of the dynamic model are shown in Table I, which also indicates the years in which the variables were measured.

The most complete set of data obtained from CCIW were temperature data. Figure 1 shows 10 years of Lake Ontario temperature data and Figure 2 shows the mean monthly temperature values over all years. Maximum temperatures usually occur in August for the epilimnion, steadily decreasing until December, when the lake is usually completely mixed. Careful examination of Figure 1 indicates some significant variability in the year-to-year temperatures, resulting in different times of stratification and fall overturn. This variability of temperature therefore becomes important to the fluctuations of phytoplankton versus time.

Figure 3 is a long term plot of corrected chlorophyll-a from 1967 to 1976, and Figure 4 shows the mean monthly chlorophyll-a over the period 1967-1976. Earlier years (1967-1971) reported only uncorrected chlorophyll-a at a 1-m depth and were corrected to a 0-17 m average [9] using the following equation:

$$Ch_{0-17} = 1.1985 + .5714 \, Ch_{ul} \tag{1}$$

where Ch_{ul} is the uncorrected chlorophyll at 1-m (μg/l) and Ch_{0-17} is the corrected chlorophyll epilimnion average (μg/l). A second adjustment was also necessary for relating surface corrected chlorophyll to 0-17 m average, and was estimated by:

$$Ch_{0-17} = 1.034 + .755 \, Ch_1 \tag{2}$$

where Ch_1 is the corrected chlorophyll (μg/l) at 1-m depth.

The time series of Figure 3 indicates considerable variability from year to year, with some years showing peak values in May (1970) and others showing peak values in July (1972). There is an apparent consistent midsummer decline. Figure 4 shows that for the monthly average over all years, peak values occur in June-July at about 5.5 μg chlorophyll-a/l, and August values decline to about 4 μg/l. A subsequent fall bloom to greater than

Table I. List of Variables and Years CCIW Cruises Sampled on Lake Ontario

Parameter	Years Measured by CCIW											
	1965[a]	1966	1967	1968	1969	1970	1971	1972	1973	1974	1975	1976
Temperature	X	X	X	X	X	X	X	X	X	X	X	X
Chlorophyll	X		X	X	X	X	X	X	X		X	X
Total P	X		X	X	X	X	X	X	X	X	X	X
$NO_2 + NO_3$ (as N)	X		X	X	X	X	X	X	X	X	X	X
NH_3 (as N)	X		X	X	X	X	X	X	X	X		X
Dissolved Orthophosphate (as P)				X	X	X	X	X	X	X	X	X
Silica	X		X	X	X	X	X	X	X	X	X	X

[a] 1965 data were surface data only and not compiled for this work; not all depths were sampled in all years.

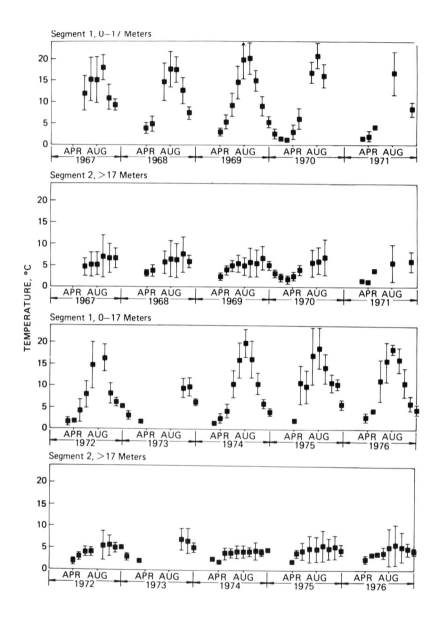

Figure 1. Temperature time series, Lake Ontario (1967-1976), mean ± 1 standard deviation (data courtesy of CCIW).

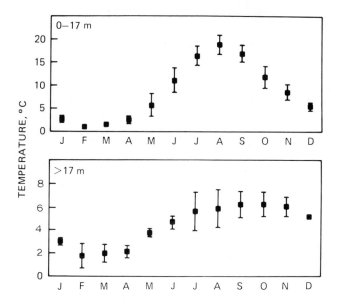

Figure 2. Mean monthly temperature over 1967-1976, mean ± 1 standard deviation (data courtesy of CCIW).

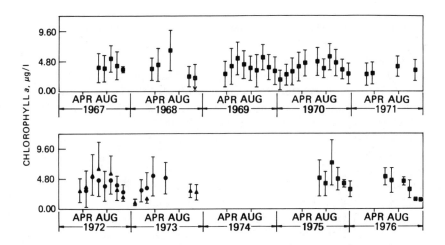

Figure 3. Chlorophyll-a time series, Lake Ontario (1967-1976), 0-17 m, mean ± 1 standard deviation (data courtesy of CCIW).

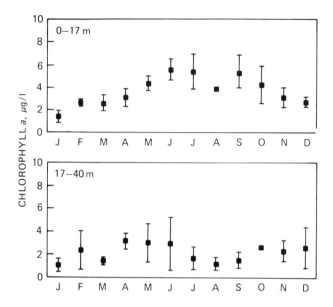

Figure 4. Mean monthly corrected chlorophyll for 1967-1976, mean ± 1 standard deviation (data courtesy of CCIW).

5 μg/l is evident in September. On the average then, Lake Ontario phytoplankton chlorophyll levels go through a relatively rapid decline and subsequent rapid growth during the period July-September. A detailed analysis of year-to-year variations in chlorophyll did not indicate any long-term temporal trends in the data of Figure 3, and one would conclude that the phytoplankton chlorophyll are at approximate equilibrium with external conditions.

The updated 1979 model (discussed more fully below) includes two phytoplankton functional groups designated as "diatoms" and "nondiatoms". Such data were not available from the U.S. EPA STORET systems, but were computed from Stoermer et al. [10] and analyzed by Shattuck [11]. Extensive analysis was done during the IFYGL (1972-1973) in classifying phytoplankton. Using Stoermer's raw data, the data were divided into the two functional groups. Table II shows the relative fractions of both groups. Diatoms appear to dominate throughout the year, with the exception of August and October, 1972. In general, an average over the 10 sampling months indicates that over 70% of all the phytoplankton in Lake Ontario are diatoms. This becomes significant since diatom growth may be limited by low silica levels. Figure 5 shows the estimated equivalent chlorophyll levels for both groups for the 1972 period. The chlorophyll levels were calculated as the relative fraction (Table II) times the 1972 chlorophyll-a

Table II. Estimated Relative Fraction of Diatoms and Nondiatoms [11] [a]

Time	Diatoms	Nondiatoms
1972		
May	0.78	0.22
June	0.68	0.32
July	0.71	0.29
August	0.26	0.74
September		
October	0.59	0.49
November	0.74	0.26
1973		
February	0.82	0.18
March	0.88	0.12
May	0.91	0.09
June	0.71	0.29
Average	0.71	0.29

[a] As derived by Shattuck [11] from Stoermer et al. [10] ; surface data only.

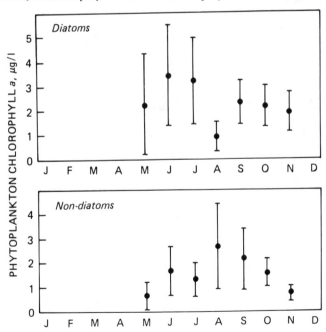

Figure 5. Estimated diatom-nondiatom phytoplankton chlorophyll for Lake Ontario (1972), Segment 1, 0-17 m (after Shattuck [1]).

values for Segment 1 (as shown in Figure 3). Figure 5 shows a diatom peak in June, with a rapid decline which may be due to silica limitation (as discussed subsequently). Nondiatom maximum growth appears to occur in August.

Figures 6 and 7 show the monthly mean total phosphorus (TP) time series and the average monthly values over all years. There does not appear to be any obvious long-term trend in TP over the data period. The range in the data for any given month or year, however, is substantial. This is particularly significant when it is recognized that changes of only several micrograms of TP can influence overall lake behavior. There is considerable variation in hypolimnion total phosphorus, and a closer analysis of TP from Figure 6. especially during 1969 and 1972, shows a marked decline in the epilimnion during the summer months. This decrease in TP results from the

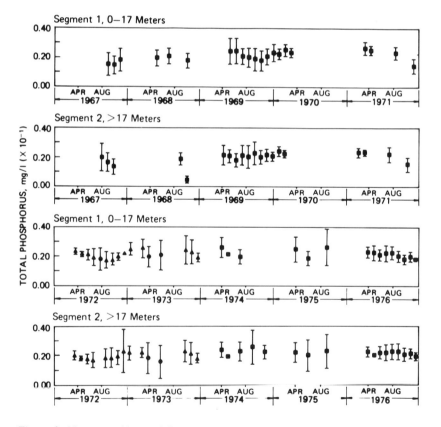

Figure 6. Mean monthly total P time series, Lake Ontario (1967-1976), mean ± 1 standard deviation (data courtesy of CCIW).

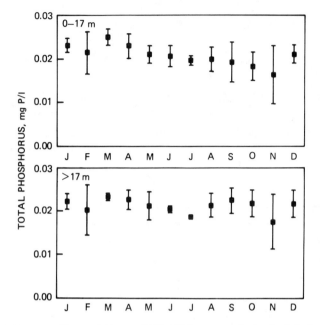

Figure 7. Mean monthly total P over 1967-1976, mean ± 1 standard deviation (data courtesy of CCIW).

generation of particulate phytoplankton from the available phosphorus pool and subsequent sinking of the plankton out of the epilimnion. This phenomenon is therefore a mechanism for checking on the net sinking velocity used in the dynamic models. The increase of nutrients due to the death of phytoplankton, plus the fall overturn, restores the loss of TP.

Figures 8 and 9 show the monthly mean values for orthophosphorus (ortho-P) and the average monthly values over all years. This form of phosphorus is considered readily available to the phytoplankton. The figures indicate high values in the early months corresponding to low phytoplankton growth, and low values in the summer due to phytoplankton uptake. In the fall, this form of phosphorus again increases, due to phosphorus regeneration and lower phytoplankton uptake.

Figure 10 shows the average monthly mean values of the ratio (ortho-P/ TP) over all years, and indicates a more stable level in the hypolimnion (> 17 m) during the year, while the upper level (0-17 m) indicates the conversion of ortho-P to phytoplankton. Therefore, from this analysis, the available phosphorus accounts for approximately 32% of the total phosphorus in the epilimnion, and about 52% of the total phosphorus in the hypolimnion.

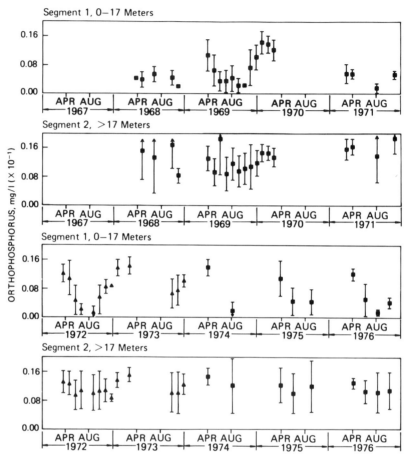

Figure 8. Mean monthly orthophosphorus time series, Lake Ontario (1968-1976), mean ± 1 standard deviation (data courtesy of CCIW).

Silica is an important measure of diatom growth in Lake Ontario. Figures 11 and 12 are the mean dissolved silica concentration (as mg SiO_2/l) versus years, and average monthly mean silica across all years. It can be seen that, for the epilimnion in Figure 12, silica is at its highest during the winter months; the subsequent decrease during the summer is due to the growth of diatoms, which use silica as a major nutrient. It should also be noted that levels of silica approach limiting levels of 0.1-0.2 mg SiO_2/l (0.05-0.1 mg Si/l) during these months. This indicates that silica limitation may be significant in controlling peak levels of diatoms in the spring.

There appear to be some year-to-year trends in the silica data. Figure 11 indicates a rise from about 0.3 mg SiO_2/l in 1967 to about 0.55 mg SiO_2/l in

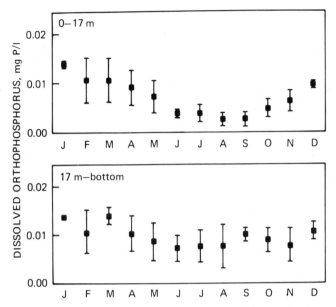

Figure 9. Mean monthly orthophosphorus over 1968-1976, mean ± 1 standard deviation.

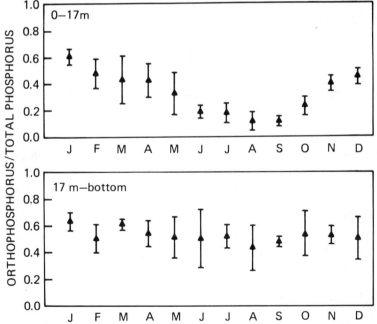

Figure 10. Mean monthly ratio of (ortho-P)/(total P) over 1968-1976, mean ± 1 standard deviation.

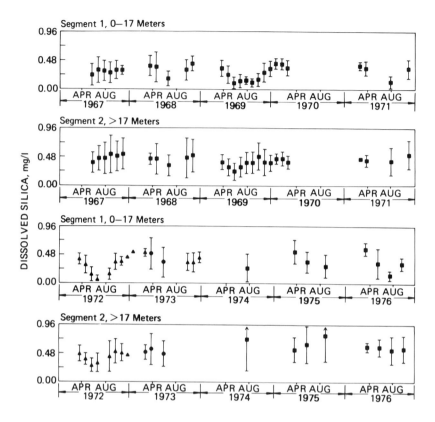

Figure 11. Mean monthly dissolved silica time series, Lake Ontario (1967-1976); mean ± 1 standard deviation (data courtesy of CCIW).

1976 for the upper segment. A similar rise is noted for the hypolimnion. The reason for this trend is not clear since data on incoming silica loads are virtually nonexistent. Also, Segna [9] has examined in detail the vertical structure of the silica and has concluded that there may be significant releases of silica from bottom sediments.

Additional compilations and analyses have been conducted of the nitrogen data. Since it appears, however, that the nitrogen cycle is not contributing significantly to the phytoplankton dynamics, such analyses are not reviewed here.

Model Calibration - Updated Kinetic Model

Since the original construction of the Lake Ontario dynamic model, additional insight has been gained in several areas; principally nutrient

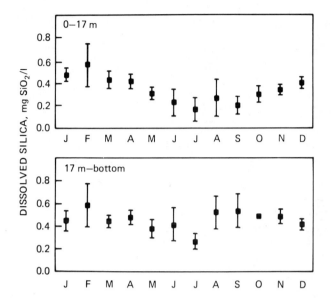

Figure 12. Mean monthly dissolved silica over 1968-1976; mean ± 1 standard deviation (data courtesy of CCIW).

recycling and the effect of silica limitation. An updated model was therefore constructed and designated as the Lake 1A kinetic structure.

The Lake 1A kinetics have been developed as part of eutrophication models constructed for the Lake Huron and Saginaw Bay system and for Lake Erie [12,13]. A systems diagram for the updated kinetics is shown in Figure 13. The additional state variables include a division of total phytoplankton chlorophyll into diatoms and "others" (or nondiatoms) chlorophyll compartments and consideration of available and unavailable silica. It should be noted that the use of the term "unavailable" in Figure 13 essentially means "slowly available" (i.e., as shown in the diagram, this form of phosphorus, silica or nitrogen is allowed to become available at a specific rate).

The principal kinetic changes therefore are as follows:

1. The use of threshold nutrient limitation, in contrast to product expressions. Therefore, the growth rate is limited by:

$$\text{Min} \left(\frac{[P_a]}{K_{sp} + [P_a]}, \frac{[N_a]}{K_{sN} + [N_a]}, \frac{[S_a]}{K_{sSi} + [S_a]} \right) \tag{1}$$

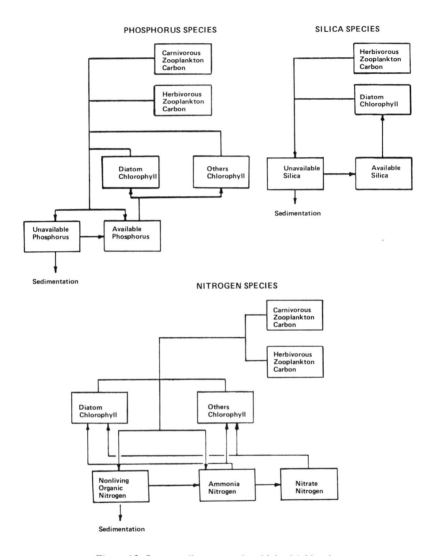

Figure 13. Systems diagram, updated lake 1A kinetics.

where K_{sp}, K_{sN}, K_{sSi} are the half-saturation constants for available phosphorus $[p_a]$, nitrogen $[N_a]$ and silica $[S_a]$, respectively.

2. Mineralization of unavailable (i.e., slowly available) to available forms through a Michaelis-Menton recycle expression with chlorophyll. Therefore, the general expression for conversion of relatively unavailable forms (p_{ua}) is:

$$R$$

$$[\text{Unavail}] \rightarrow [\text{Avail}]$$

$$\text{for } R = K\ \theta^{T-20}\ \frac{[p]}{[p]\ +\ K_{sp}}\ [p_{ua}] \tag{2}$$

where K = mineralization rate @ $20°$ C and K_{sp} = half-saturation constant for chlorophyll $[p]$ limitation.

3. Several adjustments in the basic kinetics of parameters. Table III lists the parameter values of principal interest. In addition, some vertical mixing was introduced across the boundary between segments 1 and 2, using the values of vertical dispersion estimated from a three-dimensional temperature calibration. The updated kinetics were then used to calibrate the 1972 IFYGL data set for open lake epilimnion and hypolimnion data. The results of this calibration, using the parameter values of Table III, are shown in Figures 14 and 15. The principal features of the interactive system are properly obtained by the model. The relative distribution of diatoms and "others" (nondiatoms) is approximately duplicated by the model. All chemical variable outputs were made as close to the observed data as possible, using a consistent set of coefficients. The 1972 calibrated model was then used to verify the 10 years of data discussed previously.

MODEL VERIFICATION, 1967-1976

Verification Procedure

The basic approach for the model verification was as follows:

The calibrated Lake 1 and Lake 1A models were each started in January 1966, with initial conditions specified for all variables. The Lake 1 kinetic model, as indicated previously, had been calibrated on the average of four years of data (1967-1970) and the Lake 1A model was calibrated to the 1972 IFYGL data.

Loads were estimated for phosphorus and nitrogen for each of the 10 years. The sources for these load estimates were from a variety of reports published by the IJC, and a summary report by Hydroscience [14]. The estimated TP loads for the 10 years are shown in Figure 16. The range in these loads is probably on the order of 10-20%. The silica loading was held constant at an incoming concentration of 0.6 mg Si/l over the 10-yr period because of the lack of any available loading data. For each of the nutrients, a division was made according to available and slowly available phosphorus (as well as nitrogen). Figure 16 is a plot of the phosphorus loading to Lake Ontario divided into available and relatively unavailable forms. It should be

Table III. Principal Parameter Values—Lake 1A Kinetics

Description	Value Diatoms	Others	Units
Phytoplankton Growth			
Saturated Growth Rate @ 20° C	2.1	1.6	day^{-1}
Temperature Coefficient	1.09	1.08	
Saturating Light Intensity	225.0	350	langleys/day
Half Saturation Const-P	2.0	2.0	μg P/l
Half Saturation Const-N	25.0	25.0	μg N/l
Half Saturation Const-Si	100.0		μg Si/l
C-to-Chlorophyll Ratio	100.0	100.0	μg C/μg chl-a
P-to-Chlorophyll Ratio	1.0	1.0	μg P/μg chl-a
N-to-Chlorophyll Ratio	15.0	15.0	μg N/μg chl-a
Si-to-Chlorophyll Ratio	40.0		μg Si/μg chl-a
Settling Rate for Chlorophyll			0.2 m/day
Phytoplankton Respiration			
Endogeneous Respiration Rate (20° C)	0.04	0.07	day^{-1}
Temperature Coefficient	1.08	1.08	
Available Fraction of Respired			
Phytoplankton	0.5	0.5	
Herbivorous Zooplankton			
Grazing Rate	0.07		liter/mg C/day -° C
Half Sat. Const. for Grazing Limitation	10		μg chl-a/l
Half Sat. Const. for Assimulation			
Limitation	5		μg chl-a/l
Maximum Assimilation Efficiency	0.6		
Respiration Rate @ 20° C	0.03		day^{-1}
Respiration Temperature Coefficient	1.045		
Carnivorous Zooplankton			
Grazing Rate	0.195		liter/mg C/day - ° C
Respiration Rate @ 20° C	0.007		day^{-1}
Nitrogen			
Conversion Rates of Slowly Available to			
Available P and N @ 20° C	.03		day^{-1}
Temp. Coefficient for N and P Rates	1.08		
Unavailable Si Mineralization Rate (20° C)	0.0175		day^{-1}
Half Sat. Const. for Chlorophyll			
Limitation	10		μg chl-a/l
Settling Rate of Particulate Forms			0.5 m/day

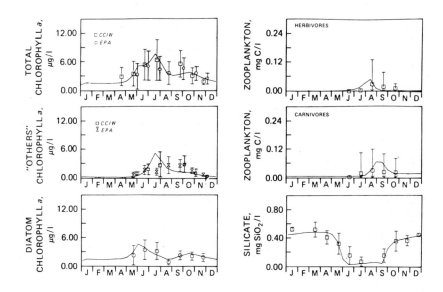

Figure 14. Lake 1A kinetic calibration, 1972.

noted again that the use of the term "unavailable" in Figure 16 does not imply that this form of phosphorus does not enter into the phytoplankton cycle. The basic model incorporates a release of this "slowly available" form to the available form at the rate indicated in Table III (0.03/day at 20° C). The rate, however, is a function of the chlorophyll level, and for the half-saturation level of 10 μg chlorophyll-a/l, the actual rate used in the calculation is somewhat less than 0.03/day, being closer to a value of about 0.005-0.01/day.

Vertical dispersion between the epilimnion and hypolimnion was computed for each year as a function of the observed temperature variations shown in Figure 1.

The models were then run for the 10-year period (1967-1976) and compared to the observed data time series. Verification statistics were computed for each of the variables, and for the model as a whole, to determine model credibility.

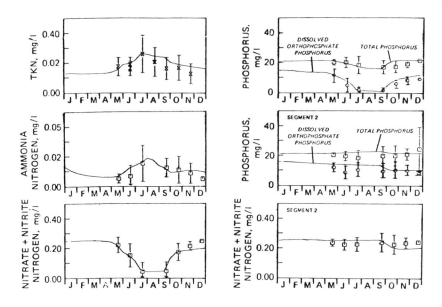

Figure 15. Lake 1A kinetic calibration, 1972 (continued).

The results of the computation indicated that the long-term behavior of the lake is particularly sensitive to the overall settling rates of the chlorophyll and particulate forms of phosphorus. For example, a computed settling velocity for the slowly available phosphorus was estimated at 0.12 m/day. This results from a particulate settling velocity of 0.5 m/day and an estimate that 24% of the slowly available phosphorus is in the particulate form. The use of 0.12 m/day, however, resulted in a loss of total phosphorus over a 10-year span slightly higher than the observed concentrations shown in Figure 6. Therefore, some adjustment was made of this net settling velocity of the slowly available form, and a resulting value of 0.11 m/day for the Lake 1A kinetics gave the best results for the phosphorus verification. This indicated that even small changes, in the order of 0.01 m/day of net settling velocity, were important in the present modeling framework.

Figures 17-20 illustrate the results of the computation for the Lake 1A kinetics (similar results were generated for the original Lake 1 kinetics). As can be seen from an inspection of these figures, the model generally duplicates the behavior of chlorophyll and the nutrients phosphorus and silica. But, in any given year, there are instances of some substantial deviations between the observed and calculated values. For example, the chlorophyll

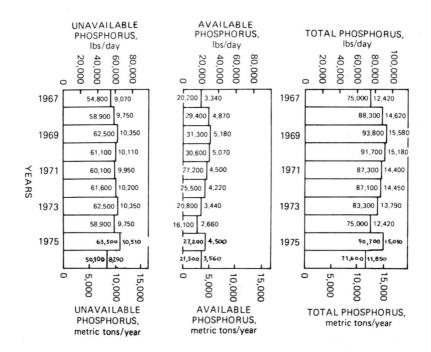

Figure 16. Assumed P loading to Lake Ontario for model verification.

comparison (Figure 17) shows some years, such as 1970, in which the model calculations are substantially below the mean value of observed chlorophyll during the spring bloom. It can also be noted that the diatom and nondiatom model, as incorporated in the Lake 1A kinetics, results in varying chlorophyll behavior from year to year, due partly to the variable dispersion and to spring stratification and fall turnover periods. Therefore, some years (such as 1971) are calculated to be essentially composed of single peak values, while other

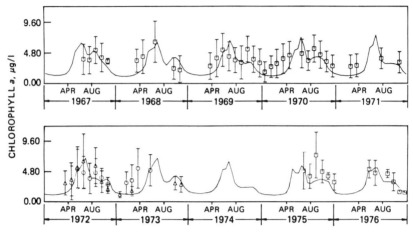

Figure 17. Comparison of observed and computed values, chlorophyll-a.

years (such as 1968) are composed of both the spring and fall bloom. The epilimnion total phosphorus illustrated in Figure 18 is an interesting comparison, since it shows the summer decline in total phosphorus due to settling of particulate forms out of the epilimnion. This seasonal variation in total phosphorus in the upper layer represents one means for calibrating the long-term settling velocity of the model. The total phosphorus behavior in the hypolimnion indicates a slight increase from the 1967 value of 20 μg/l to peak values of about 23 μg/l in 1971. This is a consequence of the peak load of 15,580 metric ton/yr in 1969 (see Figure 16). Figure 19 illustrates the seasonal and year-to-year behavior of the dissolved ortho-P (the form considered readily available). The epilimnion results indicate the late spring/early summer phosphorus decline due to phytoplankton growth. It should be noted that the model does not entirely regenerate the necessary level of the ortho-P during the spring of the latter years of the computation. This indicates that some adjustment of the recycle rate of the slowly-available to available phosphorus may be necessary to improve this condition. The silica comparison shown in Figure 20 indicates the potential for silica to limit diatom growth. It should be recognized that a constant input load for total silica was used for this computation and, hence, the annual year-to-year variations can be overestimated (as in 1969) or underestimated (as in 1976). Further data on input silica loads would improve this comparison.

Although the results shown in Figures 17-20 indicate the model can reproduce the long-term behavior of the phytoplankton-nutrient interaction in Lake Ontario, some more specific measure of model performance is

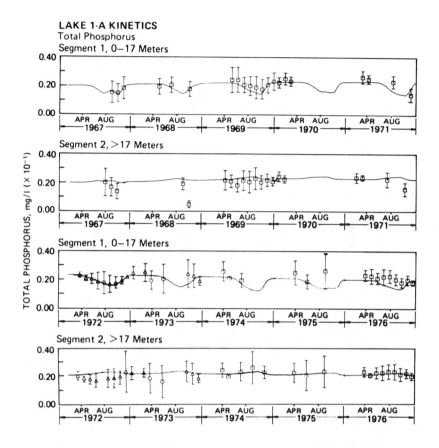

Figure 18. Comparison of observed and computed values, total phosphorus.

desirable. Therefore, extensive statistical comparisons were made of the results shown in Figures 17-20 to quantify the degree to which the model verified the observed data.

Verification Statistics, 1967-1976

A variety of simple statistical comparisons are appropriate to quantify model verification status. Such measures would be intended to supplement the qualitative comparisons shown in the preceding figures. A detailed discussion of the use of verification statistics for a three-dimensional analysis of Lake Ontario is given in Thomann et al. [3]. The three tests used to compare observed and computed values for the Lake Ontario model are:

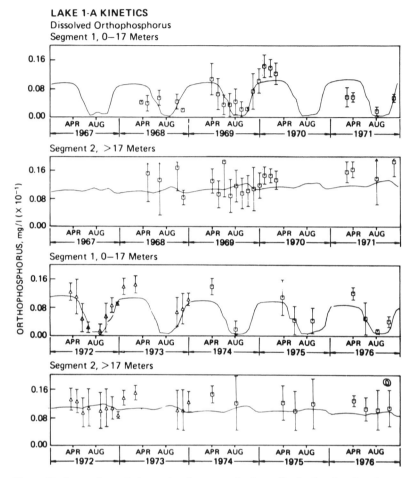

Figure 19. Comparison of observed and computed values, dissolved ortho-phosphorus.

1. comparison of means,
2. regression analyses, and
3. relative error.

These are illustrated in Figure 21.

Comparison of Means

One test of model status is to conduct a simple test of the differences between the observed mean and the computed mean. Letting $\bar{d} = \bar{x} - \bar{c}$, the test statistic distributed as a student's "t" probability density function is given by

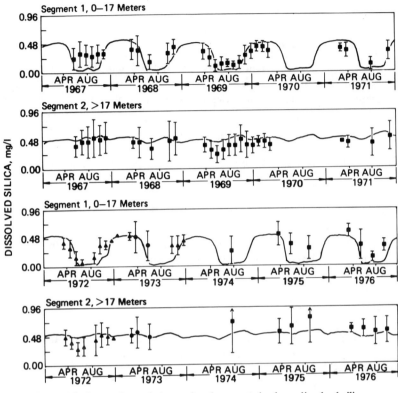

Figure 20. Comparison of observed and computed values, dissolved silica.

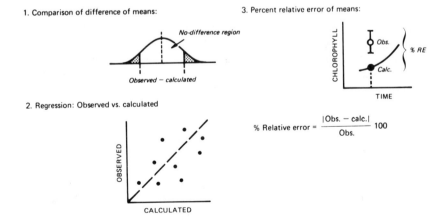

Figure 21. Three statistical tests used in comparison of model output to observed data for Lake Ontario, 1967-1976.

$$t = \frac{\bar{d} - \delta}{s_{\bar{d}}} \tag{3}$$

where δ is the true difference between model and data and $s_{\bar{d}}$ is the standard deviation of the difference given by a pooled variance of observed and model variability. If these latter quantities are assumed equal, then:

$$s_{\bar{d}} = 2 s_{\bar{x}} \tag{4}$$

where s_x is the standard error of estimate of the observed data and is given by

$$s_{\bar{x}}^2 = s_x^2 / N \tag{5}$$

Alternatively, one may simply specify the standard error of the mean as a parameter of the test. This was the procedure followed for the 10-year comparisons.

Regression Analyses

A perspective on the adequacy of a model can be obtained by regressing the calculated values with the observed values. Therefore, let the testing equation be:

$$\bar{x} = a + \beta \bar{c} + \epsilon \tag{6}$$

where a and β are the true intercept and slope, respectively, between the calculated values, $\bar{c}$, and the observed values, $\bar{x}$, and ϵ is the error of x. The regression model (Equation 6) assumes, of course, that $\bar{c}$ (the calculated value from the water quality model) is known with certainty, which is not the actual case. With Equation 6, standard linear regression statistics can be computed, including:

- the square of the correlation coefficient r^2 (the % variance accounted for) between calculated and observed values;
- standard error of estimate, representing the residual error between model and data;
- slope estimate b of β and intercept estimate a of a; and
- test of significance on the slope and intercept. The null hypothesis on the slope and intercept is given by $\beta = 1$ and $\alpha = 0$. Therefore, the test statistics $(b - 1)/s_b$ and α/s_a are distributed as student's t, with $n - 2$ degrees of freedom. The variance of the slope and intercept s_b^2 and s_a^2 are computed according to standard formulae. A two-tailed

"t" test is conducted on b and a, separately, with a 5% probability in each tail (i.e., a critical value of t of about 2 provides the rejection limit of the null hypothesis).

Relative Error

Another simple statistical comparison is given by the relative error, defined as:

$$e = \frac{|\bar{x} - \bar{c}|}{\bar{x}} \tag{7}$$

Various aggregations of this error across regions of the water body or over time can also be calculated, and the cumulative frequency of error over space or time can be computed. Estimates can then be made of the median relative error, as well as the 10% and 90% frequency of error. The difficulties with this are its relatively poor behavior at low values of $\bar{x}$ and the fact that it does not recognize the variability in the data. In addition, the statistic is poor when $\bar{x} > \bar{c}$, since under that condition the maximum relative error is 100%. As a result, the distribution of this error statistic is most poorly behaved at the upper tail. Nevertheless, if the median error is considered, this statistic is a readily understood comparison and provides a gross measure of model adequacy. It can also be especially useful in comparisons between models.

It should, of course, be recognized that the results of these statistical comparisons represent, at best, only a lower bound on the errors of predicting lake responses. The validity of any forecasts (such as those presented below), therefore, can be viewed in the light of the statistical comparisons between observed and computed values. Further, it should be noted that there have not been any substantial changes in loads to Lake Ontario during the period 1967-1976 and, as a result, the water quality of Lake Ontario has not been observed over a wide range of perturbations.

The verification scores for six-state variables, using the difference of means test, are shown in Figure 22. The standard errors of the mean used in the test are shown in the figure. With the standard errors, the Lake 1A kinetics resulted in a score of 70% (i.e., there was no statistical difference between 70% of the variables and months at which a comparison could be made). The "earlier kinetics" Lake 1 model had a score of 50%. If, however, a more stringent criteria is applied by lowering the standard errors of the mean to half the values shown, then the model performance for Lake 1A kinetics drops to 40%. This test, therefore, provides a general overview of model status.

Some verification statistics for chlorophyll are shown in Figures 23 and 24.

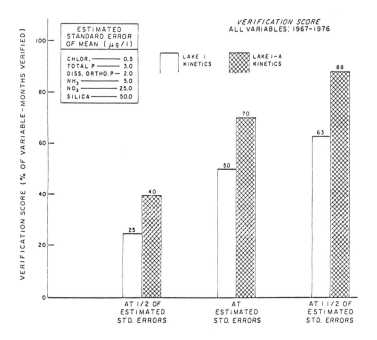

Figure 22. Verification scores, difference of means test, pooled six-state variables, 1967-1976.

For the first of these figures, the regression analyses show an improvement in the slope between Lake 1 and 1A kinetics, but no change in intercept or residual errors as measured by the square of the coefficient of correlation (a slope of unity would indicate good "tracking" of the model to data). The relative error distribution indicates an improvement in the median error (from a 42% error in chlorophyll to a 30% error), as well as in the extremes. For example, the 90% error for Lake 1 was about 100%, while for Lake 1A, the 90% error was about 55% (i.e., for 90% of the months at which a comparison between model and data could be made, the relative error was equal to or less than 55%).

The time history of the chlorophyll error is shown in Figure 24, first by year (i.e., the year-to-year relative error), and then by month over all 10 years. The year-to-year error indicates that the model error has no secular trend (i.e., where the error increases as the length of the model run increases). The average seasonal error indicates quantitatively the difficulty in accurately

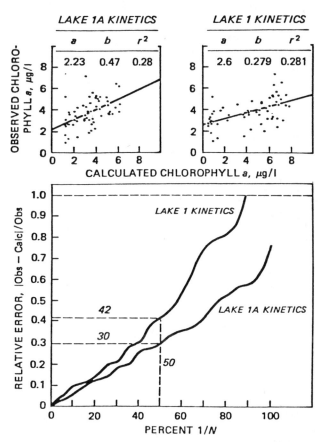

Figure 23. Chlorophyll-a verification statistics, 1967-1976, regression analyses (top) and relative error, 1967-1976.

calculating the onset of the spring bloom. Maximum errors occur in March-May, reflecting the inability to accurately compute the spring phasing. This error in phasing, however, may be misleading, since even a two-week "miss' in calculating phytoplankton growth can result in a substantial error. For most management purposes, however, the actual predicting of the seasonal onset of the bloom is not critical.

Figure 25 shows the relative error distribution for dissolved orthophosphorus (available phosphorus) and the ammonia nitrogen. The latter distribution shows the inability of the Lake 1 kinetics to describe the secular behavior of ammonia over the ten-year period. This is true principally because the original calibration for Lake 1 was to a set of data which turned out to

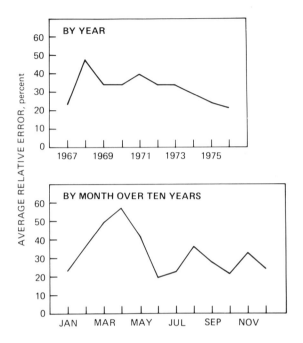

Figure 24. Annual and seasonal variation of average chlorophyll-a relative error, 1967-1976.

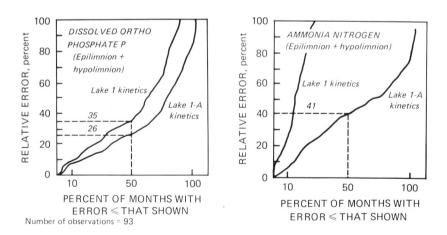

Figure 25. Dissolved orthophosphorus and ammonia nitrogen relative error distributions, 1967-1976.

have been significantly higher than later reported values of ammonia [9]. A slight improvement in the error statistics for the available phosphorus is noted in Figure 25.

The median relative errors for each state variable are shown in Figure 26,

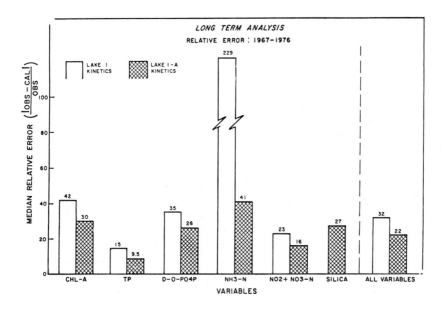

Figure 26. Relative error scores, six state variables and across all variables.

together with the median error across all variables. A general improvement is noted between the original Lake 1 kinetics and the Lake 1A kinetics. Total phosphorus errors, for example, decline from about 15% to about 10%, due principally to improvement in the seasonal behavior in the epilimnion and, concurrently, to improvement in the year-to-year behavior. Overall, with the Lake 1A kinetics, the model median relative error across all variables was about 60%.

These results provide a basis for model credibility and the usefulness of the model for making some forecasts under different land scenarios. The statistical comparisons indicate that with a consistent set of model parameters, a framework can be constructed that reproduces the 10 years of Lake Ontario data to an average of about 20%. Significant month-to-month errors, however, may occur. Further, as noted previously, the ability of the model to forecast lake behavior under substantially different loading conditions is not addressed in these verification analyses. Accordingly, short-term (one-to-three year) forecasts have been made and compared to more recent phosphorus data, and a long-term simulation under a given load scenario has also been prepared.

Short-Term Forecasts and Long-Term Simulation

Figure 27 shows the input phosphorus load distribution used in both the

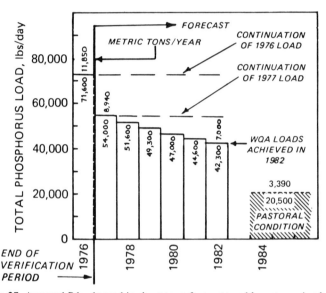

Figure 27. Assumed P loads used in short term forecasts and long-term simulations.

short-term forecast and long-term simulation. As indicated in the figure, the estimated load in 1977 of 8940 metric ton/yr is approximately equal to the load called for in the 1972 Water Quality Agreement. It was assumed for purposes of the short-term (1- to 3-yr) forecasts, that the load would decrease in equal increments to reach the proposed load of 7000 metric ton/yr in the 1972 Water Quality Agreement by 1982. As a comparison, the estimated pastoral condition of 3400 metric ton/year is shown in Figure 27. It should also be noted that these loads do not include estimates of shoreline erosion.

For the short-term forecasts, the model that was verified for the period 1967-1976 was run for an additional three years (1977-1979) under two load scenarios: (1) forecasts of 1977 to 1979 nutrient and phytoplankton conditions under a continuation of the 1976 load of 11,800 metric ton/yr; and (2) a forecast of 1977-1979 conditions under a continuation of the 1977 load of 8940 metric ton/yr. In each case, the total phosphorus load indicated in Figure 27 was considered to be made up of 30% readily available phosphorus and 70% relatively unavailable phosphorus (which becomes available in the lake at the rate of about 0.005 to 0.01/l day, depending on chlorophyll levels). For the short-term forecasts, in-lake data existed for total phosphorus only for 1977 and 1978. Results of this short-term forecast are shown in Figure 28.

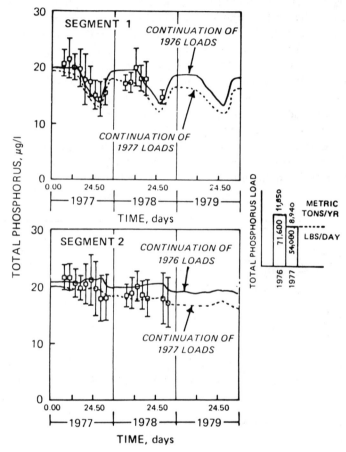

Figure 28. Short-term total phosphorus forecast, 1977-1979 using previously verified model for period 1967-1976 (data courtesy of CCIW).

It is clear from these results that the continuation of the 1976 load forecast scenario tends to overestimate the total phosphorus in the hypolimnion in 1978. The results further seem to indicate, given the assumed drop of approximately 2000 metric ton/yr between 1976 and 1977, that the 1978 lake phosphorus concentration has begun to respond to the external load reduction, and that the continuation of the 1977 loads through 1978 and 1979 appears to be a reasonable forecast over the short term. The average total phosphorus concentration for 1978 in the hypolimnion, as indicated in

Figure 28, is about 18 μg/l which, from a historical point of view (see for example, Figure 6), represents some of the lowest values of average total phosphorus recorded in the decade. It should also be noticed, however, that ± 2 standard deviations in total phosphorus is fairly substantial (on the order of ± 3 mg P/l). The variation in the epilimnion total phosphorus is properly forecasted by the model, although the model seems to be responding to the 1976 load, rather than the 1977 load. These short-term forecasts, of course, depend on the assumed input loads shown in Figure 27 and the assumption that 30% of the total is readily available phosphorus. If additional external loads, such as those given by shoreline erosion, prove to be a significant input, then sensitivity analyses with an additional 1300 metric ton/yr (divided 30% readily available) indicates a range in forecasted total phosphorus in 1977-1979 of approximately ± 1 μg P/l.

The long-term simulation and results, using the load decrease shown in Figure 27, and reaching 7000 metric tons in 1982 and remaining at that level indefinitely into the future, are shown in Figures 29-31.

SEGMENT 2 > 17 METERS

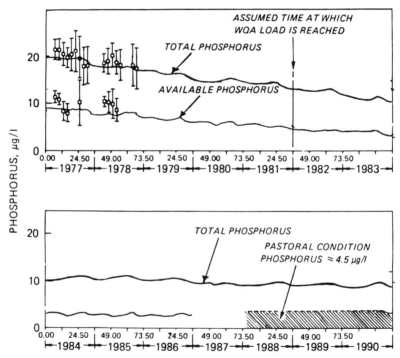

Figure 29. Long-term total and available P simulations, using input loads of Figure 27 — hypolimnion.

SEGMENT 1

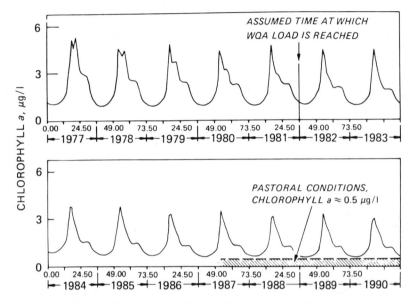

Figure 30. Long-term chlorophyll simulations–epilimnion.

As shown in Figure 29, the simulation indicates that the total phosphorus under the given load scenario would decline to an equilibrium level of about 10 μg P/l by about 1984, and a very slight decline is calculated following that period. The 10 μg P/l can be compared to an estimated total phosphorus concentration under pastoral conditions of 4.5 μg/l. On the basis of the verification statistics, the forecast error in the total phosphorus is probably about ± 4 μg P/l for any given month and somewhat less that that for an annual average. The available phosphorus simulation is also shown in Figure 29 for the hypolimnion, and it is interesting to note that with respect to 1977-1978 data, the model projects slightly lower than the observed mean values. This indicates that perhaps an additional recycling of available phosphorus, especially in the spring, would be necessary to further refine the results and achieve a better forecast of the available form. The ratio of available phosphorus to total phosphorus is also calculated to change from the historical average of 52% [9] to the predicted average of about 40%. Figure 30 shows the simulated behavior of the phytoplankton chlorophyll. The simulation indicates that the response in the phytoplankton is somewhat less than that for the phosphorus. While total phosphorus is estimated to decline from present levels of 20 μg P/l to 10 μg P/l (a 50% decline), the chlorophyll levels are calculated to decline by at most 30-40%. Further,

SEGMENT 1

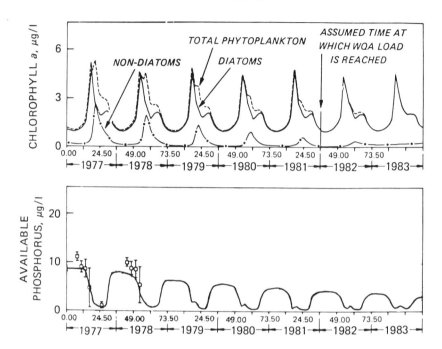

Figure 31. Long-term diatom and nondiatom chlorophyll and available phosphorus simulations–epilimnion.

the response in chlorophyll is slower and less dramatic, indicating that it may be more difficult to discuss an improvement in lake status by examining chlorophyll. The reason for this behavior in the simulation is shown in Figure 31.

Since the model consists of two phytoplankton types, diatoms and nondiatoms, the effect of a phosphorus load reduction is not necessarily equal for both types. The 10-year verification indicated that growth of the diatom group appeared to be limited by silica in the spring. As a consequence, reduction in phosphorus has no effect on the diatoms (using the minimum nutrient formulation shown previously). As shown in the simulation of Figure 31, the nondiatom group is rapidly reduced to low levels due to phosphorus reduction, and there is an effect noted in reducing the diatom group only as phosphorus is reduced below the silica limitation. This simulation would indicate the need to sample phytoplankton types in the late summer-early fall in order to detect the effect of the phosphorus reduction program.

CONCLUSIONS

The availability of a 10-year data base afforded the opportunity to test the credibility of a dynamic phytoplankton-phosphorus model of Lake Ontario. Original kinetic structure is updated to include diatom and nondiatom functional phytoplankton groups. The results of the verification of the 10 years of record indicate an overall median relative error of about 20% for the updated kinetics. The application of the model to the 1967-1976 data set appears to indicate a limitation of diatoms by silica in the spring, followed by a growth of nondiatom phytoplankton which tends to be controlled by phosphorus.

The short-term forecasts under an assumed input load scenario indicate that the 1978 hypolimnion total phosphorus levels of about 18 μg P/l in the lake are a response to a rather steep decline in the load between 1976 and 1977. The long-term simulation indicates that the total phosphorus may decline to 10 μg P/l under a 7000 metric ton/yr target load. About 40% of the projected 10 μg P/l is available phosphorus. The response in total phosphorus is approximately linear to the external load. The simulations indicated that the response in chlorophyll would not be as marked and would, in general, be difficult to detect. This is due to the calculated silica limitation of diatoms, which, therefore, do not respond to a phosphorus management program, in terms of phytoplankton, will be in the nondiatom group, which is estimated to be markedly reduced by a reduction in phosphorus load.

ACKNOWLEDGMENTS

This research was conducted under a grant from the U.S. EPA Large Lakes Research Station, Grosse Ile, Michigan under grant number R805916010. Grateful acknowledgment is made of both this financial support and the technical input of Mr. William Richardson, Dr. Vic Bierman and Mr. Nelson Thomas of EPA Grosse Ile. The insight of our colleagues at Manhattan College, Drs. Dominic DiToro and Donald J. O'Connor was particularly helpful. Thanks are also due to Ms. Cynthia O'Donnell for her patient typing of the manuscript.

REFERENCES

1. International Joint Commission. "Great Lakes Water Quality Agreement of 1978," International Joint Commission, Windsor, Ontario (1978).
2. Vallentyne, J. R., and N. A. Thomas. "Fifth-Year Review of Canada-United States Great Lakes Water Quality Agreement," Report of Task Group to Review Phosphorus Loadings (1978).
3. Thomann, R. V., A. Robertson, D. Scavia and D. M. TiToro.

"Ecosystem and Water Quality Modeling During IFYGL," (in final preparation).

4. Thomann, R. V., R. P. Winfield and D. M. DiToro. "Modeling of Phytoplankton in Lake Ontario (IFYGL)." Proc. 17th Conference Great Lakes Research, International Association of Great Lakes Research, 135-149 (1974).

5. Thomann, R. V., D. M. DiToro, R. P. Winfield and D. J. O'Connor. "Mathematical Modeling of Phytoplankton in Lake Ontario. I. Development and Verification," EPA-660/3-75-005, Environmental Research Laboratory, U.S. Environmental Protection Agency, Corvallis, OR (1975).

6. Thomann, R. V., R. P. Winfield, D. M. DiToro and D. J. O'Connor. "Mathematical Modeling of Phytoplankton in Lake Ontario 2. Simulations Using Lake 1 Model," EPA-660/3-76-065, Environmental Research Laboratory, U.S. Environmental Protection Agency, Duluth, Minnesota (1976).

7. Thomann, R. V., R. P. Winfield and D. S. Szumski. "Estimated Responses of Lake Ontario Phytoplankton Biomass to Varying Nutrient Levels," *J. Great Lakes Res.*, 3(1-2):123-131 (1977).

8. Bierman, V. J., Jr. Report to the Expert Committee on Ecosystems Aspects, International Joint Commission, on a 1976 report entitled, "Assessment of the Effects of Nutrient Loadings on Lake Ontario Using a Mathematical Model of the Phytoplankton," prepared by Hydroscience, Inc. (1977).

9. Segna, J. "Preliminary Analysis of 10 Years of Water Quality Data for Lake Ontario," Environmental Engineering and Science Program Project Report, Manhattan College, Bronx, NY (1978).

10. Stoermer, E. F. et al. "Phytoplankton Composition and Abundance in Lake Ontario," (IFYGL), EPA-660/3-75-004, U.S. Environmental Protection Agency, Environmental Research Laboratory, Duluth, MN (1975).

11. Shattuck, A. E. "A Data and Statistical Analysis on Phytoplankton Species Data of Lake Ontario," Environmental Engineering and Science Program Project Report, Manhattan College, Bronx, NY (1978).

12. DiToro, D. M., and W. F. Matystik. "Mathematical Models of Water Quality in Large Lakes. 1. Lake Huron and Saginaw Bay—Model Development, Verification and Limitations," Ecological Research Series, Environmental Research Laboratory, U.S. Environmental Protection Agency, Duluth, Minnesota (in preparation).

13. DiToro, D. M., and J. C. Connolly. "Mathematical Models of Water Quality in Large Lakes. 2. Lake Erie," Ecological Research Series, Environmental Research Laboratory, U.S. Environmental Protection Agency, Duluth, Minnesota (in preparation).

14. Hydroscience, Inc. "Assessment of the Effects of Nutrients Loadings on Lake Ontario Using a Mathematical Model of the Phytoplankton," Report prepared for International Joint Commission, Windsor, Ontario by Hydroscience, Westwood, New Jersey, for the International Joint Commission, Windsor, Ontario (1976).

CHAPTER 8

THE EFFECT OF PHOSPHORUS LOADINGS
ON DISSOLVED OXYGEN IN LAKE ERIE

D. M. Di Toro

Environmental Engineering and Science
Manhattan College
Bronx, New York

INTRODUCTION

The purposes of this chapter are (1) to summarize the methods used to analyze the present effects of phosphorus discharges on the dissolved oxygen distribution of Lake Erie; and (2) to present the results of calculations which attempt to predict the impact of phosphorus loading reductions. A complete technical report is available [1] which presents, in detail, the conceptual framework, the data employed, the mathematical formulations and their justification, the calibration and the verification calculations, and the complete results of the projection calculations. As a consequence this presentation is an overview of the methods and results without detailed technical justification.

FRAMEWORK

In the simplest terms the causal chain that links phosphorus inputs to the dissolved oxygen concentrations in Lake Erie is shown in Figure 1. Phosphorus entering the lake increases the in-lake phosphorus concentration. If the other nutrients required for phytoplankton growth are in excess supply and the population is not light-limited, the plankton will respond and an increased concentration of biomass will result. The organic carbon synthesized and the

191

LAKE ERIE

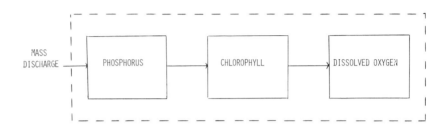

Figure 1. Simplified schematic of the causal chain linking phosphorus mass discharges to inlake phosphorus, chlorophyll and dissolved oxygen concentrations.

dissolved oxygen liberated affect the dissolved oxygen concentration—the latter directly; the former via subsequent oxidation. A quantitative assessment of these phenomena is required if projections of the effects of phosphorus loading reductions are to be attempted.

Two essentially separate phenomena must be understood—the interaction of phytoplankton and nutrients, and the interactions of the sources and sinks of dissolved oxygen. The methods employed for both of these calculations are based on mass balance equations.

CALIBRATION—PHYTOPLANKTON FOR THE WESTERN BASIN AND THE EPILIMNIA OF THE CENTRAL AND EASTERN BASINS,

A number of reviews of the general methodology employed for the phytoplankton-nutrient calculations are available and therefore, these will not be discussed in detail.

The kinetics employed are designed to simulate the annual cycle of phytoplankton production, its relation to the supply of nutrients and the effect on dissolved oxygen. The calculation is based on formulating the kinetics which govern the interactions of the biota and the forms of the nutrients and applying them to the regions of Lake Erie within the context of conservation of mass equations. The 15 variables for which these calculations are performed are:

Phytoplankton
 1. Diatom chlorophyll-a
 2. Nondiatom chlorophyll-a

Zooplankton
    3. Herbivorous zooplankton carbon
    4. Carnivorous zooplankton carbon
Nitrogen
    5. Organic nitrogen
    6. Ammonia nitrogen
    7. Nitrate nitrogen
Phosphorus
    8. Unavailable phosphorus
    9. Soluble reactive phosphorus
Silica
    10. Unavailable silica
    11. Soluble reactive silica
Carbon, Hydrogen and Oxygen
    12. Detrital organic carbon
    13. Dissolved inorganic carbon
    14. Alkalinity
    15. Dissolved oxygen

A critical requirement for each application is a complete calibration which compares the computation to the available observations. A summary of these comparisons is shown in Figure 2. The average concentration of chlorophyll-a is both computed and observed to decrease from west to east. The spring diatom bloom appears to be of comparable magnitude in all basins, whereas the magnitude of the fall nondiatom bloom progressively decreases from west to east. Shipboard ^{14}C primary production measurements are compared to the comparable kinetic expression in the calculation. The observed threefold variation from west to east is correctly reproduced. Total phosphorus concentrations are observed and calculated to decrease from west to east. The higher observed concentrations in the late fall and early spring are attributed to wind-driven resuspension of sedimentary phosphorus due to high winds during this period. These effects do not appear to persist into the productive period of the year.

The calculated and observed orthophosphorus, nitrate nitrogen (the ammonia concentrations are quite small), and silica concentrations are shown in Figure 3. The depletion of silica terminates the spring diatom bloom, whereas the exhaustion of nitrogen primarily in the western basin and phosphorus in the central and eastern basin terminates the nondiatom bloom. Again the progressively decreasing concentrations from west to east are suggested but not as pronounced as in the previous figure. The somewhat scattered behavior of silica in the western basin is unexplained at present.

One of the principal purposes of calibration is to demonstrate that the calculation can reproduce the major features of the seasonal distribution of phytoplankton and nutrients over a range of observed concentrations.

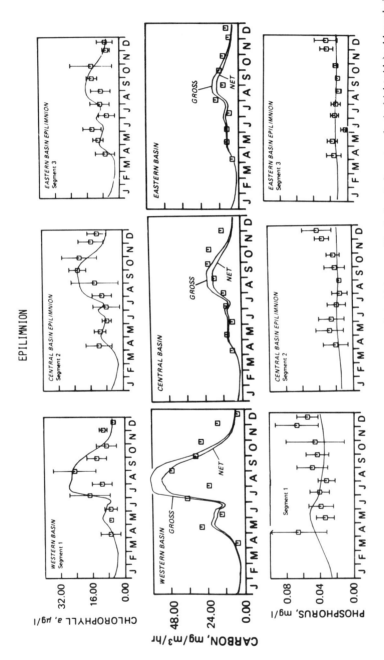

Figure 2. Lake Erie calibration results, 1970 epilimnion. Western basin (left side), central basin (center), eastern basin (right side); chlorophyll α μg/l, (top), C^{14} shipboard primary production mg $C/m^3/hr$ (middle), total phosphorus mg/l (bottom). Symbols: mean ± standard deviation; lines are the computations.

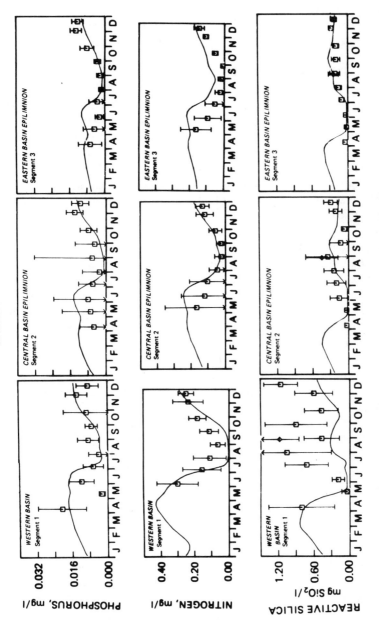

Figure 3. Lake Erie calibration results, 1970 epilimnion. Western basin (left side), central basin (center), eastern basin (right side); orthophosphorus (top), nitrate nitrogen (middle) reactive silica (bottom). Symbols: means ± standard deviation; lines are the computation.

The fact that the western, central and eastern basin distributions are all reasonably well-reproduced using the same kinetic structure and coefficients suggests that the calculation has a certain generality and can reproduce conditions as distinct as those in the western and eastern basin.

CALIBRATION–DISSOLVED OXYGEN (DO)

The second phenomena of concern are those that influence the DO balance in the basins. Since the comparatively shallow western basin is essentially completely mixed vertically and remains nearly saturated throughout the year, it is not of concern, although it is part of the oxygen balance calculation. Three regions are represented in each basin: the epilimnion, the hypolimnion and the surface sediment layer. The principal sources and sinks of DO are associated with the phytoplankton and detrital organic carbon in these segments. Although nitrification is included in the calculation, its effect is negligible. The principal aerobic reactions which affect DO are primary production and respiration of phytoplankton, and the oxidation of detrital organic carbon via bacterial synthesis and respiration. The primary production reaction liberates DO and decreases total CO_2. The aerobic respiration reactions consume DO and liberate CO_2. Thus both DO and total inorganic carbon mass balances are considered. This provides an additional variable for the calibration, giving a total of four principal state variables of concern as shown in Figure 4. Of these, three are directly observed. As discussed below, BOD measurements provide an indirect measurement of the fourth variable, the detrital organic carbon. The concentrations of concern are affected by two classes of transport phenomena: (1) dispersion, which exchanges matter across the segment interfaces; and (2) advection, which, in the vertical direction, is associated with settling of the particulates. These are shown in Figure 4 together with the principal reactions of concern. If all the transport fluxes and reaction rates were directly observed, then the mass balance calculation could be accomplished by simple summations. These could then be compared to the observed concentrations as a check on the reaction and transport fluxes. Unfortunately, neither all the variables nor all the fluxes are observed, as indicated in Figure 4. These must be estimated either by using tracers or from the calibration itself.

Air-water interface exchange is assigned based on wind velocity and empirical relationships. Thermocline exchange is estimated using temperature as a tracer. Hypolimnion-interstitial water exchange is estimated using chloride as a tracer. Settling velocities of phytoplankton and detrital carbon are assigned based on literature values and the calibration. Sedimentation settling velocity is estimated using pollen tracers and sedimentation trap results.

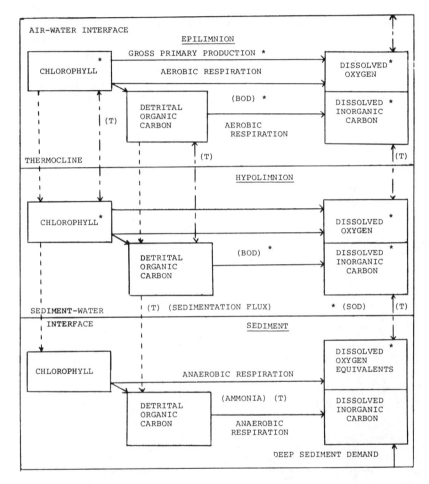

Figure 4. Schematic diagram of the dissolved oxygen mass balance computation. The epilimnion, hypolimnion (for central and eastern basins) and surface sediment segments as indicated. Directly observed concentrations and reaction rates are indicated (*). Transport rates estimated using tracers are indicated (T).

Certain reaction rates are also directly observed; for example, primary production. The aerobic respiration rate of oxygen is directly observed as biochemical oxygen demand (BOD). Since this rate is the sum of both detrital oxidation and phytoplankton respiration, it provides an indirect measurement of the detrital carbon concentration.

For the sediment segment, direct observations of the DO equivalents in the interstitial water are available. However, due to the complex chemistry associated with the anaerobic respirations the total CO_2 is not easily computed. The interstitial ammonia concentration provides a tracer which is liberated as the anaerobic respirations proceed, so that this reaction rate can be estimated.

As previously discussed, the purpose of the calibration is to estimate the individual kinetic coefficients based on the observed state variable concentrations and composite reaction rates, and to demonstrate their applicability in differing regions of the lake. The results for the central basin are shown in Figure 5. Dissolved oxygen, while remaining essentially saturated in the

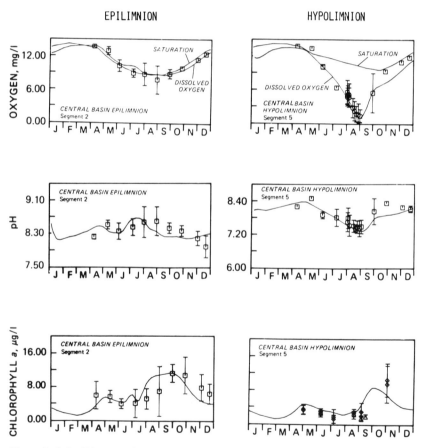

Figure 5. Lake Erie central basin calibration for dissolved oxygen and related variables. Epilimnion (left side), hypolimnion (right side); dissolved oxygen, mg/l (top), pH (middle), chlorophyll a =g/l (bottom). The hypolimnion chlorophyll a data is 1973-1974, as no 1970 data are available. Symbols: mean ± standard deviation; lines are computations.

epilimnion, declines after stratification to a minimum of 1.0 mg/l before overturn. The lack of fit after overturn is due to an inaccurate estimate of the vertical dispersion. The pH comparison is a direct calibration of the total CO_2 computation since the pH is determined by changes in total CO_2, the alkalinity being essentially constant. There is a dramatic difference between the epilimnion pH reflecting the net total CO_2 loss which is occurring and the hypolimnion pH reflecting the net total CO_2 production. The chlorophyll concentrations also reflect the production in the surface layer, the settling into the hypolimnion and the respiration and other losses. Figure 6 gives the

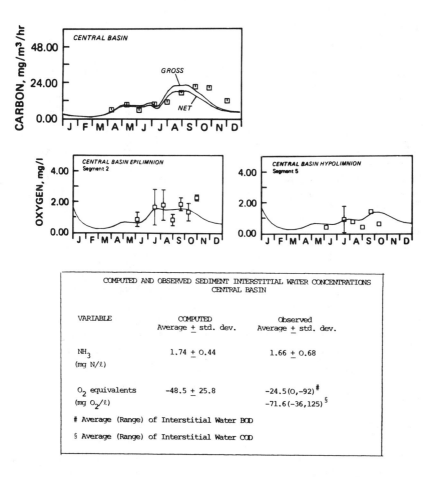

Figure 6. Lake Erie central basin calibration. Epilimnion (left side), hypolimnion (right side), shipboard C^{14} primary production (top). No hypolimnion data available, BOD_5 (middle). This is 1967 data as no 1970 data are available.

comparisons to epilimnion primary production and BOD in both layers. The larger quantity of BOD in the epilimnion reflects the larger quantity of phytoplankton and detrital organic carbon as well as the higher temperature. The table presents the observed and computed sediment interstitial water concentrations. The ammonia and oxygen equivalents are within the range of observations. The deep sediment flux of oxygen equivalents is chosen to match both the observed oxygen equivalent concentrations and also the observed oxygen equivalent flux which is observed as the sediment oxygen demand at the sediment-water interface. Similar comparisons for the eastern basin have also been made with essentially comparable results.

A summary of the computed sources and sinks of dissolved oxygen in the central basin hypolimnion during the summer stratified conditions is presented in Table I. The aqueous reactions: phytoplankton respiration

Table I. Sources and Sinks of Dissolved Oxygen Central Basin Hypolimnion—1970

Reaction[a]	Average Volumetric Depletion Rate (mg O_2/l-day)	Dissolved O_2 Change Day 180-240 (mg O_2/l)
Phytoplankton Respiration	−0.020	−1.20
Detrital Organic Carbon Oxidation	−0.0646	−3.88
Sediment Oxygen Demand	−0.0597	−3.58
Thermocline Transport	+0.0305	+1.83
Total O_2 Change		−6.83

[a]Source (+); sink (-).

and detrital organic carbon oxidation, consume a total of 5.08 mg O_2/l during the period of stratification. The oxygen decrease attributed to sediment oxidation is 3.58 mg O_2/l. These sinks are balanced by the only significant source considered in the computation: transport of DO through the thermocline, which increases the concentration by 1.83 mg O_2/l. The net result is a decrease of 6.83 mg O_2/l during the 60-day period analyzed in this table. These sources and sinks combine to produce the result shown in Figure 5.

VERIFICATION

The DO distribution in the Central Basin in 1975 was quite different from that in previous years, and this provides an opportunity to verify the computation. The result is shown in Figure 7. The principal difference between 1975 and the previous four years was the abnormally shallow depth at which the thermocline developed. This increased the volume of oxygen

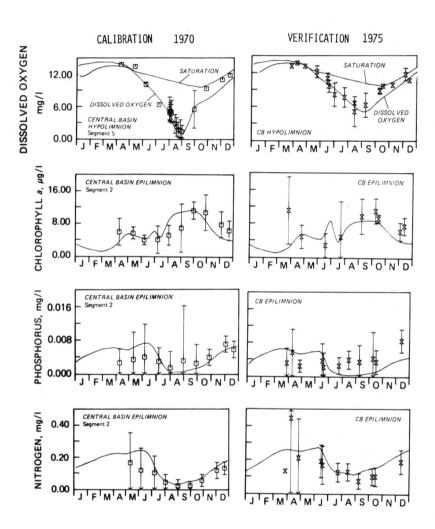

Figure 7. Comparison of 1970 calibration (lefthand side) and 1975 verification (right-hand side) dissolved oxygen (top), chlorophyll *a* (upper middle), orthophosphorus (lower middle), nitrate nitrogen (bottom). Symbols: mean ± standard deviation; lines are computations.

trapped below in the hypolimnion. It is interesting to note that if oxygen depletion were due to purely volumetric consumption, this would have no effect on the DO distribution, since the extra trapped DO is exactly balanced by the extra trapped volumetric DO demand, assuming the production of phytoplankton and detrital carbon is the same. However, if the oxygen depletion were due only to areal oxygen demand at the sediment water inter-face, then the effect of essentially a doubling of the hypolimnetic volume would be to halve the DO depletion. In fact, what was observed and calcula-ted is the result of the combination of these two sinks of DO. The fact that the computations are able to verify the 1975 DO distribution implies that the proportions of volumetric and areal oxygen demand in the calcula-tion are correct.

Chlorophyll, orthophosphorus and nitrate nitrogen are also shown. Note that slightly less chlorophyll is calculated in 1975 than in 1970 due to a relatively small (12%) phosphorus loading reduction achieved between 1970 and 1975. Althought the effect is slight in the chlorophyll, it is calculated and observed in the nitrate distribution. Slightly less nitrate uptake was observed and calculated in 1975. Admittedly, this effect is small but encouraging.

PROJECTIONS

A comprehensive calculation of this sort can be used to calculate the effects of phosphorus loading reductions on the DO, in particular, in the central basin. An immediate question regarding these calculations is: "How accurate are these projections likely to be?" Although there is no firm methodology available at present to answer this question, it is certainly the case that the projections are not likely to be any more accurate than the calibration and verification. An analysis of the residuals, the observed minus calculated DO concentration for the central basin is shown in Table II. Resi-dual standard deviations of $\sim$ 0.5-1.0 mg/l imply that the prediction error standard deviation is likely to be at least this large for any specific time during the computation. In terms of median relative error ($\sim$ 5-15%) the calibration and verification of the central basin is among the more precisely calibrated of DO models [2].

The critical question concerning the central basin is "At what phosphorus mass discharge rate will the anoxia in the central basin be eliminated?" For the mass balance calculation, the entire hypolimnion is represented by effectively one segment. However, the anoxic area that develops is confined to a region close to the sediment water interface. Therefore it is necessary to relate the volume average hypolimnetic DO concentration to the minimum DO in the hypolimnion. The results are shown in Figure 8 which is a plot of

Table II. Dissolved Oxygen Calibration–Residual Analysis, 1970

| Central Basin | Year | Residual[a] | | Median[b] Relative Error (%) |
		Mean (mg/l)	Standard Deviation (mg/l)	
Epilimnion	1970	−0.13	0.60	4
	1973-74	−0.16	1.26	7
	1975	−0.24	0.95	6
Hypolimnion	1970	0.56	0.77	12
	1973-74	0.63	1.08	15
	1975	0.29	0.93	8

[a]Residual = Calculated-Observed

[b]Relative Error = | Calculated-Observed | /Observed.

both these observations. It is clear from these data that at a hypolimnetic average oxygen concentration of 4.0 mg/l, anoxia begins in the basin.

The projection computations are also shown. The short-term effect is the projected concentration five years after an abrupt reduction in phosphorus loading as specified in the abscissa. The ultimate effect assumes that the deep sediment oxygen demand has responded to the reduction of the influx of organic carbon to the sediment in proportion to the reduction computed in the surface sediment segment. The solid line corresponds to point source reductions, the dotted line to diffuse source reduction which is less effective. A reduction of between 9500 and 12,000 metric ton/yr of total phosphorus loading is projected to eliminate anoxia in the central basin. This range is due to the uncertainty of the long-term sediment response.

A further uncertainty is due to the prediction error. If one assumes that the prediction error in DO is equal to the calibration error, then with a linear relationship between loading and concentration (Figure 8), the standard deviation of the projected loading is linearly related to the prediction standard deviation. For a predicted loading of 10,000 metric ton/yr and using a range of 0.5-1.0 mg/l for the prediction standard deviation, the predicted loading standard deviation is 1000-2000 metric ton/yr, so that the prediction error due to error in calibration is estimated to be in the range of 10-20%.

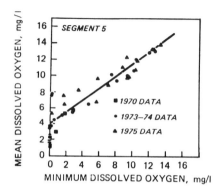

MINIMUM CENTRAL BASIN HYPOLIMNION DISSOLVED OXYGEN
(VOLUME AVERAGED) vs. LAKE TOTAL PHOSPHORUS

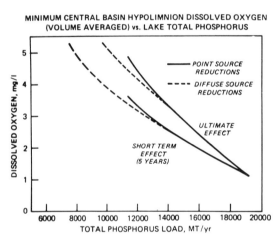

Figure 8. Observed hypolimnion mean dissolved oxygen vs minimum observed hypolimnion dissolved oxygen (top). Predicted mean hypolimnion dissolved oxygen just prior to overturn vs lake total P loading (bottom).

ACKNOWLEDGMENTS

The paper is based on a report [1] coauthored by John Connolly whose contribution is acknowledged. The support and contribution of the members of the EPA Large Lakes Research Station—Nelson Thomas, William Richardson and Victor J. Bierman, and other members of our research group—Robert V. Thomann and Donald J. O'Connor are greatly appreciated.

REFERENCES

1. DiToro, D. M. and J. Connolly, "Mathematical Models of Water Quality in Large Lakes: Part II. Lake Erie." EPA Ecological Research Series (in press).
2. Thomann, R. V. "Measures of Verification," presented at the National Workshop on Verification of Water Quality Models, West Point, NY, March 1979. Sponsored by EPA, Environmental Research Laboratory, Athens, GA, and Hydroscience Inc., Westwood, NJ.

CHAPTER 9

THE PHOSPHORUS LOADING CONCEPT
AND GREAT LAKES EUTROPHICATION

R. A. Vollenweider
Canada Centre for Inland Waters
National Water Research Institute
Burlington, Ontario

W. Rast
International Joint Commission
Washington, DC

J. Kerekes
Canadian Wildlife Service
Environment Canada
Halifax, Nova Scotia

It is now well accepted that eutrophication of lakes depends on excessive loads or inputs of phosphorus and nitrogen to the lakes. Eutrophication control programs are frequently based on controlling the inputs of these aquatic plant nutrients, especially phosphorus, to water bodies. Attention is given to phosphorus inputs because this nutrient has most often been found to limit aquatic plant productivity in lakes, in the sense of Liebig's Law of the Minimum [1]. Further, controlling phosphorus loads to water bodies is both technically sound and economically feasible for many water bodies [2,4].

This approach has led to the development of the nutrient loading concept in limnology. This concept implies that a quantifiable relationship exists between the quantity of phosphorus entering a water body and its response to this phosphorus input, as measured through some kind of trophic scale index or water quality parameter. A trophic index would allow for expressing the response in the classical limnological categories of oligotrophy, mesotrophy and eutrophy, while the use of some type of water quality parameter(s) will allow the response to be expressed in terms with a practical meaning in present water quality management programs (*e.g.*, chlorophyll levels, water transparency).

The nutrient loading concept has been known on a qualitative level for several decades. Rawson [5,6], Sawyer [7], Ohle [8], Edmondson [9] and Sakamato [10] have all presented some expression of the effects of nutrient loads on the trophic conditions of water bodies. Sawyer [7] was among the first to use the concept of nutrient loading with his observations on lakes in Madison, Wisconsin, when he reported that the lakes which received the greatest amount of nitrogen and phosphorus per unit surface area of the lake experienced the most severe algal blooms during the growing season. He also established critical threshold levels for nutrients in these water bodies, above which algal nuisance conditions could be expected in a lake during the growing season. None of these investigators, however, were able to present definitive quantitative conclusions concerning nutrient loading levels and expected trophic conditions in water bodies.

Vollenweider [3] made the first attempt to formulate loading criteria for phosphorus and nitrogen by defining a boundary level between oligotrophic and eutrophic water bodies, taking into account nutrient loadings relative to mean depth (a measure of lake volume) as the principal parameters. Vollenweider [11,12] has subsequently refined his approach through several stages of development. Other investigators have also subsequently developed models based on the nutrient loading concept for predicting and assessing various responses of lakes to nutrient inputs [13-16]. These latter models, as well as the refinements of Vollenweider, focus on phosphorus loads as the major nutrient influencing the eutrophication responses of water bodies. This is because of the technical and economic considerations indicated earlier, as well as the factors which complicate accurate measurements of nitrogen loads to water bodies, such as nitrification, denitrification, and fixation of nitrogen from the atmosphere by blue-green algae. Accordingly, further discussion is limited to phosphorus model formulations.

CONTINUITY EQUATION AND MASS BALANCE

An underlying concept developed in this report is that of a lake as an open system, which can in turn be expressed in the form of a mass balance equation, as follows:

$$\frac{\Delta M}{\Delta t} = I - O - (S - R) \tag{1}$$

where   I   =   external load
               O =   outflow loss
               S   =   loss to sediments
               R   =   sediment regeneration (internal load)
       $\Delta M/\Delta t$ =   storage gain, loss of substance M over time Δt

This equation will be applied to phosphorus in following sections, although in principle, Equation 1 would be applicable to any substance involved in lake dynamics. As such, Equation 1 is an accountability model, concerned only with the balance of substance M between its sources and sinks.

The values of the terms in Equation 1 can generally be obtained by direct measurement. (S - R) is usually difficult to measure directly. It can be obtained, however, by difference from the more easily measured terms of Equation 1. It is noted that it is also possible to split this equation into two separate equations, one referring to the epilimnion, and the other to the interaction between the epilimnion and hypolimnion [12].

Equation 1 is the basis for any model development, whether it be the empirical approach described here, or the dynamic models of the type exemplified by Thomann et al. [17,18]. It should be noted that the dimensional correctness of any derived equation is not a prerequisite as long as it is understood why the dimensional correctness has been given up. In relation to some of the formulae derived in this report, it is important to keep this thought in mind, since some of the terms in the models have, for practical reasons, been derived statistically.

The steady state condition for Equation 1 is simply a particular form of the equation, determined by setting $\Delta M/\Delta t = 0$. In the real world, no lake is in a steady state condition over shorter (and sometimes even longer) periods of time. Lakes with nutrient loads which oscillate around a relatively constant mean value over prolonged periods of time (years) can, however, be characterized as being in a repetitive steady state. This steady state can be defined, for example, as the storage content measured at spring overturn or the yearly average storage content. Most empirical models have been used under the assumption of steady state conditions.

MODEL DEVELOPMENT

As with virtually all current phosphorus models, the model development presented herein is based on the principle of conservation of mass, in this case phosphorus. A basic hypothesis of this model development is that a lake is considered to be a completely mixed reactor (i.e., the phosphorus load is instantaneously and completely mixed throughout the lake water). In reality, this assumption is only partially true for most lakes. In some cases, it might be more appropriate to treat the lake as a sequence of coupled mixed reactors (e.g., lakes with morphometric characteristics such that distinct trophic conditions exist from one end of the lake to the other).

No one model is entirely satisfactory for the large variety of water bodies existing in the world. On the other hand, the multistage reactor concept indicated above, because of its normally extensive data requirements, is not practical for comparing the eutrophication status or responses of large numbers of water bodies. Therefore, Vollenweider restricted his approach to basically the mixed reactor concept, although augmenting the application of his model in data elaboration, and interlake comparison of trophic conditions, with statistical techniques in some cases. As a result, the relationships derived in this report are quantitative methodologies for describing the average statistical behavior of a large population of lakes, rather than for describing the specific behavior of a particular lake. It is possible, however, from the knowledge of the statistical behavior of a large population of lakes to obtain a measure of how much a particular water body will deviate from the responses and conditions normally observed in a large number of water bodies.

Mass Balance Equation

At any given time, the instantaneous change of phosphorus in a lake can be derived from the general mass balance formula (Equation 1), as follows:

$$\frac{d\,M\lambda,t}{dt} = Q_{j,t}\,[M]_{j,t} - Q_{w,t}\,[M]_{w,t} - A_e\,F_e\,(M) + A_s F_s\,(M) \qquad (2)$$

where $Q_{j,t}$ = volume of the ith inflow at time t (m^3/sec)
$[M]_{j,t}$ = concentration of the ith inflow at time t $(mg\ P/m^3)$
$Q_{w,t}$ = outflow at time t (m3/sec)
$[M]_{w,t}$ = concentration of the outflow at time t $(mg\ P/m^3)$
A_e = area of epilimnion
A_s = area of sediments
$F_e\,(M)$ = positive and negative fluxes of substance M through the epilimnion $(mg\ P/m^2$-sec$)$
$F_s\,(M)$ = positive and negative fluxes of substance M through the sediments $(mg\ P/m^2$-sec$)$

This equation can be solved through appropriate measurement of each term in the equation. Note that Equation 2 is simply a more detailed reiteration of Equation 1.

A few assumptions and simplifications make Equation 2 more easily managed. Rather than considering the various instantaneous processes,

Equation 2 can be written as a difference equation over time Δt, which has been assumed to be 1 year for our purposes. When this is done, Q becomes the total water discharge over one year. To obtain the phosphorus mass involved, the concentration [M] can than be approximated with a term denoting an "average concentration." The same procedure can be applied to other terms in Equation 2.

It is noted particularly that $A_e F_e$ (M) may also be approximated. This may be done with the hypothesis $A_e F_e$ (M) = $A_e \bar{V} [\bar{M}]_\lambda$, in which $[M]_\lambda$ is an average lake concentration, and $\bar{V}$ is an apparent average settling rate (m/t) for phosphorus. The average settling rate is substantially different from the instantaneous settling rate.

General Form of Calculation Model

Several definitions and simplifications can be made, as follows:

$$[\bar{M}]_w \equiv [\bar{M}]_\lambda$$

$$\frac{\Sigma \, Q \, [M]}{\Sigma \, Q} \equiv [M]_j$$

$$\Sigma \, Q_j \equiv \Sigma \, Q_w \equiv \bar{Q}$$

Using these definitions, Equation 2 can be rewritten as:

$$\frac{\Delta M}{\Delta t} = \bar{Q} [\bar{M}]_j - \bar{Q} [\bar{M}] - A_e \, \bar{V} \, [\bar{M}] + A_s F_s \, (M) \tag{3}$$

Both sides of Equation 3 can then be divided by the lake volume V_0 as follows.

$$\frac{1}{V_0} \frac{\Delta M}{\Delta t} = \frac{\bar{Q}}{V_0} [\bar{M}]_j - \frac{\bar{Q}}{V_0} [\bar{M}]_\lambda - \frac{A_e}{V_0} \, \bar{V} [\bar{M}]_\lambda + \frac{A_s}{V_0} \, F_s \, (M) \tag{3a}$$

When steady state conditions are assumed, then one may write an expression for average lake concentration, as follows:

$$[M]_\lambda \approx \frac{\dfrac{\bar{Q}}{V_0} [\bar{M}]_j + \dfrac{A_s}{V_0} \, F_s \, (M)}{\dfrac{\bar{Q}}{V_0} + \dfrac{A_e}{V_0} \, \bar{V}} \tag{4}$$

Examination of Equation 4 shows that the average lake concentration of substance M (phosphorus) is a function of the relationship between the total volumnar load to the lake (including both external load to the lake and internal regeneration from the sediments), as expressed in the numerator, and the elimination of the substance from the lake (including outflow from the lake and settling to the sediments), as expressed in the denominator.

Specific Forms of Calculation Model

Several specific models have been derived from the general model expressed in Equation 4. Several additional definitions and simplifications are necessary prior to considering these specific models. First, by definition, V_O/Q_O is equivalent to the filling time, or hydraulic residence time τ_w. V_O and Q_O refer to the outflow volume and discharge, respectively. Further, A_e and A_s both approximate the lake surface area (A_O), while V_O/A_e and V_O/A_s both approximate the mean depth (Z).

Further items of definition include:

$$\bar{Q}/A_O = q_a \quad = \text{ hydraulic load (m/yr)}$$

$$[M]_j \cdot q_a = L\,(M) = \text{ areal load of substance M (phosphorus to the lake}$$

$$L\,(M)/\bar{z} \quad = 1\,(M) = \text{ volumnar loading of substance M to the lake}$$

$$\tau_w \quad\quad = \text{ reciprocal of flushing rate/yr = hydraulic residence}$$
$$\text{time (m/yr)}$$

It is noted that the volumnar loading 1 (M) is identitical to the "potential loading" defined by Edmondson [9].

Because these definition terms are interchangeable, Equation 4 may assume several different forms, each having identical content. For example, if the numerator and denominator of Equation 4 are multiplied by $V_O/\bar{Q}_O$, assuming $A_e \approx A_s \approx A_O$, then Equation 4 becomes:

$$[M]_\lambda \approx \frac{[M]_j + \tau_w/\bar{Z} \cdot \dot{F}\,(M)}{1 + \tau_w/\bar{Z}\ \bar{V}}$$

Noting that the above assumption was somewhat arbitrary, Equation 4 can be generalized to:

$$[M]_\lambda \approx \frac{[M]_j + F\,(M)\ f_1\,(\tau_w,\ \bar{Z})}{1 + \dfrac{}{\bar{V}}\ f_2\,(\tau_w,\ \bar{Z})} \tag{4a}$$

When the internal loading of phosphorus is insignificant (i.e., $f_2 = 0$), this constitutes an important limiting case of Equation 4a. Accordingly,

$$[M]_\lambda \approx \frac{[M]_j}{1 + \bar{V} f (\tau_w, \bar{Z})} \qquad (4b)$$

$\bar{V} f (\tau_w, \bar{z})$ can then be obtained statistically, using the equivalency expression.

$$\frac{[M]_\lambda}{[M]_j} = \pi_r = \frac{1}{1 + \bar{V} f (\tau_w, \bar{Z})}$$

This latter expression defines the ratio between average lake concentration $[M]_\lambda$, and average inflow concentration $[M]_j$, and usually leads to nonlinear terms in τ_w and Z. Details concerning the derivation of π_r and implications in relating the residence time of phosphorus to the residence time of water are provided by Vollenweider [12] and Rast and Lee [16].

Vollenweider [11,12,19] and others [2,13] have applied Equation 4b to phosphorus. Vollenweider's [19] original equation, in which he first considered the impact of the flushing rate (as expressed in the hydraulic residence time term) as a principal variable in affecting how a water body responds to phosphorus inputs, was as follows:

$$[P]_\lambda = \frac{L (P)}{\bar{Z} (1/\tau_w + \sigma)}$$

where: $[P]_\lambda$ = average lake concentration of phosphorus
σ = phosphorus sedimentation coefficient
L (P) = annual areal phosphorus load

This equation can be expressed in the same form as Equation 4b:

$$[\bar{P}]_\lambda = \frac{[\bar{P}]_j}{1 + \sqrt{\tau_w}} \qquad (5)$$

where $[P]_j$ = average inflow phosphorus concentration. The same form applies to Vollenweider's [12] more recent modification:

$$[\bar{P}]_\lambda = \frac{[\bar{P}]_j}{1 + \sqrt{\tau_w}} \tag{6}$$

The dimensional meaning of σ in Equation 5 was equivalent to $\overline{V}/\overline{Z}$, with $\overline{V}$ (settling rate) having a value of 10-20 m/yr. If one assumes $\sigma = 1/\sqrt{\tau_w}$ [14], then Equation 6 is equivalent to Equation 5. Both relationships have been derived from statistical analyses of data.

Larsen and Mercier [14] and Dillon and Rigler [13] have both derived other versions of Equation 4, both involving the phosphorus retention coefficient. The phosphorus retention coefficient R_c represents the portion of the phosphorus load which is retained in the lake (i.e., that portion which is not lost by outflow from the lake), and is usually equivalent to that portion which ultimately undergoes sedimentation. The phosphorus retention coefficient can be defined as:

$$R_c = 1 - \frac{[\bar{P}]_o}{[\bar{P}]_j}$$

where $[\bar{P}]_o$ = average outflow phosphorus concentration. When one assumes that the outflow concentration is equal to the average lake concentration (i.e., $[\bar{P}]_o = [\bar{p}]_j$), then the lake phosphorus concentration can also be derived as:

$$[\bar{P}] = [\bar{P}]_j \, (1 - R)$$

Larsen and Mercier [14], using approximations similar to those of Vollenweider [11,12], did a statistical analysis of several relationships for calculating R_c and concluded that the most accurate expression was:

$$R_c = \frac{1}{1 + \sqrt{1/\tau_w}} \tag{7}$$

Examination of Equation 7 shows that it is equivalent to Equation 6.

Dillon and Rigler [13], by contrast, did not attempt to define R_c independently. Rather, they used the expression

$$R_c = 1 - \frac{[\bar{P}]_o}{[\bar{P}]_j}$$

to calculate the value of σ in Equation 5, as:

$$\sigma = \frac{R_c \, \rho_w}{1 - R_c}$$

where $\rho_w = 1/\tau_w$. From this calculation, Equation 5 becomes:

$$[P]_\lambda = \frac{L\,(P)\,(1 - R_c)}{\bar{Z}\,\rho_w} \tag{8}$$

Note that in this form, Equation 8 does not predict the lake phosphorus concentration from the phosphorus load. Rather, the right hand side of Equation 8 equals the outflow phosphorus concentration. Thus, Equation 8 is a test for the hypothesis that the outflow phosphorus concentration from a lake equals the in-lake phosphorus concentration. Equation 8, therefore, cannot be used for predictive purposes. Kircher and Dillon [20], however, solved this problem by statistically elaborating R_c as a function of the hydraulic load, as follows:

$$R_c = 0.426 \cdot \exp\,(-0.271\ q_a) + 0.574\ \exp\,(-0.00949 \cdot q_a) \tag{9}$$

Because this is an independent estimate of R_c, when inserted in Equation 8, the equation may then be used to predict the in-lake phosphorus concentration from the phosphorus load.

Chapra [21] and Reckhow [22] have formulated other modifications of Equation 4b. Reckhow [22] has introduced uncertainty analysis to assess the statistical validity of the various parameters used in the models. The uncertainty terms derived by Reckhow can be used to restate trophic state predictions in terms of a probability statement which reflects the uncertainty involved in the parameters.

Reckhow [23] has also attempted to separate water bodies into three classes, using specific characteristics of each class, as well as a quasi-general solution for distinguishing lake classes. The three classes are: (1) lakes in which the hypolimnion becomes anoxic during the period of thermal stratification; (2) lakes in which the hypolimnion remains oxic and $q_a < 50$ m/yr; and (3) lakes in which the hypolimnion remains oxic and $q_a > 50$ m/yr. It remains to be seen how far Reckhow's approach can be applied to lakes of deviating nature. It is noted, however, that Reckhow's methodology is of importance in relation to further development of empirical solutions for Equation 4b.

Inclusion of Biological Parameters

The principles involved in quantifying the relationship between in-lake phosphorus concentrations and biological responses (e.g., chlorophyll levels) is a matter of straightforward statistical analyses. Sakamoto [10], Dillon and Rigler [24] and Jones and Backmann [15], have all provided examples of such analyses in predicting chlorophyll levels from either spring overturn or growing season in-lake phosphorus concentrations. While determination of the relationships between phosphorus loads and biological responses is less straightforward, the principles are the same as those for predicting lake phosphorus concentrations from loads. Vollenweider [12] and Rast and Lee [16] have provided examples of relationships between chlorophyll levels and phosphorus loads.

Thomann [25,26] and Schindler [27] have demonstrated that the same principles may apply to phosphorus load-chlorophyll response relationships as those which apply to phosphorus load-phosphorus concentration response relationships. Thomann [25], for example, indicated that empirical solutions to models such as Equation 4 are simply special cases of more complex dynamic phytoplankton models and that their predictive capabilities depend on whether the relationship between the phosphorus content of phytoplankton and the total phosphorus concentration in the water body remains relatively constant. Schindler [27,28] and Rast and Lee [16] have demonstrated that this is the general case.

Equation 4b therefore represents a reasonable predictor for chlorophyll, if the appropriate ratio between phytoplankton phosphorus and total phosphorus is used, and assuming that phosphorus is the controlling factor. Based on these modifications, Equation 4b can be altered to allow for the prediction of chlorophyll, as follows:

$$[\text{Chlorophyll}] \approx \frac{\alpha \ [P]_j}{1 + \bar{V} \ f \ (\tau_w \ \bar{Z})} \tag{10}$$

Vollenweider [12] and Rast and Lee [16] have demonstrated this relationship holds within certain boundaries, using the standard predictor $1/(1 + \sqrt{\tau_w})$. Modifications of Equation 10 are also applicable for prediction of primary productivity in water bodies.

PHOSPHORUS LOADING CRITERIA

Phosphorus criteria for water bodies can be derived from Equations 4a and 10. Although these equations are initially meant to relate phosphorus loads to qualitative lake conditions as defined by the classical lake trophic categories, it is also possible now to use these relationships in the sense of a sliding scale, with which one may arbitrarily or purposefully choose any desired water quality goal or water body response. The models can then be used to determine and evaluate the phosphorus load which must be met in order to achieve the desired goals in the water body.

Vollenweider [3], as noted earlier, made the first attempt to formulate phosphorus and nitrogen loading criteria by defining the boundary loading levels between oligotrophic and eutrophic water bodies, using phosphorus loading relative to mean depth (a measure of lake volume) as the primary criterion in affecting water body response. His initial information for determining the critical phosphorus loading level was:

$$L\ (P)_c\ \ (mg\ P/m^2\text{-}yr)\ =\ (25\ to\ 50)\ \ \bar{Z}^{0.6}$$

where: $L\ (P)_c$ = critical areal phosphorus load
  25 = factor for oligotrophic-mesotrophic boundary loading level
  50 = factor for mesotrophic-eutrophic boundary loading level.

Vollenweider noted, however, that the effect of mean depth as a sole reference parameter for determining the transition range for critical phosphorus loads was only a first approximation, and that other parameters would also have to be considered in refining the general applicability of the phosphorus loading concept. Some of these factors include the extent of the littoral zone, internal phosphorus regeneration, seasonal variation in phosphorus loads, and water renewal time (hydraulic residence time).

In order to make his phosphorus loading criteria more general, Vollenweider [11,12] subsequently modified his initial formulation to include the hydraulic residence time as an essential parameter in the loading criteria [i.e., $L\ (P)_c = f\ (\bar{Z}, \tau_w)$].

Accordingly, it is logical to postulate that the critical phosphorus load for a given lake (i.e., the tolerance of the lake for assimilating phosphorus inputs without producing excessive algal growths) is directly proportional to its volume (as manifested in the mean depth term $\bar{Z}$ = lake volume/lake surface area) and to its flushing rate. Since the hydraulic residence time is the reciprocal of the flushing rate, the loading tolerance is therefore also inversely proportional to the hydraulic residence time.

In expressing this relationship, Vollenweider [11,12] derived two versions of the formulation:

$$L\ (P)_c\ (mg\ P/m^2 \cdot yr) = [\bar{P}]_{\lambda \cdot c}\ \left(\frac{\bar{Z}}{\tau_w}\right)\ (1 + k\tau_w)$$

and

$$L\ (P)_c\ (mg\ P/m^2 \cdot yr) = [\bar{P}]_{\lambda \cdot c}\ \left(\frac{\bar{Z}}{\tau_w}\right)\ (1 + \sqrt{\tau_w})$$

where $[P]_{\lambda \cdot c}$ = critical phosphorus concentration for separating oligo-trophic and eutrophic water bodies. Assumptions would include a well-mixed lake, outflow phosphorus concentration equal to in-lake concentration, equivalent inflow and outflow rates, no net phosphorus loads from sediments and phosphorus sedimentation proportional to in-lake phosphorus concentration. $[\bar{P}]_{\lambda \cdot c}$ is usually taken for simplicity as 10 and 20 mg P/m^3 at spring overturn, with the 10 mg/m^3 defining the boundary between oligotrophic and mesotrophic conditions and 20 mg P/m^3 defining the mesotrophic-eutrophic boundary conditions [12]. The second version of the above equation is more specific and eliminates choosing values for k.

Using the above formulations as guides, Equation 4b may be rewritten for the purpose of defining phosphorus loading criteria as either:

$$[P]_{j \cdot c} \approx [\bar{P}]_{\lambda \cdot c}\ [1 + \bar{V}\ f\ (\tau_w)] \tag{11}$$

or

$$L\ (P)_c \approx [\bar{P}]_{\lambda \cdot c}\ \left(\frac{\bar{Z}}{\tau_w}\right)\ [1 + \bar{V}\ f\ (\tau_w, \bar{Z})] \tag{12}$$

Imboden [29] and Snodgrass and O'Melia [30] have produced similar expressions for phosphorus using two-layered phosphorus models.

One may use the same relationships for deriving phosphorus loading criteria for achieving specific biomass (chlorophyll) levels, in addition to their use above for achieving critical phosphorus concentrations. This allows for a more flexible consideration of phosphorus loading criteria, since the desired goal is then a level of biomass compatible with water use objectives. This is a desirable advancement of the phosphorus loading concept in that it allows water quality managers to base phosphorus loading objectives on a more readily appreciated water quality parameter related to eutrophication in lakes, namely biomass as measured by chlorophyll. Obviously, for different water uses, such phosphorus loading objectives may vary considerably. The criteria for establishing phosphorus loading objectives relating to the use of a lake as a drinking water supply, for example, will be much more stringent than criteria relating only to the recreational use (e.g., fishing, swimming) of a water body.

APPLICATION OF THE PHOSPHORUS LOADING CONCEPT TO THE GREAT LAKES

The previous sections dealt with the theoretical development of the phosphorus loading concept in limnology. This section presents a practical application of this concept in assessing phosphorus loads to the Great Lakes. Since the specific forms of the Vollenweider models used in establishing the new phosphorus loading objectives for the Great Lakes are discussed in other sections of these Proceedings (see the report by Bierman) and elsewhere [31], they will not be repeated in this section. Rather, an application of the phosphorus loading models developed by Rast and Lee [16] is summarized here to illustrate how such models may be used on a practical level to assess Great Lakes phosphorus inputs.

Introduction

The eutrophication of the Great Lakes continues to be a major concern to both the U.S. and Canada. The 1972 Great Lakes Water Quality Agreement [32] focused considerable attention on the control of phosphorus as a means of controlling Great Lakes eutrophication. Since 1972, governmental agencies in the U.S. and Canada have taken various steps to control the input of phosphorus to the Great Lakes, including effluent phosphorus limitations for municipal wastewater treatment plants discharging to the Great Lakes or their tributaries, and regulation of detergent phosphate in a number of Great Lakes states and the Province of Ontario.

In 1978 a new Great Lakes Water Quality Agreement [33] was signed, which has as one of its primary objectives, the control of eutrophication through further reducing phosphorus inputs to Lakes Erie and Ontario. New phosphorus loading objectives have been proposed in the 1978 Agreement, based on a number of revised water quality goals for the Great Lakes. These goals include restoration of year-round aerobic conditions in the bottom waters of the central basin of Lake Erie, and a substantial reduction in the present levels of the algal biomass in Lake Erie and in Saginaw Bay. The goal for Lake Ontario is reduction in the present algal biomass levels to below nuisance conditions. For Lakes Superior, Michigan and Huron (excluding Saginaw Bay), the objective was the maintenance of currently acceptable oligotrophic conditions and algal biomass levels.

The basic approach to setting the proposed target loads presented in the 1978 Agreement was to establish desirable in-lake conditions. These conditions were then related to the external phosphorus loads to the various lakes. A number of programs were also outlined for attaining the proposed target loads, including further reductions in effluent phosphorus concentrations from 1 to 0.5 mg P/l for municipal wastewater treatment plants [33] in the Lake Erie and Lake Ontario basins and reductions in phosphorus inputs from nonpoint sources (mainly agricultural and urban runoff).

The models and procedures used to derive the proposed phosphorus target loads, as well as the data base used in the process, are presented in Chapter 10 and elsewhere [31].

Coincident with these developments has been the international eutrophication study being conducted by the Organization for Economic Cooperation and Development (OECD), which focuses on quantifying and assessing nutrient load-eutrophication response relationships for water bodies. The international OECD eutrophication study is a multinational cooperative study initiated in the early 1970s for the purpose of developing quantitative models for assessing and predicting eutrophication responses of lake and impoundments to nutrient inputs. The OECD study was conducted within four regional projects, as follows:

1. Alpine Project—Swiss, German, French, Austrian and Italian Alpine lakes;
2. Nordic Project—Scandinavian lakes;
3. Shallow Lakes and Reservoirs Project—German, Dutch, Belgian, English. Spanish, Japanese and Australian lakes and impoundments; and
4. North American Project—U.S. and Canadian lakes and impoundments.

The overall study involved over 200 water bodies, almost half of which were located in the U.S. and Canada. The study concentrated on determination of nutrient load and resultant eutrophication responses as expressed by a variety of chemical and biological parameters. The data gathered in the Alpine, Nordic, Shallow Lakes and Reservoirs, as well as the Canadian portion of the North American Project, are currently being reviewed. The U.S. data have been extensively analyzed by Rast and Lee and a final report covering the

U.S. portion of the North American Project has been published by the U.S. EPA [16]. Reports on the individual lakes in the U.S. study were also compiled by the U.S. EPA [34].

The OECD eutrophication program provides a potential basis by which predictions can be made of the changes in water quality that will occur as a result of changes in the phosphorus loads to phosphorus-limited waterbodies, such as the changes brought about by implementing the proposed loading objectives put forth in the 1978 Agreement [33]. Rast and Lee [16] and Lee et al. [35] have shown that the results of the U.S. portion of the OECD eutrophication modeling program are applicable to the Great Lakes. These results include empirical models which relate phosphorus loads to several common water quality parameters, including chlorophyll concentrations, Secchi depths and hypolimnetic oxygen depletion. These models are based on the empirical approaches developed earlier by Vollenweider (e.g., see Equations 4b, 10 and 12), and represent expansions of his phosphorus loading concept for predicting additional water quality responses in water bodies. The U.S. OECD modeling approach is the basis for the model development, with a Great Lakes data base being used to derive the specific models presented herein.

The following section presents an evaluation of the expected responses of the Great Lakes to the proposed phosphorus target loads in the 1978 Great Lakes Water Quality Agreement [32]. Loads in these target loads are as follows: Lake Superior, 3400 metric tons/yr; Lake Michigan, 5600 metric tons/yr; Lake Huron (including Saginaw Bay), 4360 metric tons/yr; Lake Erie, 11,000 metric tons/yr; and Lake Ontario, 7000 metric tons/yr.

Great Lakes Empirical Phosphorus Models

Because the approaches used in this section are based on the U.S. OECD eutrophication models, a brief description of these models is presented.

Phosphorus Load - Chlorophyll-a Model

As noted earlier, Vollenweider [12] developed a relationship (Equation 10) between algal biomass (expressed as chlorophyll-a) and phosphorus load. In practical terms, the phosphorus load is incorporated in the phosphorus loading term

$$\frac{L\ (P)/q_a}{1 + \sqrt{\tau_w}} \tag{13}$$

where $L(P)$ =   annual areal phosphorus load (mg P/m^2-yr)

              q_a =   hydraulic load (m/yr) = $\overline{Z}/\tau_w$

      and τ_w =   hydraulic residence time = lake volume/annual outflow volume (m^3/yr)

Vollenweider [12] noted the above phosphorus loading term is equivalent to the mean in-lake steady-state phosphorus concentration. Vollenweider's approach thus becomes one of relating predicted phosphorus concentrations and resultant chlorophyll levels, as demonstrated previously by Sakamoto [10] and Dillon and Rigler [24]. A detailed derivation of this relationship was provided by Vollenweider [12] and Rast and Lee [16]. Figure 1 shows the relationship between mean steady-state phosphorus concentration, as developed by Rast and Lee [16], based on data from the U.S. OECD water bodies. The U.S. OECD results support the usefulness of this model

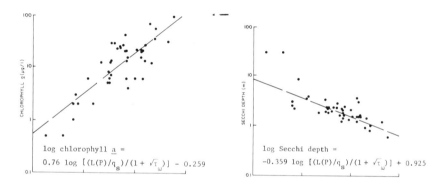

$$\log \text{ chlorophyll } \underline{a} = 0.76 \log \left[(L(P)/q_s)/(1 + \sqrt{\tau_\omega}) \right] - 0.259$$

$$\log \text{ Secchi depth} = -0.359 \log \left[(L(P)/q_s)/(1 + \sqrt{\tau_\omega}) \right] + 0.925$$

Taken from Lee et al. [39]

$$\log \text{ areal hypolimnetic oxygen depletion} = 0.467 \log \left[(L(P)/q_s)/(1 + \sqrt{\tau_\omega}) \right] - 1.07$$

$$\frac{L(P)/q_s}{1 + \sqrt{\tau_\omega}} \quad (\text{mg P/m}^3)$$

$L(P)$ = areal annual P load (mg P/m^2 · yr)

q_s = mean depth (m) ÷ hydraulic residence time (yr)

t = hydraulic residence time (yr)

Figure 1. U.S. OECD empirical phosphorus loading models [39].

as a quantitative tool for predicting expected mean summer chlorophyll-a levels as a function of phosphorus load.

Phosphorus Load-Hypolimnetic Oxygen Depletion Relationship

The use of Secchi depth (a measure of water clarity) as an indicator of algal biomass and overall water quality has been proposed by several investigators [36,37]. An additional feature of this parameter is that increased water clarity is one variable easily perceived by the public as indicative of improved water quality.

Several investigators have demonstrated an inverse, nonlinear relationship between chlorophyll content and Secchi depth in a waterbody [36,38]. Rast and Lee [16] further verified this relationship using literature values. Using chlorophyll-a as the common parameter, they developed a model which directly related Secchi depth and phosphorus load (Figure 1). Details of this model development are provided by Rast and Lee [12] and Lee et al. [35].

With this relationship it is possible to predict changes in Secchi depth resulting from altering phosphorus loads. As noted by Rast and Lee, this approach is applicable only to water bodies in which water clarity is primarily a function of phytoplankton content. It would not be applicable in its present form for water bodies with excessive amounts of inorganic turbidity or color. Further, it would be inappropriate for use in water bodies whose excessive phosphorus loads are manifested principally in excessive macrophyte or attached algal growths.

Phosphorus Load - Hypolimnetic Oxygen Depletion Relationship

Hypolimnetic oxygen depletion is of concern because of its implications for producing anoxic conditions in hypolimnetic waters. The consequences of anoxic hypolimnetic waters on cold water fisheries which frequently populate this region of a water body are obvious. Excessive hypolimnetic oxygen depletion can also produce chemically reducing conditions in the hypolimnion, causing the release of phosphorus and other materials from lake sediments back into the water column.

Rast and Lee [16] further extended Vollenweider's basic relationship to include the impact of phosphorus load on hypolimnetic oxygen depletion. The rationale for this approach is that hypolimnetic oxygen depletion is largely (though not exclusively) a function of bacterial decomposition of dead algal cells and other organic matter settled from the epilimnion into the hypolimnion. Since the growth of algae in the epilimnion is a function of the water body's phosphorus load, a relationship is also expected between phosphorus load and hypolimnetic oxygen depletion. The relationship developed by Rast and Lee is included in Figure 1. Although some scattering of data exists, a positive relationship between the phosphorus loading term and the hypolimnetic oxygen depletion is evident.

Modification of U.S. OECD Models for Analysis
of Great Lakes Water Quality

Because the U.S. OECD nutrient load-eutrophication response relationships (Figure 1) were based primarily (though not exclusively) on data from inland lakes, Rast and Lee modified their models to reflect Great Lakes conditions. The same techniques used to derive the U.S. OECD models (namely, least squares linear regression techniques applied to log transformations of the appropriate data) were applied to exclusively Great Lakes data. The resulting models are presented in Figures 2 and 3. Generally, the regression lines of best fit for the Great Lakes data exhibited a similar slope, but lower intercept than the regression line developed with the U.S. OECD data sets. It is noted that the Great Lakes regression line showed a remarkably close fit to the data sets derived for the various lakes. Part of the difference between the U.S. OECD and Great Lakes regression lines reflects the fact that some of the data upon which the Great Lakes' mean values were based were not from the summer period. This would produce average Great Lakes chlorophyll values somewhat lower than the average summer values typically used for the U.S. OECD models, and conversely would produce Great Lakes Secchi depth mean values somewhat higher than the U.S. OECD summer average values. In the case of the phosphorus load-hypolimnetic oxygen depletion model, Great Lakes data were only available for the central basin of Lake Erie. Consequently, the U.S. OECD regression line was used as the basis for evaluating the hypolimnetic oxygen depletion expected from the Lake Erie phosphorus loading objective. Development of the appropriate data base describing present water quality conditions in the Great Lakes is described in detail elsewhere [39].

It is noted that 1972-1973 chlorophyll data and 1976 phosphorus loading and Secchi depth values were used for Lake Ontario. This was done because the 1976 chlorophyll data for Lake Ontario were more limited than the 1972-1973 IFYGL data base, as well as to lend comparability to the various estimated loading changes and target load values.

Point A in Figure 2 represents the phosphorus load-chlorophyll relationship using the 1976 phosphorus load and 1972-1973 chlorophyll data; it does not fit the regression line. This anomalous fit for Lake Ontario was likely due to the lake being in a period of adjusting to phosphorus loads, which have been reduced each year since the 1972 Great Lakes Water Quality

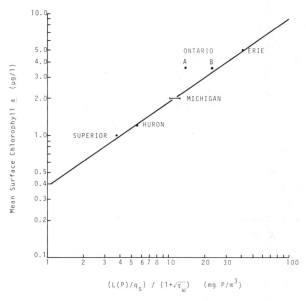

A - based on 1976 normalized base load and 1972-73 IFYGL chlorophyll
      data.
B - based on 1972 IFYGL load **and 1972-1973 chlorophyll data**

Figure 2. Phosphorus load–chlorophyll-a relationship for the Great Lakes [39].

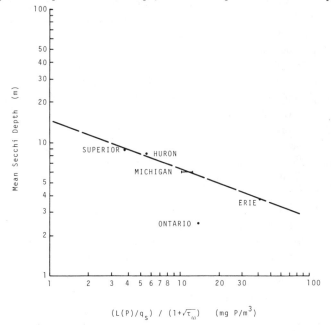

Figure 3. Phosphorus load–Secchi depth relationship for the Great Lakes [39].

Agreement. All of the Great Lakes will in fact experience a similar lag. However, the magnitude of the phosphorus load reductions in the upper Great Lakes (i.e., Superior, Michigan and Huron) called for in the 1978 Water Quality Agreement is small, relative to the present total phosphorus loads to these lakes. Therefore the "base year" (1976) loading data presented in the 1978 agreement can be used as a basis for correlating water quality to phosphorus loads. The fact that Lake Erie fits the regression line with the upper Great Lakes, even though its phosphorus loads have also been significantly reduced, can be explained on the basis of a rapid adjustment to new phosphorus loads because of its relatively short phosphorus residence time of approximately 0.3 yr [40]. Consequently, Lake Erie should adjust rapidly to changes in phosphorus loads and reach a new equilibrium condition in less than a year following an alteration in its phosphorus load (i.e., a period equal to approximately three times the phosphorus residence time). For Lake Ontario, however, the phosphorus residence time is on the order of 1.5 yr. In accord with the Sonzogni et al. model [40], Lake Ontario should essentially reach a new equilibrium about 4.5 years after a load reduction, approximately three times its phosphorus residence time. The concept of lake response time is discussed by Vollenweider [11,12,19] and Sonsogni et al. [40].

Because of its anomalous fit, Lake Ontario was not included in computation of the regression lines for the phosphorus load-chlorophyll-a and phosphorus load-Secchi depth models for the Great Lakes. However, it is important to note that when the 1972 phosphorus load is correlated with the 1972-1973 chlorophyll data (point B in Figure 2), Lake Ontario does follow the same load-response pattern as the other Great Lakes. As seen in Figures 2 and 3, there is a good correlation between the phosphorus loading expression and both the mean summer chlorophyll and Secchi depth values. These relationships appear to be appropriate for predicting the changes in water quality to be expected from the phosphorus loading objectives in the 1978 Great Lakes Water Quality Agreement.

In all the models, changes in water quality were determined by moving the data point parallel to the regression line of best fit, in response to phosphorus load reductions for given scenarios, including reduction to the proposed target loads. The present phosphorus loads and water quality conditions were used as the initial data point for each lake. The appropriateness of this procedure is discussed by Rast and Lee [16].

Effect of 1978 Water Quality Agreement Phosphorus Loading Objectives on Water Quality

Based on the models described above, plus the Great Lakes data base described by Lee et al. [39], the water quality expected to result from the implementation of the proposed phosphorus loading objectives in the 1978 Great Lakes Water Quality Agreement has been calculated. The results are summarized on the right side of Table I. Examination of Table I shows there

Table I. Comparison of Predicted Water Quality Based on
Objective and that Based on Achieving Target Loads [a]

Lake	Predicted Water Quality Based on Phosphorus Objective		Predicted Water Quality Based on Target Loads	
	Chlorophyll $(\mu g/l)$[b]	Secchi Depth (m)[b]	Chlorophyll $(\mu g/l)$	Secchi Depth (m)
Superior	1.2	8	0.9	9.3
Michigan	1.7	6.6	1.8	6.5
Huron	1.2	8	1.1	8.3
Erie	2.4	5.2	3.3	4.7
Ontario	2.6	5	1.7	6.7

[a]Taken from Lee et al. [39].

[b]Phosphorus objective data taken from International Joint Commission Science Advisory Board [41].

should be no discernible change (relative to the present conditions) in either the chlorophyll-a or Secchi depths in Lakes Superior and Huron upon achievement of the loading objectives. A slight, though still small, charge in chlorophyll-a and Secchi depth is predicted for Lake Michigan. This is not unexpected since the target loads for these three lakes were set so as to maintain present water quality. The predicted predictions in chlorophyll-a levels for Lakes Erie and Ontario are more substantial, with the Lake Erie mean chlorophyll-a level predicted to decrease from about 5.0 to 3.3 mg/l upon achievement of the target load. Corresponding changes in the average Secchi depths are also predicted for these two lakes.

It is of interest to compare the predicted water quality based on the U.S. OECD modeling approach as applied to the target loads, with the predicted water quality based on the phosphorus water quality objectives developed by the Scientific Basis for Water Quality Criteria Committee of the International Joint Commission's Science Advisory Board [41]. The predicted water quality based on the phosphorus objectives is also summarized on the left-hand side of Table I. Examination of Table I shows that there is reasonably good agreement between the water quality attainable under target load conditions and that based on the water quality objectives. From an overall point of view, it appears that the two different approaches (loads vs concentrations) yield comparable results for chlorophyll-a and Secchi depth.

In the case of hypolimnetic oxygen depletion, there was not a reliable "average" hypolimnetic oxygen depletion rate analogous to the average values established for chlorophyll-a and Secchi depth. Consequently, the range of oxygen depletion rates reported by the International Joint Commission's Water Quality Board [42] were used as the initial or "present" values. This range was from a minimum of 0.43 g O_2/m^2/day in 1970 to a maximum of 0.75 g O_2/m^2/day in 1976. Lee et al. [39] felt these minimum and maximum values were reasonable estimates of the range of depletion rates indicative of recent conditions in Lake Erie. The present phosphorus load estimate was then applied to this range, and the phosphorus load-hypolimnetic oxygen depletion model (Figure 1) was used to predict the hypolimnetic oxygen depletion expected upon achievement of the Lake Erie target load. The procedure was identical to that used for the chlorophyll-a and Secchi depth models (Figures 2 and 3).

Based on the U.S. OECD modeling approach, it was predicted that achievement of the target load for Lake Erie would reduce the hypolimnetic oxygen depletion rate from its initial range of 0.43-0.75 g O_2/m^2/day to 0.32-0.52 g O_2/m^2/day.

To translate these areal hypolimnetic oxygen depletion rates into expected oxygen levels in hypolimnetic waters at the end of the thermal stratification period, it is necessary to know certain characteristics of the Lake Erie hypolimnion. The hypolimnetic oxygen depletion depends on several factors, including hypolimnion volume, area, temperature, epilimnetic primary production, etc. Based on the U.S. OECD approach, the Lake Erie phosphorus loading objectives may or may not prevent the hypolimnion of Lake Erie from becoming anoxic, depending on the characteristics of the hypolimnion. The hypolimnetic volume (and corresponding oxygen reservoir) and the period of thermal stratification appear to be especially important parameters for Lake Erie. Examination of several sources [31,43-45] indicates the mean depth of the Lake Erie central basin hypolimnion ranges from 2.5 to 4.0 m, with a corresponding range of hypolimnetic volumes.

Using the Lake Erie loading objective of 11,000 metric tons/yr and the predicted maximum oxygen depletion rate of 0.52 g O_2/m^2/day, based on the loading objective, depletion of oxygen to undetectable levels would occur by the end of the average 110-day thermal stratification period in Lake Erie, regardless of whether a 2.5-, 3.3- or 4-m mean hypolimnetic depth and corresponding volume were used. If the same load and the minimum depletion rate of 0.32 g O_2/m^2/day were used, complete oxygen depletion would occur with a mean hypolimnetic depth of 2.5 m, but not with a mean depth of 3.3 or 4m. Using the U.S. OECD model, it is predicted that an oxygen depletion rate of approximately 0.3-0.4 g O_2/m^2/day must not be exceeded to maintain oxygen in Lake Erie's central basin hypolimnion through the end of the thermal stratification period. The approach of Cornett and Riglar [46] leads basically to the same conclusion.

It is noted that the target load for Lake Erie is based in part on long term "average" tributary flows. As noted above, it was not possible to define an average hypolimnetic oxygen depletion rate, because of the sensitivity of the depletion rate to hypolimnetic volume—depth, as well as the length of the thermal stratification period and climatic conditions. Based on the U.S. OECD relationship, it is estimated that to maintain oxic conditions (i.e., dissolved oxygen greater than 0 mg/l) in Lake Erie's central basin hypo-limnion on a consistent year-to-year basis, the Lake Erie phosphorus load would have to be reduced to the order of 2500-6500 metric tons/yr depending on hypolimnetic volume, climatic conditions, etc. Although it is noted that different hypolimnetic characteristics were assumed in the various relationships used to derive the Lake Erie target load [31], Vollenweider [47] has independently estimated that the Lake Erie load would have to be re-duced to less than 4000 metric tons/yr to maintain at least 6 mg O_2/l in the hypolimnion throughout the summer thermal stratification period. Based on this observation, Vollenweider [47] and others [39,45] have concluded that Lake Erie's central basin has probably always exhibited some degree of depleted oxygen conditions since colonization of the Lake Erie basin. Since the basis for the development of the Lake Erie target load in the 1978 Water Quality Agreement is maintenance of oxic conditions in Lake Erie's central basin hypolimnion, and since the OECD approach predicts that oxic conditions in these waters will not be achieved on a consistent year-to-year basis, it is suggested that a review of the basis for the Lake Erie target load is in order. Such a review may indicate that the target load for this lake should be based on surface water characteristics, as was done with all the other Great Lakes, rather than on hypolimnetic oxygen conditions.

Overall Conclusions

The previous several sections have demonstrated the usefulness of the U.S. OECD empirical phosphorus loading modeling approach for predicting the water quality of the open waters of the Great Lakes, based on their present phosphorus loads and the proposed target loads of the 1978 Great Lakes Water Quality Agreement. The potential utility of these relatively simple water utility relationships as management tools is evident. It is of interest to note that Simons and Lam [48], in discussing seasonal vs long-term behavior in water bodies, have compared the predictive capabilities of a simple input-output model, based on Vollenweider's mass balance approach, with a more elaborate, two-layer, two-component model. They have concluded that both models appear to have an equivalent predictive capability in terms of the overall response of a lake to changing phosphorus loading conditions.

Caution should, of course, be employed in the use of any model. The validity of the predictions of a model are based on the validity of its conceptual framework, the parameters, variables and assumptions included in the model, and the adequacy and accuracy of the initial conditions and verification data base. Deficiencies in any of these components will affect the validity and accuracy of model predictions. Biswas [49] has provided a good summary of what constitutes a "good" model. In the case of the empirical models presented here, the conditions under which the models should not be used in their present form have been summarized by Rast and Lee [16]. For example, Vollenweider developed criteria for assessment of recreational water quality in the open waters of phosphorus-limited water bodies, as manifested in excessive planktonic algal growths. It would, therefore, be inappropriate to use these criteria to assess the quality of water bodies for drinking water supplies, or for assessing the eutrophication status of water bodies whose excessive algal productivity is manifested in excessive growths of attached algae. These criteria will have to be considered in relation to the intended use of the water. As another example, the use of the phosphorus load-Secchi depth relationship (Figure 3) would be inappropriate for assessing the eutrophication status of water bodies with excessive quantities of inorganic turbidity or color, since the relationship was developed on the premise that decreased transparency in a water body was due principally to increased levels of planktonic algae in the water body. The reader is referred to Vollenweider [11,12] and Rast and Lee [16] for discussion of the appropriate conditions under which these empirical relationships should and should not be used.

Based on the results presented in this report, it is concluded that the empirical phosphorus loading relationships originally developed by Vollenweider are, and will continue to be, valuable quantitative methodologies for establishing critical phosphorus loading levels, based on desired in-lake conditions, and for predicting expected water quality changes resulting from alterations in phosphorus loads to water bodies. The application of these models for assessing nearshore water quality, and the development of such models for evaluating explicitly the effects of the biologically available fraction of the total phosphorus load to water body, would appear to be some of the logical avenues of future development of empirical modeling approaches.

REFERENCES

1. Odum, E. P. *Fundamentals of Ecology* (Philadelphia: W. G. Saunders Co., 1971), pp. 106-139.

2. Vollenweider, R. A. and P. J. Dillon. "The Application of the Phosphorus Loading Concept to Eutrophication Research," NRC No. 13690, NRC Associate Committee on Scientific Criteria for Environmental Quality, National Research Council Canada, (1974) p. 42.
3. Vollenweider, R. A. "Scientific Fundamentals of the Eutrophication of Lakes and Flowing Waters, with Particular Reference to Nitrogen and Phosphorus as Factors in Eutrophication," Technical Report DAS/CSI/68.27, OECD, Paris, France, (1968) p. 159.
4. Lee, G. F. "Role of Phosphorus in Eutrophication and Diffuse Control," *Water Res.* 7:111-128 (1973).
5. Rawson, D. S. " Some Physical and Chemical Factors in the Metabolism of Lakes," *Am. Assoc. Adv. Sci.* 10:9-26 (1939)
6. Rawson, D. S. "Morphometry as a Dominant Factor in the Productivity of Large Lakes," *Verh. Internat. Verein. Limnol.* 12:164-175 (1955).
7. Sawyer, C. N. "Fertilization of Lakes by Agricultural and Urban Drainage," *New England Water Works Assoc.* 61:109-127 (1947).
8. Ohle, W. "Bioactivity, Production and Energy Utilization of Lakes," *Limnol. Oceanog.* 1:139-149 (1956).
9. Edmondson, W. T. "Changes in Lake Washington Following An Increase In Nutrient Income," *Verh. Internat. Verein. Limnol.* 14:167-175 (1961).
10. Sakamoto, M. "Primary Production of Phytoplankton Community in Some Japanese Lakes and Its Dependence on Lake Depth," *Arch. Hydrobiol.* 62:1-28 (1966).
11. Vollenweider, R. A. "Input-Output Models, with Special Reference to the Phosphorus Loading Concept in Limnology," *Schweiz. Z. Hydrol.* 37:53-84 (1975).
12. Vollenweider, R. A. "Advances in Defining Critical Loading Levels for Phosphorus in Lake Eutrophication," *Mem. 1st. Ital. Idrobiol.* 33:53-83 (1966).
13. Dillon, P. J., and F. H. Rigler. "A Test of A Simple Nutrient Budget Model Predicting the Phosphorus Concentration in Lake Water," *J. Fish. Res. Bd. Can.* 31:1771-1778 (1974).
14. Larsen, D. P., and H. T. Mercier. "Phosphorus Retention Capacity of Lakes," *J. Fish. Res. Bd. Can.* 33:1742-1750 (1976).
15. Jones, J. R., and R. W. Bachmann. "Prediction of Phosphorus and Chlorophyll Levels in Lakes," *J. Water Poll. Control Fed.* 48:2176-2182 (1976).
16. Rast, W., and G. F. Lee. "Summary Analysis of the North American (U.S. Portion) OECD Eutrophication Project: Nutrient Loading-Lake Response Relationships and Trophic State Indices," U.S. EPA Report No. EPA-600/3-78-008, U.S. EPA Environmental Research Laboratory, Corvallis, OR (1978) pp.455.
17. Thomann, R. V., D. M. DiToro, R. P. Winfield and D. J. O'Connor. "Mathematical Modeling of Phytoplankton in Lake Ontario I. Model

Development and Verification," U.S. EPA Report No. EPA-660/3-75-005, U.S. EPA Environmental Research Laboratory, Corvallis, OR (1975) 177 pp.

18. Thomann, R. V., R. P. Winfield, D. M. DiToro and D. J. O'Connor. "Mathematical Modeling of Phytoplankton in Lake Ontario. II. Simulations Using Lake I Model," U.S. EPA Report No. EPA-660/3-76-065, U.S. EPA Environmental Research Laboratory, Duluth, MN (1976) 87 pp.

19. Vollenweider, R. A. "Possibilities and Limits of Elementary Models Concerning the Budget of Substance in Lakes," *Arch. Hydrobiol.* 66:1-36 (1969).

20. Kirchner, W. B., and P. J. Dillon. "An Empirical Method of Estimating the Retention of Phosphorus in Lakes," *Water Resources Res.* 11: 182-183 (1975).

21. Chapra, S. C. "Total Phosphorus Model for the Great Lakes," *J. Envir. Eng. Div., Am. Soc. Civil Engr.* 103(EE2):147-162 (1977).

22. Reckhow, K. H. "Uncertainty Analysis Applied to Vollenweider's Phosphorus Loading Criterion," *J. Water Poll. Control Fed.* 51(8): 2123-2128 (1979).

23. Reckhow, K. H. " Lake Quality Discriminant Analysis," *Water Res. Bull.* 14:856-869 (1978).

24. Dillon, P. J., and F. H. Rigler. "The Phosphorus-Chlorophyll Relationship in Lakes," *Limnol. Oceanog.* 19:767-773 (1974).

25. Thomann, R. V. " Comparison of Lake Phytoplankton Models and Loading Plots," *Limnol. Oceanog.* 22(2):370-373 (1977).

26. Thomann, R. V. "Reply to Comment by D. W. Schindler," *Limnol. Oceanog.* 23(5):1082-1083 (1978).

27. Schindler, D. W. "Predictive Eutrophication Models," *Limnol. Oceanog.* 23(5):1080-1081 (1978).

28. Schindler, D. W., E. J. Fee and T. Ruszczunski. "Phosphorus Inputs and Its Consequences for Phytoplankton Standing Crop and Production in the Experimental Lakes Area and in Similar Lakes," *J. Fish. Res. Bd. Can.* 35:190-196 (1977).

29. Imboden, D. M. "Phosphorus Model of Lake Eutrophication," *Limnol. Oceanog.* 19:297-304 (1974).

30. Snodgrass, W. J., and C. R. O'Melia. "Predictive Tool for Phosphorus in Lakes," *Envir. Sci. Tech.* 9:937-944 (1975).

31. Vallentyne, J. R., and N. A. Thomas. "Fifth Year Review of Canada-United Staes Great Lakes Water Quality Agreement," Report of Task Group III, A Technical Group to Review Phosphorus Loadings (Windsor, Ontario: International Joint Commission, Great Lakes Regional Office, 1978), 86 pp.

32. International Joint Commission."Great Lakes Water Quality Agreement with Annexes and Terms of Reference, Between the United States and Canada," signed at Ottawa, Ontario, April 15, 1972 (Windsor, Ontario: Great Lakes Regional Office, 1972).

33. International Joint Commission. "Great Lakes Water Quality Agreement of 1978," Agreement with Annexes and Terms of Reference, Between the United States of America and Canada, signed at Ottawa, Ontario, November 22, 1978 (Windsor, Ontario: Great Lakes Regional Office, 1978).

34. Seyb, L., and K. Randolph. "North American Project—A Study of U.S. Water Bodies," U.S. EPA Environmental Research Laboratory, Corvallis, OR, (1977) 537 pp.

35. Lee, G. F., W. Rast and R. A. Jones. "Eutrophication: New Insights For An Age-Old Problem," *Envir. Sci. Tech.* 12(8):900-908 (1978).

36. Edmondson, W. T. "Nutrients and Phytoplankton in Lake Washington," in *Nutrients and Eutrophication: The Limiting Nutrient Controversy Special Sym., Vol. I, Limnol. Oceanog.*, G. E. Likens, Ed. (1972) pp. 172-193.

37. Shapiro, J., G. J. Lundquist and R. E. Carlson. "Involving the Public in Limnology—An Approach to Communication," *Verh. Internat. Verein. Limnol.* 19:866-874 (1975).

38. Carlson, R. E. "A Trophic State Index for Lakes," *Limnol. Oceanog.* 22(2):361-368 (1977).

39. Lee, G. F., W. Rast and R. A. Jones. "Use of the OECD Modeling Approach for Assessing Great Lakes Water Quality," Occasional Paper No. 42, Environmental Engineering, Colorado State University, Fort Collins, CO (1979).

40. Sonzogni, W. C., P. D. Uttormark and G. F. Lee. "The Phosphorus Residence Time Model," *Water Res.* 10:429-435 (1976).

41. International Joint Commission. "Annual Report to the Research Advisory Board," International Joint Commission (Windsor, Ontario: Great Lakes Regional Office, 1978), pp. 39-52.

42. International Joint Commission. "Great Lakes Water Quality," Appendix B: Sixth Annual Report of Surveillance Subcommittee to the Implementation Committee, Great Lakes Water Quality Board (Windsor, Ontario: Great Lakes Regional Office, 1978), p. 32.

43. Burns, N. M., and C. Ross. "Oxygen-Nutrient Relationships Within the Central Basin of Lake Erie," in *Project Hypo,* Paper No. 6, N. M. Burns and C. Ross, Eds. (Burlington, Ont.: Canada Centre for Inland Waters, pp. 85-119.

44. Burns, N. M. "Oxygen Depletion in the Central and Eastern Basins of Lake Erie, 1970," *J. Fish. Res. Bd. Can.* 33(3):512-519 (1976).

45. Charlton, M. N. "Hypolimnetic Oxygen Depletion in Central Lake Erie: Has There Been Any Change?" Scientific Series Institute, Canada Centre for Inland Waters, Burlington, Ontario (1979), 24 pp.

46. Cornett, R. J., and F. H. Rigler. "Hypolimnetic Oxygen Deficits: Their Prediction and Interpretation," *Science* 205:580-581 (1979).

47. Vollenweider, R. A. Memorandum to Members of Task Group on Phosphorus Loadings for the Re-Negotiation of the U.S.-Canada Agreement, Canada Centre for Inland Waters, Burlington, Ontario

48. Simons, T. J., and D. C. L. Lam. "Water Quality Similations for Lake

Ontario," National Water Research Institute, Canada Centre for Inland Waters, Burlington, Ontario. Unpublished (1979) 53 pp.

49. Biswas, A. K. "Mathematical Modeling and Water Resources Decision-Making," in *Systems Approach to Water Management,* (Highstown, NJ: McGraw-Hill, Inc., 1976), A. K. Biswas, Ed., pp. 398-414.

CHAPTER 10

A COMPARISON OF MODELS DEVELOPED FOR PHOSPHORUS MANAGEMENT IN THE GREAT LAKES

V. J. Bierman, Jr.

Large Lakes Research Station
U.S. Environmental Protection Agency
Grosse Ile, Michigan

INTRODUCTION

The 1978 Great Lakes Water Quality Agreement (WQA) between the Governments of the United States and Canada was signed on November 22, 1978. As part of this agreement, the U.S. and Canada established total phosphorus loading objectives for each of the Great Lakes. These objectives were based on the recommendations of Task Group III, a bilateral technical working group established under the aegis of the U.S. Department of State and the Canadian Department of External Affairs.

The general criterion used in developing the objectives was inteference with water use by man. In contrast to the approach used in the 1972 WQA, the 1978 WQA objectives were based on the resulting water quality corresponding to the phosphorus loads in each basin. In developing the 1972 objectives, the principal emphasis was on the technical feasibility of point source phosphorus control.

The basic problem in developing objectives for phosphorus loading is the identification of underlying cause-effect relationships in the lake system. External phosphorus loads are a cause, and lake responses, in terms of phosphorus, phytoplankton and dissolved oxygen concentrations, are effects. To develop loading objectives, quantitative estimates must be made of lake responses to phosphorus loads that are different from the present loads.

To estimate lake responses to changes in phosphorus loads, the Task Group (TG) used five different mathematical models for the major basins in the Great Lakes. The models used ranged from simple, empirical correlations between total phosphorus loads and several primary and secondary response parameters, to extremely complex mechanistic models which involved dynamic calculations for the major physical, chemical and biological processes that actually occur in the lakes. A feature common to all of the models was that they were based on the principle of conservation of mass. At least three of the models were used on each of the major basins. Comparisons were made among the abilities of the models to describe present conditions and among predictions of the models under changes in phosphorus loads.

The purpose of this chapter is to compare and contrast the five mathematical models used by the TG in developing the 1978 WQA phosphorus loading objectives. The problems associated with comparing the results of different mathematical models will be discussed. Special attention will be given to reconciling some of the apparent differences among model results in terms of the different assumptions used in constructing each model.

SCOPE

Only the technical aspects of the relationships between phosphorus loads and in-lake responses are presented in this paper. Discussion of the rationale for determining desirable water quality conditions in each basin is contained in the final report of the TG [1] and in Thomas et al. [2].

Results of model comparisons are presented only for Saginaw Bay, Lake Erie and Lake Ontario. These are the most highly enriched of the major basins in the Great Lakes. Although Saginaw Bay lies entirely within the U.S. border, the outflow form Saginaw Bay constitutes a substantial source of phosphorus loading to Lake Huron.

The primary water quality indicator used was total phosphorus concentration. Other indicator parameters were utilized where necessary to address specific issues in more detail in certain basins. Dissolved oxygen (DO) concentration was used in the central basin of Lake Erie. The anoxic conditions which occur in the central basin hypolimnion have a deleterious effect on certain fish species and lead to substantial phosphorus release from the sediments. Phytoplankton biomass data for different functional groups of algae were used in Saginaw Bay. Severe taste and odor problems at the principal water intake plant on Saginaw Bay were found to be statistically correlated with the blue-green component of the total phytoplankton crop.

All references to phosphorus loads and phosphorus concentrations in this chapter should be taken to mean total phosphorus unless otherwise specified. No attempt was made by the TG to separate total phosphorus into components which were available or unavailable for algal growth. It was decided by the TG that the existing data base was not sufficient to allow a meaningful distinction between these components in all of the basins. Some of the models used for the lower lakes do include an explicit distinction between available and unavailable phosphorus forms.

Only the water quality conditions in the open-water zones of the basins were considered. Nearshore water quality problems, such as excessive *Cladophora* growth, were not directly addressed. Although *Cladophora* is believed to be related to the general level of phosphorus enrichment, there is presently little scientific understanding of the dynamics of *Cladophora* in the Great Lakes. More research is needed before effective control strategies can be developed.

MODEL DESCRIPTIONS

This chapter contains only brief descriptions of the mathematical models used. The principal characteristics of each model are summarized in Table I. For detailed information, refer to the primary references cited for each model.

Vollenweider

One version of the Vollenweider model used by the TG was based on empirical correlations between total phosphorus load and in-lake concentrations of total phosphorus and chlorophyll-a [3]. This model is generally referred to as the Vollenweider loading plot model. The correlations were a function of lake depth and hydraulic detention time. The correlations in this version were developed using data from 60 temperate-zone lakes representing a range of conditions from oligotrophic to eutrophic. Another version of the Vollenweider loading plot model used included correlations developed using data from only Lakes Erie and Ontario [4].

Conceptually, the Vollenweider loading plot model originated from the solutions of a simplified mass balance model for a mixed reactor [5,6]. In practice, the Vollenweider loading plot model is based on the steady-state solutions of this mixed reactor model. Hence, it cannot give information on the response time of a system to a change in phosphorus load. The loading plot model can only give an estimate of in-lake conditions after equilibrium has been reached. Vollenweider [5] has developed other models besides the loading plot model which are time-variable and can be used to estimate such response time.

Table I. Summary of Principal Model Characteristics

Characteristic	Vollenweider (All Basins)	Chapra (All Basins)	Thomann/ DiToro (Lakes Ontario and Huron)	DiToro (Lake Erie)	Bierman (Bay)
Time Dependence					
Dynamic		X	X	X	X
Steady-State	X				
Spatial Segmentation					
None	X				X
Horizontal		X	X	X	
Vertical			X	X	
Primary Variables					
Phosphorus	X	X	X	X	X
Nitrogen			X	X	X
Silicon				X	X
Total forms only	X	X			
Available and Unavailable forms			X	X	X
Secondary Variables					
Chlorophyll	X	X	X		
Diatom/Nondiatom chlorophyll				X	
Multiclass biomass					X
Zooplankton			X	X	X
Dissolved Oxygen	X	X		X	
Direct Calculation			X	X	X
Empirical Calculation	X	X			
Input Requirements					
External loads for primary variables	X	X	X	X	X
Depth	X	X	X	X	X
Volume	X	X	X	X	X
Hydraulic detention time	X	X			
Temperature			X	X	X
Light			X	X	X
Water circulation rates			X	X	X
Sediment nutrient release rates				X	

Chapra

Chapra [7] has used the time-variable approach by Vollenweider [5] as the basis for a simple dynamic mass balance model with total phosphorus concentration as the primary variable. Total phosphorus is considered to be a nonconservative substance which does not undergo any transformation in the water column, but which is lost from the water column via an apparent settling velocity. This velocity corresponds to the net flux of total phosphorus from the water column into the sediments.

The Chapra model is a dynamic model. Given phosphorus load, volume, depth and hydraulic detention time, the model calculates in-lake phosphorus concentrations as a function of time. This model can be used to estimate response times to changes in phosphorus loads.

The Chapra model contains an empirical component which involves correlations between the primary variable, total phosphorus concentration, and the secondary variables, chlorophyll-a and DO concentrations.

The Chapra model is different from the Vollenweider loading plot model in that it involves a dynamic calculation for total phosphorus concentration. The Chapra model and the Vollenweider loading plot model are similar in that both models use empirical correlations between phosphorus concentration and chlorophyll-a and DO concentrations.

Thomann

The Thomann model was originally applied to the Great Lakes on Lake Ontario [8,9]. The Thomann model is a dynamic mass balance model which includes direct calculations for available and unavailable forms of phosphorus and nitrogen, chlorophyll-a concentration and zooplankton concentrations. Phytoplankton chlorophyll-a is a function of temperature, light and nutrient concentrations. Zooplankton concentration is a function of temperature and phytoplankton chlorophyll-a concentration.

The Thomann and Chapra models are similar in that they both include dynamic calculations for phosphorus as a primary variable. The Thomann model differs from the Chapra model in that it also includes direct dynamic calculations, as opposed to empirical correlations, for the secondary variables chlorophyll-a and zooplankton. In addition, the Thomann model also includes nitrogen as a second primary variable.

DiToro

DiToro extended the basic conceptual framework of the Thomann model and developed dynamic mass balance models for the Lake Huron-Saginaw

Bay system [10] and Lake Erie [11]. The DiToro model for Lake Huron-Saginaw Bay is essentially the same as the Thomann model for Lake Ontario. The DiToro model for Lake Erie is based on a more advanced conceptual framework.

The DiToro model for Lake Erie differs from the Thomann model in that it includes two different types of phytoplankton: diatoms and nondiatoms, direct calculation of DO concentration, and sediment nutrient release under anaerobic conditions.

Bierman

The Bierman model is similar to the Thomann and DiToro models in that it is a dynamic mass blance model which includes direct calculation of available and unavailable nutrient forms, phytoplankton and zooplankton [12,13]. The Bierman model differs from these models in that phytoplankton biomass is partitioned into five functional groups: diatoms, greens, N_2-fixing blue-greens, non-N_2-fixing blue-greens and "others". In addition, somewhat more detailed kinetic mechanisms are used to describe phosphorus and nitrogen dynamics. Saginaw Bay is the only system for which the Bierman model has been used in a predictive mode.

APPLICATIONS TO THE GREAT LAKES

Space and Time Scales

Careful consideration of spatial segmentation is an important factor in comparing the results of different models. Apparent disagreement can occur among models where there are significant spatial gradients in water quality and where the models being compared do not have the same spatial segmentation.

The Bierman model contains no spatial segmentation and was applied only to the inner portion of Saginaw Bay. The Vollenweider and Chapra models include horizontal spatial segmentation, but no vertical spatial segmentation. The Thomann and DiToro models include both horizontal and vertical spatial segmentation.

The Vollenweider, Chapra, DiToro and Bierman models were all applied to approximately the same spatial segment on Saginaw Bay. The Vollenweider, Chapra and DiToro models were all applied to the same horizontal spatial segmentation for the western, central and eastern basins of Lake Erie. The DiToro model included vertical segmentation for the central and eastern basins. The Vollenweider and Chapra models implicitly included vertical segmentation for the central basin by using hypolimnetic DO concentrations for their empirical correlations. The Vollenweider and Chapra models did not include any spatial segmentation for Lake Ontario. The Thomann model inlcuded vertical segmentation for Lake Ontario, but no horizontal segmentation.

All model results were compared for equilibrium conditions. In all cases it was assumed that reductions in phosphorus loads were instantaneous in time. The dynamic models were run until steady-state phosphorus concentrations were obtained with the new phosphorus loads. All results were expressed either as annual averages or averages over shorter time scales, depending on the particular variable and basin. Refer to Vallentyne and Thomas [1] for a discussion of response times to equilibrium.

Comparisons with Existing Data

All of the models were calibrated to existing data for the Great Lakes. The references cited for each model contain the calibration results. A complete discussion of these results is beyond the scope of the present paper.

It should be stressed that none of the models used have been tested for phosphorus loads in the Great Lakes other than the present loads. This is because there have been no significant changes in phosphorus loads during the period of time for which comprehensive in-lake data are available for comparison with model output (approximately 1967 to present). The responses predicted by these models to phosphorus load reductions are strictly best estimates and not absolute guarantees of future conditions.

The Vollenweider loading plot model [3] was used for phosphorus concentration on Saginaw Bay and chlorophyll concentrations on Lakes Erie and Ontario. This version of the loading plot model was not developed specifically for the Great Lakes. The Vollenweider results for phosphorus concentration on Lakes Erie and Ontario and DO concentration on Lake Erie were taken from Vollenweider [4]. These results were based on revised correlations developed specifically for Lakes Erie and Ontario.

The Chapra model was applied to 10 major basins in the Great Lakes as a coupled system of completely mixed reactors. The results were based on a recent calibration to historical data for the period 1970-1976 [14].

The Thomann results for Lake Ontario were taken from Thomann et al. [5]. These results were based on a calibration of the Lake 1 model to data for the period 1967-1970. Originally, phosphorus load reduction simulations were developed for three different kinetic hypotheses with the Thomann model. Subsequently, it was shown that the so-called "optimistic" kinetic assumption was consistent with the most recent data [16]. Accordingly, only the Thomann results for this assumption were used.

The DiToro results for Lake Erie were taken from DiToro and Connolly [11]. The DiToro model was calibrated to a set of field data for 1970. Subsequently, the calibrated model was successfully verified to an independent set of field data for 1975. Although the phosphorus loadings were not significantly different for these two years, the meteorological conditions were significantly different, and this led to a marked reduction in the area of anoxia of the Central Basin hypolimnion in 1975.

The Bierman results for Saginaw Bay were taken from Bierman et al. [17] and Dolan [18]. These results were based on calibration to field data for 1974 on Saginaw Bay.

Output Formats

For some of the models, a range of results was presented for a given phosphorus load. All results for the Chapra model were expressed as ranges with the extreme values corresponding to different assumptions on phosphorus feedback rates from the sediments. The Vollenweider and DiToro results for Lake Erie spanned a range of phosphorus loads, reflecting their similar uncertainties in sediment phosphorus release. The Bierman results for Saginaw Bay spanned the range between two different assumptions on boundary conditions between the inner and outer portions of the bay.

All results were expressed in terms of base year loads. Base year loads were developed by combining 1976 point source loads with 1976 diffuse source loads normalized to historical average tributary flows. This was done to account for the large year-to-year variability in the diffuse source loads caused by flow variations.

RESULTS

Saginaw Bay

Figure 1 contains the model results for phosphorus concentration in the inner portion of Saginaw Bay. Given the extremely complex and dynamic nature of this system, there was a good agreement among the different models.

Results for chlorophyll concentration were not presented because this measurement does not directly address the principal issue of taste and odor in the municipal water supplies on Saginaw Bay. In addition, the relationship between phosphorus load and chlorophyll concentration in the bay does not follow the same pattern as that observed in other parts of the Great Lakes. The Vollenweider loading plot model predicts that summer average

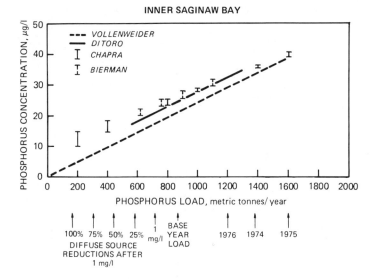

Figure 1. Relationship between P concentration and P load in Inner Saginaw Bay for the Vollenweider, DiToro, Chapra and Bierman models.

chlorophyll concentrations are approximately 25-30% of the total phosphorus concentrations in most lakes. In Saginaw Bay, this fraction is approximately 50%.

The results of the Bierman model were used for phytoplankton concentration in Saginaw Bay. Figure 2 contains results of this model for biomass of diatoms and blue-green algae relative to 1974 conditions. A statistically significant correlation has been established between measurements for taste and odor and blue-green algae concentrations at the principal water intake on Saginaw Bay [19]. Further, a statistically significant correlation has been established between measured blue-green algae concentrations in the inner portion of the bay and measured taste and odor at the intake site [18]. These results were used to relate the output of the Bierman model to taste and odor at the intake.

The phosphorus loading objective recommended by the TG for Saginaw Bay was based on the average total phosphorus results for all four models and on the results of the Bierman model for taste and odor interferences to the principal municipal water supply.

Lake Erie

Model results for Lake Erie are contained in Figures 3-10. The results for phosphorus concentration indicated reasonably good agreement among the

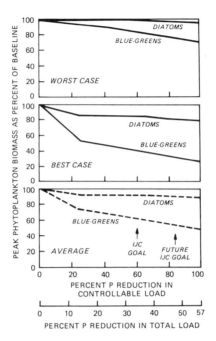

Figure 2. Relationship between phytoplankton biomass concentrations and P load reductions in Inner Saginaw Bay for the Bierman model. Reductions are relative to 1974 conditions.

Vollenweider, Chapra and DiToro models. The DiToro phosphorus concentrations showed a tendency to be slightly lower than the Chapra phosphorus concentrations, especially at lower phosphorus loads. This occurred because the DiToro results were for summer average epilimnion concentrations and the Chapra results were for annual average whole-basin concentrations.

In general, agreement among model results for chlorophyll-a concentration was not as good as the agreement for phosphorus concentration. The DiToro results for chlorophyll-a concentration in the western basin were much higher than the corresponding results for the Vollenweider and Chapra models. There was better agreement in the central and eastern basins.

The most important water quality indicator for Lake Erie was DO concentration in the central basin hypolimnion (Figure 9). The DiToro results for DO were approximately 2 mg O_2/l higher than the Vollenweider results in the loading range from 8000 to 10,000 metric ton/yr. DO results for the DiToro and Chapra models were in agreement at phosphorus loads less than 12,000 metric ton/yr; however, results for these two models progressively diverged as phosphorus loads increase from 12,000 metric ton/yr to the base year load. At the base year load, the DiToro DO concentration was approximately 2-3 mg O_2/l higher than the Chapra dissolved oxygen concentration.

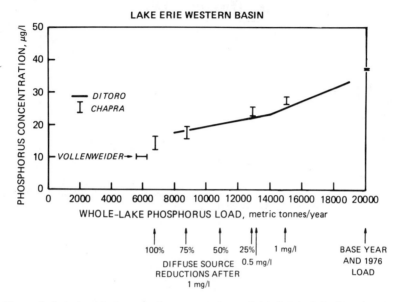

Figure 3. Relationship between P concentration and whole-lake P load in the western basin of Lake Erie for the Vollenweider, DiToro and Chapra models.

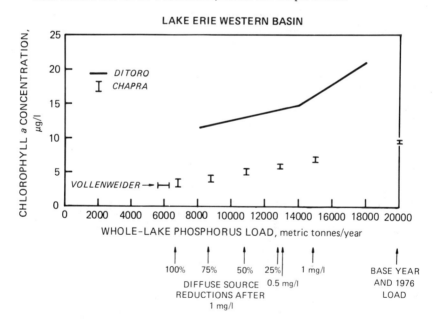

Figure 4. Relationship between chlorophyll *a* concentration and whole-lake P load in the western basin of Lake Erie for the Vollenweider, DiToro and Chapra models.

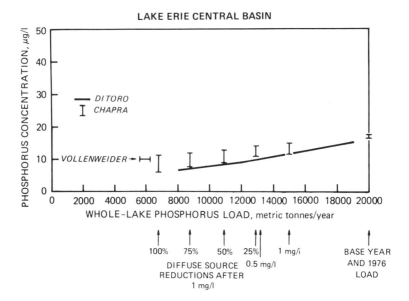

Figure 5. Relationship between P concentrations and whole-lake P load in the central basin of Lake Erie for the Vollenweider, DiToro and Chapra models.

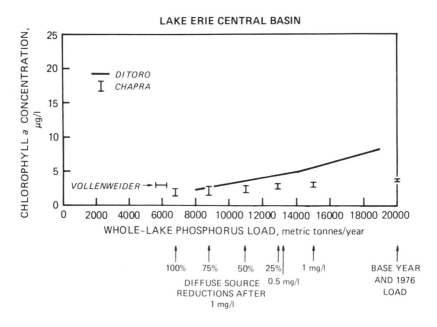

Figure 6. Relationship between chlorophyll *a* concentration and whole-lake P load in the central basin of Lake Erie for the Vollenweider, DiToro and Chapra models.

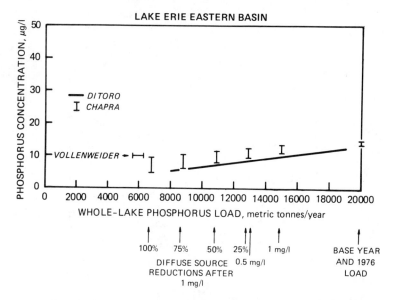

Figure 7. Relationship between P concentration and whole-lake P load in the eastern basin of Lake Erie for the Vollenweider, DiToro and Chapra models.

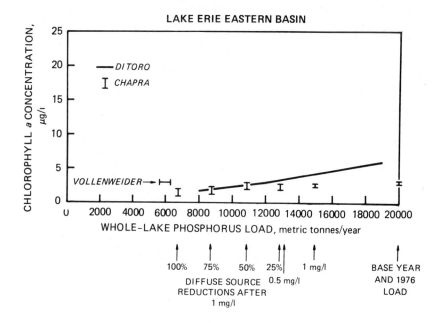

Figure 8. Relationship between chlorophyll *a* concentration and whole-lake P load in the eastern basin of Lake Erie for the Vollenweider, DiToro and Chapra models.

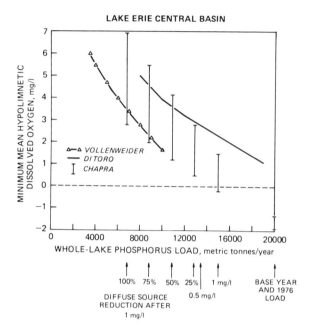

Figure 9. Relationship between minimum mean hypolimnetic dissolved oxygen concentration and whole-lake P load in the central basin of Lake Erie for the Vollenweider, DiToro and Chapra models.

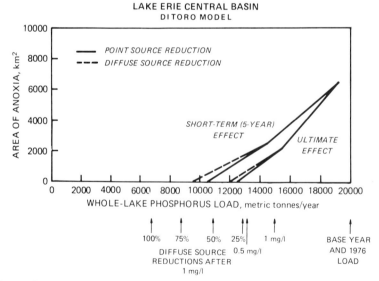

Figure 10. Relationship between area of anoxia and whole-lake P load in the central basin of Lake Erie for the DiToro model.

Differences in the definition of the hypolimnetic volume in the central basin were the probable cause of apparent discrepancies between the DO results of the DiToro model and the DO results of the Vollenweider and Chapra models. The hypolimnion in the latter two models was operationally defined by the data used in their empirical correlations with phosphorus loads. DO results for the Vollenweider and Chapra models represented average concentrations in an assumed 3.3-m hypolimnion at the end of summer stratification. DO results for the DiToro model represented average concentrations in the entire volume of water below a depth of 17 meters in the central basin at the end of the stratification period. The volume of the Vollenweider and Chapra hypolimnion thus defined was approximately 38 km^3, while the volume of the DiToro hypolimnion thus defined was approximately 51 km^3. It is expected that the DiToro results should be higher than results for the Vollenweider and Chapra models because waters at the bottom of the hypolimnion near the sediment-water interface become more oxygen-depleted than waters at the top of the hypolimnion near the 17-m depth.

To relate his results more directly to actual conditions in the central basin, DiToro correlated his model output for DO concentration to area of anoxia at the end of summer stratification (Figure 10). This correlation was developed using measured average DO concentrations in the hypolimnetic volume and the corresponding areal extent of the individual sampling stations that reported anoxic values. DiToro and Connolly [11] have shown that when the average DO concentration for the hypolimnetic volume below 17 m decreases to 4 mg O_2/l, individual sampling stations in the central basin begin to reach zero values for DO concentration. The areas encompassed by such stations were correlated with the simultaneous average DO concentrations for the overlying hypolimnetic volume. Thus, the connection was made between the DiToro model results for volumetric DO concentration and area of anoxia in the central basin hypolimnion.

Note that in the DiToro model, an average hypolimnetic DO concentration of 1 mg O_2/l corresponded to an anoxic area of 6500 km^2 (compare Figures 9 and 10). This result was in agreement with field observations and was not necessarily inconsistent with the Vollenweider and Chapra results. In comparing DO results for the Lake Erie central basin, a clear distinction must be made between volumetric conditions and areal conditions. The DiToro DO concentrations were more representative of volumetric conditions and the Vollenweider and Chapra DO concentrations were more representative of areal conditions.

The different response trajectories for short-term effect and ultimate effect in the DiToro results for area of anoxia (Figure 10) corresponded to different assumptions on phosphorus release from the deep sediment layer.

The different response trajectories for point and diffuse source reductions reflected the fact that a higher proportion of the total phosphorus load from diffuse sources was in a form which was not immediately available for algal growth. There was approximately a 10% difference in response to phosphorus loading between these two trajectories.

The phosphorus loading objective recommended by the TG for Lake Erie was based primarily on the results of the DiToro model. This was done because only the DiToro results were expressed in terms of volumetric as well as areal conditions. Volumetric results can be related to optimum requirements for fish species and areal results can be related to the extent of sediment phosphorus release under anaerobic conditions. In the judgement of the TG, there were no fundamental inconsistencies among the results of the Vollenweider, Chapra and DiToro models. The decision to rely primarily on the DiToro model was strengthened by the fact that this model was successfully verified to an independent set of data for Lake Erie.

Lake Ontario

Model results for Lake Ontario are contained in Figures 11-14. Results were presented for two cases: first, the case where phosphorus load reductions were assumed to occur only in the Lake Ontario basin and Lake Erie input was held constant; second, the case where phosphorus load reductions were assumed to occur simultaneously in the Lake Ontario and Lake Erie basins.

For both of the preceding cases, the Thomann results for phosphorus concentrations were consistently lower than the phosphorus concentrations for the Vollenweider and Chapra models. Bierman [16] has shown that during the period 1968-1974, there was a dynamic equilibrium between an average epilimnetic phosphorus concentration of 20.5 ± 3.2 μg/l, and an average phosphorus load of 14,000 ± 2190 metric ton/yr for Lake Ontario. Using data for the same period, Chapra (personal communication) determined that the average phosphorus concentration for the whole lake was 21.3 ± 2.5 μg/l. Upon extrapolating the results in Figure 11 to a phosphorus load of 14,000 metric ton/yr, it appeared that the Chapra results were closest to the actual data and that the Vollenweider and Thomann results were near the upper and lower ranges, respectively, of the standard deviations in the data.

Model results for chlorophyll-a concentration showed more scatter than the results for phosphorus concentration. This situation was similar to the corresponding results for Lake Erie.

The primary water quality indicator used for Lake Ontario was total phosphorus concentration. The phosphorus loading objective recommended by the TG for Lake Ontario was based on the average results for the

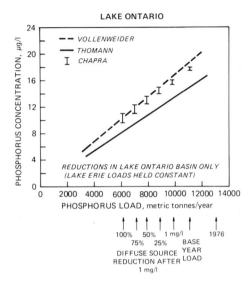

Figure 11. Relationship between P concentration and P load in Lake Ontario for the Vollenweider, Thomann and Chapra models for the case where loading input from Lake Erie is held constant.

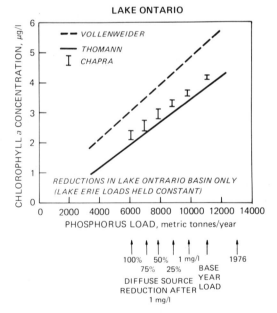

Figure 12. Relationship between chlorophyll *a* concentration and P load in Lake Ontario for the Vollenweider, Thomann and Chapra models for the case where loading input from Lake Erie is held constant.

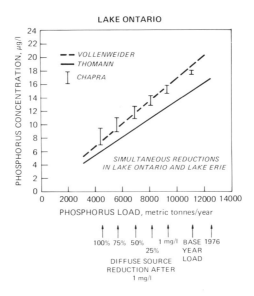

Figure 13. Relationship between P concentration and P load in Lake Ontario for the Vollenweider, Thomann and Chapra models for the case where simultaneous load reductions occur in Lakes Erie and Ontario.

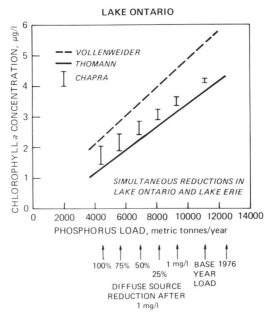

Figure 14. Relationship between chlorophyll *a* concentration and P load in Lake Ontario for the Vollenweider, Thomann and Chapra models for the case where simultaneous load reductions occur in Lakes Erie and Ontario.

Vollenweider, Chapra, and Thomann models. In the judgement of the TG, all three models adequately described the existing data.

DISCUSSION

The development of the 1978 WQA phosphorus loading objectives was the first occasion in which the results of different mathematical models were synthesized and used as a basis for management recommendations. The success of this synthesis reflects the fundamental underlying unity of these models, in spite of the fact that they were developed independently and under different sets of assumptions. This approach provided the strongest possible basis for scientific credibility of the loading objectives.

It should be re-emphasized that none of the models used have been tested for phosphorus loads in the Great Lakes other than the present load. The loading objectives should be considered best estimates based on state-of-the-art research and not absolute guarantees of future conditions. In perspective, the models have provided a quantitative framework for organizing and interpreting the existing data. This framework is a far more logical basis for extrapolating from present conditions than mere intuition.

An important unresolved research issue not addressed by the TG is the question of phosphorus availability. It has been argued that control of unavailable phosphorus forms is unnecessary because these forms cannot be directly utilized for algal growth. This argument fails to consider that a certain fraction of the unavailable portion of the total phosphorus load will become available in the water column through chemical and biological transformation reaction and the net settling flux of unavailable phosphorus to the sediments before it can undergo transformation. In this context, it is not meaningful to consider only the ratio of available to unavailable phosphorus forms in the total phosphorus load at the delivery point to a system.

A clear distinction should be made between resolution of the scientific questions concerning phosphorus availability and the implications of a resolution to these questions in a mangement context. Recall that the Vollenweider and Chapra models included only total phosphorus and the Thomann, DiToro and Bierman models included explicit distinctions between available and unavailable phosphorus forms in the external loads as well as in the water column. In spite of uncertainties in describing the dynamics of unavailable phosphorus forms, these latter models successfully described the present data for available and unavailable concentrations in the lakes. Further, all five models agreed well with each other for total phosphorus concentrations under present conditions and under conditions of reduced phosphorus loads.

It is the author's opinion that a final resolution of the phosphorus availability issue will not have a significant impact on the whole-lake phosphorus loading objectives. This issue, although it represents an important scientific question, should not be a cause for delay in implementing the 1978 WQA phosphorus loading objectives for the Great Lakes.

REFERENCES

1. Vallentyne, J. R., and N. A. Thomas, "Fifth Year Review of the Canada-United States Great Lakes Water Quality Agreement," report of Task Group III, a Technical Group to Review Phosphorus Loadings, U.S. Department of State, (1978) 86 pp.
2. Thomas, N. A., A. Robertson and W. C. Sonzogni. "Review of Control Objectives, New Target Loads, and Input Controls," Chapter 4, this volume.
3. Vollenweider, R. A. "Advances in Defining Critical Loading Levels for Phosphorus in Lake Eutrophication," *Mem. Ist. Ital. Idrobiol.* 33:53-83 (1976).
4. Vollenweider, R. A. "Memorandum to Members of Task Group III on Phosphorus Loadings for the Re-negotiation of the U.S.-Canada Agreement." (July 7, 1977).
5. Vollenweider, R. A. "Possibilities and Limits of Elementary Models Concerning the Budgets of Substances in Lakes," *Arch. Hydrobiol.* 66:1-36. (1969).
6. Vollenweider, R. A. "Input-Output Models with Special Reference to the Phosphorus Loading Concept in Limnology. *Schweizer. Z. Hydrol.* 37: 53-84. (1975).
7. Chapra, S. C. "Total Phosphorus Model for the Great Lakes," *J. Environ. Eng. Div., Am. Soc. Civil Eng.* 103(EE2):147-161 (1977).
8. Thomann, R. V., D. M. DiToro, R. P. Winfield and D. J. O'Connor. Mathematical Modeling of Phytoplankton in Lake Ontario. 1. Model Development and Verification," U.S. Environmental Protection Agency Ecological Research Series, EPA-660/3-75-005, (1975), 177 pp.
9. Hydroscience Inc. "Assessment of the Effects of Nutrient Loadings on Lake Ontario Using a Mathematical Model of the Phytoplankton," Report prepared for the Surveillance Subcommittee, Water Quality Board, International Joint Commission, Windsor, Ontario, (1976), 116 pp.
10. DiToro, D. M., and W. Matystik, Jr. "Mathematical Models of Water Quality in Large Lakes. Part I: Lake Huron and Saginaw Bay Model Development, Verification, and Simulations," U.S. Environmental Protection Agency Ecological Research Series (in press).
11. DiToro, D. M., and J. F. Connolly. "Mathematical Models of Water Quality in Large Lakes. Part II. Lake Erie," U.S. Environmental Protection Agency Ecological Research Series (in press).

12. Bierman, V. J., Jr., and D. M. Dolan. "Mathematical Modeling of Phytoplankton Dynamics in Saginaw Bay, Lake Huron," in *Environmental Modeling and Simulation, Proceedings* of a conference sponsored by the U.S. Environmental Protection Agency held at Cincinnati, Ohio, April 19-22, 1976. pp. 773-779 (1976).

13. Bierman, V. J., Jr., D. M. Dolan, E. F. Stoermer, J. E. Gannon and V. E. Smith. "The Development and Calibration of a Multi-class, Internal Pool, Phytoplankton Model for Saginaw Bay, Lake Huron," U.S. Environmental Protection Agency Ecological Research Series (in press).

14. Chapra, S. C. "Effect of Phosphorus Load Reductions in the Great Lakes," report prepared for the Large Lakes Research Station, U.S. Environmental Protection Agency, Grosse Ile, Michigan (1978).

15. Thomann, R. V., R. P. Winfield, D. M. DiToro and D. J. O'Connor. "Mathematical Modeling of Phytoplankton in Lake Ontario II. Simulations Using Lake 1 Model," U.S. Environmental Protection Agency Ecological Research Series, EPA-600/3-76-065 (1976) 87 pp.

16. Bierman, V. J., Jr. "Evaluation of Hydroscience Lake Ontario report to the Surveillance Subcommittee, Water Quality Board, International Joint Commission," prepared for the Expert Committee on Ecosystems Aspects, International Joint Commission, Windsor, Ontario, (1977) 21 pp.

17. Bierman, V. J., Jr., W. L. Richardson and D. M. Dolan. "Responses of Phytoplankton Biomass in Saginaw Bay to Changes in Nutrient Loadings," report prepared for the International Reference Group on Upper Lakes Pollution, International Joint Commission, Windsor, Ontario (1975).

18. Dolan, D. M. "Memorandum on Whitestone Point Odor Reductions as a Function of Total Phosphorus Loads," Large Lakes Research Station, U.S. Environmental Protection Agency, Grosse Ile, Michigan (September, 1977).

19. Bratzel, M. P., M. E. Thompson and R. J. Bowden, Eds. "The Waters of Lake Huron and Lake Superior. Vol. II (Part A). Lake Huron, Georgian Bay, and the North Channel," report to the International Joint Commission by the Upper Lakes Reference Group. Windsor, Ontario. (1977), 292 pp.

SECTION III

AVAILABILITY OF PHOSPHORUS TO AQUATIC LIFE

CHAPTER 11

AVAILABILITY OF PHOSPHORUS TO PHYTOPLANKTON AND ITS IMPLICATIONS FOR PHOSPHORUS MANAGEMENT STRATEGIES

G. F. Lee and R. A. Jones

Department of Civil Engineering
Colorado State University
Fort Collins, Colorado

W. Rast

International Joint Commission
Washington, D. C.

INTRODUCTION

One of the questions associated with the development of a cost-effective phosphorus management strategy for a waterbody concerns whether the control program should be based on total phosphorus or on algal available phosphorus entering the waterbody. Until now, eutrophication control programs have been based largely on the control of any form of phosphorus that was amenable to control, irrespective of whether the phosphorus was in a form which could support algal growth.

Thus far, with few exceptions eutrophication control programs in the Great Lakes and other areas have been largely directed toward the control of phosphorus inputs from domestic wastewaters. These programs have included complete elimination of phosphorus input from this source via diversion of wastewaters to another waterbody, limitation of wastewater treatment plant effluent P concentrations to 1 mg P/l and limiting the phosphorus in detergent formulations. Major wastewater diversions have

259

taken place in the Madison, Wisconsin lakes, from Lake Washington in Seattle, Washington and others. Dramatic improvements in eutrophication-related water quality have been found in several waterbodies where wastewater diversion has occurred, the most notable example being Lake Washington.

A second method for P control that has been attempted in several locations is the limitation of the P content in detergent formulations. These limitations have ranged from a few percent by weight allowable phosphorus, such as in Ontario, Canada (2.2% P allowed in formulations), to "complete" detergent P bans where the P content is limited to "trace" amounts. "Trace" is generally translated to a maximum of 0.5% P in the formulation. In the early 1970s the complete bans, such as passed in Indiana and Dade County, Florida, resulted in a decrease in phosphorus content of domestic wastewaters of about 50 to 60%. Today, with the voluntary decrease in the phosphorus content in detergent formulations that has taken place across the U.S., a complete phosphorus ban would be expected to reduce the phosphorus concentration in domestic wastewater treatment plant influents where no P limitations are already in effect, by 30 to 35%. Several years ago, Michigan enacted a detergent P ban. It has been found by comparison of P concentrations in Michigan domestic wastewater treatment plant influents before and after initiation of the ban, that the phosphorus concentration in domestic wastewaters generally decreased by 20-35%, although some cities such as Detroit showed no apparent decrease in domestic wastewater P concentration after initiation of the ban. The smaller percent concentration decrease was generally found for cities with a high percentage of the total flow to the municipal system coming from industrial wastewater sources. Attempting to control eutrophication by detergent P bans is generally of questionable value because of the relatively small reduction in phosphorus input that can be achieved by this approach. Thus far there have been no documented cases where there have been measurable improvements in water quality in lakes or impoundments because of detergent P limitation.

In the Great Lakes region, the primary approach to eutrophication control has been to establish limitations on phosphorus loads from domestic waste-water treatment plants based on what could be readily removed by chemical treatment processes (iron, alum or lime addition). Several years ago, when domestic wastewaters contained around 10 mg/l P (late 1960s to early 1970s), it was found that 90% removal of phosphorus could be attained without greatly increasing the cost of domestic wastewater treatment. This reduction translated into a 1-mg/l P effluent concentration. With the decrease in the phosphorus content of domestic wastewaters because of voluntary reductions in the phosphorus content of detergent formulation, domestic wastewaters today typically contain on the order of 5-7 mg/l P. It is generally found today that P removal to 1 mg/l P can be achieved at a maximum total cost on the order of 0.25¢ per person per day for the population served, for treatment plants serving 10,000 people or more. The treatment of domestic wastewaters for phosphorus removal has not yet produced a dramatic improvement in Great Lakes water quality, although it is expected that some measureable improvement will be found in the Lower Great Lakes and some parts of the Upper Great Lakes as a result of achievement of the 1 mg/l P limitation in domestic wastewaters. It is generally agreed that this approach has halted Great Lakes eutrophication-related water quality deterioration, but that measures will have to be taken to reduce further the P load to see a significant improvement in eutrophication-related water quality in the lower Great Lakes.

Somewhat fortuitously, the phosphorus control programs that have been initiated have focused on what are generally considered to be available forms of phosphorus, since the forms of phosphorus in effluents from domestic wastewater treatment plants not practicing phosphorus removal are believed to be in chemical forms largely available to support algal growth (i.e., algal-available phosphorus). With attention being given to reducing the phosphorus load from other sources as well as additional removal from domestic wastewaters, increasing emphasis must be placed on assessing the algal availability of phosphorus from controllable sources in order to develop cost-effective management strategies. This assessment is of particular significance to the Great Lakes because of the 1978 U.S.-Canada Water Quality Agreement [1] committing the U.S. and Canada to further reductions in P loads to the Great Lakes. This paper reviews the current state of information on the algal availability of phosphorus from major sources contributing P to the Great Lakes, methods for assessing the algal availability of phosphorus and the importance of developing phosphorus management strategies based on algal-available phosphorus rather than on total phosphorus. Although the focal point of the paper is the Great Lakes, the results are equally applicable to waterbodies throughout the world.

PREVIOUS STUDIES ON AVAILABLE P INPUTS TO THE GREAT LAKES

The amount of algal available phosphorus in an aquatic system is a function of a complex set of physical, chemical and biological processes. Attempts have been made to assess the amounts of sediment and soil-associated phosphorus available for aquatic plant growth, using both chemical fractionation and biological procedures. Although a variety of biological techniques have been used, the preferable method involves the use of a standardized batch algal culture test, such as that recommended by the U.S. EPA [2,3]. In this algal culture test (bioassay), all conditions needed to support algal growth are optimal for a particular alga, except for the phosphorus concentration. The phosphorus needed for algal growth in the test culture must be derived from the total P present in the water sample being investigated. A set of standard algal cultures is prepared in which various levels of available P are added to culture media and inoculated with algae. A calibration curve is developed, based on the standards, to relate the available phosphorus in the cultures to the algal biomass present on the plateau of the resulting growth curve. The amount of algal biomass present in the test water sample at the end of a 1- to 2-week incubation period (i.e., on the growth plateau) is compared to the calibration curve to estimate the amount of phosphorus in the sample that became available to the algae during the incubation period. Other biological tests have been employed by some investigators in order to short-cut the several-week incubation period. As discussed in subsequent sections of this paper, a number of these procedures have significant limitations in their ability to estimate algal available phosphorus in a sample.

The chemical extraction techniques have their basis in the soils literature, where empirical correlations have been developed between the amount of phosphorus extracted from soils with various reagents under certain laboratory conditions and the growth of terrestrial crops on these soils. NaOH is believed to extract from particulates phosphorus associated with iron and aluminum (sometimes called "nonoccluded" P). These forms have been generally found to be available to terrestrial crops. Dilute HCl extracts some calcium phosphates-apatite, which are largely unavailable or very slowly available to support the growth of terrestrial vegetation. Although a number of investigators have applied these procedures directly to aquatic systems, few have investigated the appropriateness of these soil extraction techniques for determining the algal availability of suspended particulate, atmospheric, or sediment-associated P. There may be difficulty in applying procedures designed for the measurement of root uptake of P from soils to the estimation of algal uptake of P from aqueous systems since the conditions of uptake are somewhat different.

Tributary Inputs

One of the early studies of availability of particulate P for algal growth examined the availability of phosphorus associated with sediments from certain Dutch lakes. They indicated that most (approximately 90%) of the P entering these lakes is stored in their sediments. Since primary production in these lakes was high, they suggested that either phosphorus is rapidly regenerated by algae, or there is a regeneration of the sediment-associated phosphorus. As a result of algal assays run in which sediment was the only P source, it was found that 7-30% of the sediment P became available for algal growth during the 3- to 4-week assays. From chemical analyses, they determined that the extractable portion of the sediment phosphorus decreased during the assay by an amount approximately equal to the amount taken up by algal cells.

Golterman [5] found that only a small amount of the phosphate in Frisian marine sediments (total P content 0.9 mg P/g) was available for growth of *Scenedesmus*. He estimated the amount available was equivalent to that portion extracted with NTA (0.01 N) which is said to be on the order of magnitude of the sum of calcium and iron phosphates bound in the sediments. Golterman [5] indicated that recently sorbed phosphate is biologically and chemically different from structural phosphate molecules, in that it tends to become available for algal growth. The P firmly bound to clay or humic compounds is not readily available to support algal growth.

Golterman [5] and Golterman et al. [4] have indicated that "$FePO_4$" and hydroxyapatite are available to stimulate algal growth, whereas $Ca_3 (PO_4)_2$ shows little availability. This is not in agreement with data reported by the IJC Pollution From Land Use Activities Reference Group (PLUARG) with respect to hydroxyapatite [6]. In algal assays, PLUARG found that in cultures receiving soluble orthophosphate as a P source, there was an immediate and rapid increase in primary production. Cultures receiving only erosional materials from Lake Erie shortline bluffs, in which most of the P was in the form of insoluble apatite, showed a maximum of 20% of the average response to soluble ortho-P spikes. They concluded that there was not enough release of P from the bluff material to support high levels of productivity. This point is discussed further in subsequent sections.

One of the most comprehensive studies concerned with assessing algal available P was conducted in 1972-1973 by Cowen and Lee [7,8] and Cowen [9], who studied the algal availability of particulate P in stormwater runoff from Madison, Wisconsin and tributary waters to Lake Ontario. They evaluated a variety of chemical and physical extraction techniques (acid, base and anion exchange resin) for estimating the fraction of P that becomes available to algae during an 18-day period, using a number of bioassay techniques. Bioassays were run on the separated particulates (without drying) in Algal Assay Procedure (AAP) medium (minus P), on the filtered tributary water after autoclaving or on the particulates in AAP (-P) medium after autoclaving. A variety of other procedures were also used, including chloroform treatment followed by incubation, and short and long dark incubations of tributary water and lakewater with tributary particulates added with anion exchange resin to determine microbial and organic mineralization contributions to available P. Dark incubations of tributary water and lakewater with tributary particulates added were conducted with and without anion exchange resin to determine the effect of soluble P concentration on P release from particulates.

Urban runoff samples were collected by Cowen and Lee [7,8] and Cowen [9] during 12 runoff events between August 1972 and March 1973 from eight locations in Madison, Wisconsin receiving runoff from residential, commercial and urban construction land areas. Acid-extractable inorganic particulate P (extraction with concentrated $(0.083N)HCl + H_2SO_4$) ranged from 13 to 60% of the total particulate phosphorus (PP_T) in the individual samples; means for the variety of urban land-use fell into the 33-46% PP_T range. Between 9 and 49% PP_T was extracted using a base (NaOH) treatment. Mean values for various land uses ranged from 22 to 27% PPT. The anion exchange resin extraction was designed to measure that fraction of inorganic P (P_i) involved in solid-solution exchange. From 2 to 28% of PP_T was extracted from the urban runoff samples collected, with mean values for different urban land areas varying from 13 to 17% PP_T. There was no discernible difference in percent extractable P_i in particulates from the urban land areas evaluated. With the particulates from runoff as the only source of P for *Selenastrum* in P-free AAP medium, between 8 and 55% of the PP_T became available for growth (based on comparison of amount of algae present after 18 days in test water with that in the AAP standard assays). While the amount of P from Madison urban runoff extracted with acid was greater than that extracted with NaOH, which was greater than that with anion exchange resin, the overall means did not vary greatly (between 15 and 38% PP_T). The mean percent P available from PP_T as measured by *Selenastrum* growth was 30%, which is in the midst of that range. In 6 of the 10 samples tested by base extraction, the percent PP_T extracted and the percent PP_T used by *Selenastrum* for growth were within 10%; in 7 of the 13 samples tested using anion exchange resin extraction, and in 4 of the 13 samples extracted with acid, the extraction and bioassay results varied by less than 10% of the PP_T. This indicates that the NaOH and anion exchange resin extractions were, for the samples evaluated, a better estimator of the amount of PP_T available for *Selenastrum* growth than acid extractions. As discussed by Cowen and Lee [8], specific conclusions regarding the appropriateness of particular chemical extractions for predicting availability of P could not be made on the basis of their data; a host of other factors influence the eventual availability of particulate P in various aquatic systems. The effective availability would, because of these factors, likely be lower than that predicted by *Selenastrum* growth and would likely be better predicted for these samples by anion exchange resin extraction. In general, there was little evidence of mineralization of organic particulate P in the urban runoff. Release of P from particulates was apparently controlled by physical-chemical sorption-desorption and precipitation-dissolution reactions.

Data were collected for Cowen and Lee [8] and Cowen [9] for two urban-residential areas in the Genesee River Basin. Eight samples were collected from both stations between October 1972 and June 1973. The mean percents acid extractable PP_i of PP_T were 30 and 48%, which were comparable to the mean 33 to 46% found for Madison, Wisconsin urban runoff. The apparent available P from 18-day *Selenastrum* bioassays on the 16 samples from the two Genesee River Basin sites ranged from < 1 to 34% PP_T, typically on the order of the apparent availability found for Madison, Wisconsin runoff.

Cowen and Lee [8] and Cowen [9] also collected data on seven water samples from three streams in the Genesee River Basin draining crop, brush and pastureland, collected between November 1972 and June 1973. In a separate sampling program they collected 34 tributary water samples from 4 major New York tributaries to Lake Ontario (Black, Oswego, Genesee and Niagara Rivers) between August 1972 and June 1973. The samples from the Genesee River Basin draining nonurban areas showed about 20-35% PP_T extracted with acid, 13-18% PP_T extracted with NaOH, and 6 to 17% PP_T extracted with anion exchange resin. Apparent availability to *Selenastrum* of the particulate-associated P in these samples ranged from about < 3 to 10%, but as high as 20% when the samples wre autoclaved prior to inoculation and assay. The four Genesee River mouth water samples collected as part of the Lake Ontario, New York tributary sampling program showed that 21-79% PP_T was base-extractable, and 6-31% PP_T was extractable with anion exchange resin. These results were comparable to the other set of Genesee River Basin samples. The other New York tributary samples were not evaluated for chemically extractable forms, but they were evaluated using bioassays. The whole river water samples from the New York tributaries evaluated showed very low P availability, typically < 5% of the total PP_T. The difference between P chemically extracted and that taken up by *Selenastrum* may have been due, according to Cowen and Lee, to competition for P by native algae and bacteria present in the sample at the time of collection. When samples were autoclaved prior to inoculation and incubation, availability of particulate P increased to 26-57% PP_T.

Cowen and Lee proposed that the short-term resin extraction incubation carried out in the dark may give a better estimate of readily available PP_T than bioassay of the natural PP_T alone, since the resins would take up all the P that would normally be taken up by the algae and bacteria. This amount could readily be determined using this technique. They also suggested, without further substantiation, that by autoclaving river water particulates in AAP (−P) (without filtration) a better estimate of potentially available P can be obtained.

Cowen and Lee [7] concluded that in the absence of site-specific data, an upper-bounds estimate could be made of the available P in urban runoff or tributary waters using the following equation:

$$\text{Available P} = \text{TSP} + 0.3 \ PP_T \tag{1}$$

where TSP is the total soluble P, or where there is a large soluble organic P contribution, soluble reactive P (SRP); and PP_T is the total particulate P. Subsequently it has been found by one of the authors (Lee) that due to factors affecting P availability in the receiving waters, namely particle residence time in the photic zone, and the optimum nature of algal assay tests, a more appropriate estimate of the ultimate availability of tributary water phosphorus would be:

$$\text{Available P} = \text{SRP} + 0.2 \ PP_T \qquad (2)$$

A number of recent investigators have cited the work of Sagher [10] to ascribe a major significance to NaOH extractable inorganic phosphorus (P_i) in assessing the algal availability of tributary particulate phosphorus. Sagher [10] proposed a method for assessing the "availability" of particulate P, in which NaOH extractable P_i + dissolved P_i are determined in a soil/water system (to which P-starved algae have been added) at the beginning of the assay and after 48-hr incubation. The difference between the NaOH-P_i + dissolved P_i at those two times supposedly gives an estimate of the amount of P "available" to algae. The evaluation of this method was based on two sets of experiments. One set involved correlating NaOH-extractable P_i and dissolved P_i to algal numbers during 30-day bioassays, and the other involved measuring the change in concentrations of the various P_i forms over a 48-hr period in a laboratory sediment-soil-algae system.

Sagher's "growth-related" P uptake experiments involved measuring the amounts of NaOH and HCl extractable P before and during 30-day bioassays, using three soil horizon samples from a Wisconsin Miami silt loam soil and *Selenastrum* cultures. NaOH-extractable P_i generally showed decreasing trends in the bioassays in which the soil was the only P source, indicating to Sagher that the NaOH-P_i fraction was highly mobile and was the source of replenishment of phosphorus for the algae. In one of the three experiments the NaOH-P_i fraction increased. It was assumed that organic P contribution to available P would be negligible, but that aspect was not investigated by Sagher [10]. Sagher found that, in assays not receiving supplemental available P over the 30-day bioassay period, the total P_i decreased by 2, 37 and 54% in the three soil samples. These decreases corresponded to decreases in the NaOH extraction P of –50 (i.e., increased by 50%), 82 and 65% respectively. These numbers were mathematically manipulated to yield what he termed microbial P to estimate what had been taken up and used by algae.

Correlating P_i concentrations with corresponding cell numbers over the culture period (at 0, 15 and 30 days), as Sagher did as part of this experiment, is not appropriate if the goal is to assess the capability of chemical procedures to estimate the amount of P available for algal growth. Using the initial levels, it should be possible, if the procedure is valid, to predict some endpoint algal biomass. While the correlation between "microbial P" decrease and microbial P based on cell numbers at the end of the incubation period made by Sagher was often not strong, there was an apparent relationship between the initial NaOH-P_i and the number of *Selenastrum* cells present at the end of the incubation period which was not discussed by Sagher. It should be noted again, however, that the correlations made between algal numbers and NaOH extractable P_i were based on samples of three horizons on one soil. As discussed by Sagher, in order to make any generalized statements about such relationships, a much wider variety of samples needs to be evaluated using this procedure.

Sagher [10] also conducted a series of experiments on the < 20-μ fraction of six soil samples from two watersheds in Wisconsin, in which dissolved P, NaOH-P_i and HCl-P_i fractions were measured over 48-hr periods in soil-water-*Selenastrum* bioassays to determine shorter-term availability or uptake by algae. He found that after 48-hr incubation, generally 60-70% of the initial NaOH-P_i plus dissolved P_i were removed from solution. According to Sagher, this translated into an algal assimilation of 70-90% of the initial NaOH-P_i plus dissolved P_i.

Sagher [10] conducted the same type of experiments on simulated runoff, in which an equivalent of 4.6 in. (11.7 cm) rainfall were applied to two plots of agricultural land and runoff was collected. The percentage of NaOH-extractable plus dissolved P_i available from these unfractionated samples was 10-30% higher than that typically found by Sagher using the fractionated soil samples and the fraction of NaOH-P_i that became available. The dissolved P_i concentrations in the runoff were comparatively high. This suggests that this approach, if truly applicable to land runoff, may be highly specific to certain types of land.

While Sagher's recommended procedures appear to have some applicability to the algal availability of the phosphorus in shoreline erosion such as occurs from the bluffs of Lake Erie, and from certain types of soils, it may not be appropriate for assessing the availability of tributary particulate P at the point where the tributary empties into a waterbody, such as one of the Great Lakes, because of the differences in the nature of the materials. This point will be discussed further in a subsequent section. Further, there are several aspects of Sagher's methodology which could distort the results obtained. The soil samples were dried prior to processing which would likely change their sorption/desorption character for representing tributary suspended matter. In the soil preparatory step to fractionate particle size, the soil was leached for about six hours, and (although not specified in the procedure, it was ultimately recommended) the suspension was stored for several days. These steps alter the "initial" concentrations of the various P_i forms and likely do not result in a consistent error across a variety of soil samples. Sagher used a very limited number of soil samples from a restricted geographical area. The general applicability of his procedures must be evaluated. As discussed by Sagher, his method of separating "dissolved P_i" may have not separated out some finely divided particulates/colloidal forms which were included in the "dissolved" fraction.

Sagher [10] also proposed an indirect chemical method for assessing available P_i which only required analysis of the soil suspension. This was based on the "fact that essentially all of the soil runoff NaOH-P_i (0.1 N NaOH extraction for one hour)" assumed from the context of Sagher's thesis to include dissolved P_i, "present in the bioassay system was assimilated by P-limited *Selenastrum* after (assumed to be *during*) the 48-hr incubation period." Examination of the data base presented on five soil samples from two watersheds shows that "essentially all available" ranged from 66 to 94% available (with a mean of 79% available) in those assays not spiked with P; those spiked showed much less NaOH-P_i utilization. This procedure would not likely be applicable to soils with high organic content since it does not take into account the mineralization of organic P that would take place over time in a lake. From an overall point of view, the work of Sagher alone, as indicated by Sagher as well, is not adequate justification for use of the NaOH extractable P_i as a measure of biologically available particulate P at the tributary/lake interface. Therefore, those studies discussed below which have relied on the Sagher [10] work for justifying the use of NaOH extraction as a measure of available P must be viewed in light of this potentially significant deficiency.

Sagher et al. [11] conducted four-week bioassays and chemical fractionation studies on seven samples of Wisconsin lake sediments to determine the availability of sediment P to algae. P-starved algae were added to mixtures of wet sediment and algal assay medium with and without added P. Chemically, availability was assessed in terms of a decrease in total P_i concentration (HCl extractable P_i plus dissolved P_i concentration), which was mathematically converted to increases in microbial P. They indicated that typically on the order of 50-95% (average of about 75%) of the total HCl extractable P_i was available to algae (i.e., was converted to microbial P). Most of that which became available was from the NaOH extractable P rather than from apatite-P. They found that on the average of about 74% of the NaOH-P_i plus dissolved P_i (50-93% range) was removed during the course of the 3- to 4-week bioassays. Sagher [10] found that on the average approximately 79% of NaOH extractable P_i + dissolved P_i in soils was removed by algae in short-term (48-hr) uptake studies. One of the major differences between these two sets of results [10,11] was in the rate of decrease in extractable P_i levels. Sagher [10] reported that essentially all of the NaOH-P_i (actually approximately 65-75%) was removed by algae within a 48-hr period, while Sagher et al. [11] reported that the rapid decline in dissolved P_i in one of their sediment-PAAP (Provisional Algal Assay Procedure medium) systems in the first few days was likely a function of P_i sorption by the sediment particles. Sagher et al. found that uptake of sediment P_i generally gradually occurred over one to two weeks of exposure.

Sagher et al. [11] did not provide any substantial discussion of the use of this approach for predicting the significance of sediment P_i in affecting algal growth in natural water systems, or the relationship between initial P_i levels and resulting algal biomass. Further, there was generally a poor agreement between microbial P based on *Selenastrum* counts and that based on the chemical procedure, pointing to deficiencies in one or both methods of assessing microbial P uptake. The amount of uptake of P_i also appeared to have depended to some extent on the type of algae assayed. Based on the limited data provided by Sagher et al., there appears to be some relationship between the log of cell numbers present at the end of the bioassay system. The slope of this straight line relationship appeared to be the same as that between the log of the number of cells after 30 days vs. initial dissolved P_i plus NaOH-P_i from treated soils from the work of Sagher [10]. Greater growth per unit concentration of NaOH-P_i + dissolved P_i was found than per unit total P_i which would be expected, since in general, the sequential HCl extractable (i.e., Ca-P) fraction has been found to be considerably less available than NaOH-extractable P_i.

Sagher et al. [11] pointed out that one of the potential limitations to their approach was its inability to account for available P present as mineralizable P_O (organic P) in sediments. In general, the results of this study of sediment available P by Sagher et al. should not be directly applicable to assessing the availability of phosphorus associated with suspended sediments in a tributary river to algae in receiving waters, mainly because of the general difference in character of the two types of particulates (i.e., sediment and river suspended particulates).

Huettl et al. [12] used five of the Wisconsin soils tested by Sagher [10] to evaluate the use of aluminum-saturated strong-acid cation exchange resin-extractable P as a measure of the availability of soil particulate P to aquatic algae. While they concluded that for 24-hr exposure periods the resin extractable P averaged 98% of that found to be "available" to algae by bioassay, it is unclear how those authors assessed algal uptake. It appears that they may have used Sagher's [10] proposed method of taking the difference between the sum of dissolved P_i plus NaOH-P_i before and after incubation with algae, or perhaps they used Sagher's results for the soils he evaluated.

Basically Huettle et al. [12] have shown that, for several soils from central Wisconsin, an aluminum-treated cation exchange resin yields about the same extractable P as the decrease in NaOH extractable P and dissolved P' from an algae-containing suspension. Using the same type of particulate material, but longer term bioassays in which algal biomass was measured, Cowen [9] and Cowen and Lee [8] found essentially the same general relationship. It is therefore concluded that with these types of samples, the results of all of these procedures are essentially the same. It is, however, important to emphasize that these relationships would not necessarily hold for other soil-suspended river particulate samples. Also, for other systems, tests on soils and on suspended particulates may not yield similar results. Further, Huettl et al. have presented an inappropriate critique of previous studies involving growth bioassays when they state, "Because several days were allowed for cell growth, the relationships between test results (i.e., those of Chiou and Boyd [13], Cowen and Lee [8], and Porcella et al. [14]) and P availability may be invalid for estimating short-term availability. Short-term availability of P is important in surface waters because the time that eroded soil particles are present in the photic zone prior to settling may be relatively short." Studies on the aqueous environmental chemistry of phosphorus have shown that the P present in the photic zone of a waterbody can be of minor significance in determining the overall phytoplankton biomass in the waterbody. The photic zone may be a relatively thin band of water compared to the mixed layer above the thermocline or bottom. Wind or other current-induced mixing is normally sufficiently great in waterbodies so that available P present in that portion of the mixed layer below the photic zone must be included in any assessment of the phytoplankton growth that will occur in a waterbody. Further, as shown by Stauffer and Lee [15], in some waterbodies even available P below the thermocline can, as a result of thermocline migration, play an important role in regulating phytoplankton growth within waterbodies. Most importantly, it is well known that there is appreciable release of P from sediments under both oxic and anoxic conditions. This was demonstrated by Lee et al. [16] for Lake Mendota, Wisconsin. Therefore, in general, the short-term bioassays recommended by Huettl et al. [12] and Sagher [10] can produce an erroneous estimate of the available P associated with a particular sample. Even the extended bioassays of the type used by Cowen and Lee [8] may be too short to measure properly the amount of P from particulate matter entering a waterbody that would typically become available for algal growth. The insufficient duration of the bioassays done by Huettl et al. [12] and Sagher [10] points out the importance of designing the bioassay system so that it properly reflects the environmental chemistry of P in aquatic systems.

Li et al. [17] evaluated the uptake of various forms of sediment P_i by the aquatic macrophyte *Myriophyllum spicatum* by determining changes in the concentration of total inorganic P, nonoccluded (NaOH extractable) P_i, exchangeable P_i using radioisotopes of P and plant tissue P after exposure to eight Wisconsin lake sediment samples. Wet sediment was mixed with P-free liquid growth medium and allowed to settle for one week. *M. Spicatum* were added, and when roots appeared (one week later) an additional 9 liters of P-free medium were added. After three more weeks the macrophytes were harvested, 9 liters of medium were removed and replaced with fresh medium. Two more growth cycles were conducted on the same sediment in the same manner. The growth response was proportional to the mass of sediment P added. Results from addition of 43.4 mg sediment P/test culture were used for comparison across sediments since addition of additional P resulted in no biomass increase but did result in luxury uptake of P. The uptake of dissolved phosphorus was not discussed. It should be noted, however, that in light of the contact time between the sediments and water prior to and during the assays, there would likely have been, at least in the first run, considerable dissolved P_i in the surrounding waters, which perhaps could have been taken up by the plants. Li et al. reported that the results of the three runs on the sediment sample were variable but comparable. Some mineralization of the sediment organic P occurred, generally less than 10% of the total P_o.

The percentage plant uptake of total sediment P_i, 13-17%, was considerably below the 60-95% microbial uptake of sediment P_i found by Sagher et al. [11] in sediments from the same lakes that Li et al. examined. For calcareous sediments tested, 33-50% of the nonoccluded P_i was taken up by the plants, considerably less than the 60-85% found by Sagher et al. [11]. From 43 to 85% of the total exchangeable P_i was available to the macrophytes. For noncalcareous sediments, 18-24% of the nonoccluded P_i was available, whereas 33-44% of the exchangeable P_i was available. Nonoccluded P_i appeared more available in calcareous than noncalcareous sediments. Since such differences were not observed for exchangeable P_i availability, it was concluded that radioisotopically exchangeable P_i is more representative of the available fraction.

Williams et al. [18] investigated the algal availability of various forms of phosphorus associated with 5 Lake Ontario bluff erosional material samples, 3 tributary suspended solids samples from Lake Ontario tributaries, 2 samples from Lake Erie tributaries, and 12 samples collected from the top few centimeters of lake sediments from the Lake Ontario basin. Although the availability of the tributary suspended solids is being given particular emphasis in this chapter, Williams et al. presented only a generalized discussion in their synopsis paper of the results of all samples. Bioassays were conducted in which about 100 mg dry solid equivalent of wet sediment was mixed with 500 ml culture media and algae and incubated for 12 days. It was concluded that apatite P is generally not available to algae, and that algal uptake of P was linearly related to the nonapatite inorganic P (NAIP) present. Uptake averaged about 75% of the NAIP added; in no case was all the NAIP utilized. Algal P uptake was even more closely related to the NaOH extractable P_i, which they found by correlation to be about 70% of the NAIP. The amount of algal uptake of P was not related to the total P content (8-50% of the total P was taken up); however, it was found that the higher percentage uptakes were in samples having higher total P content.

Williams et al. pointed out that P utilized under laboratory conditions should be considered as only potentially available, as those optimum conditions would not likely prevail in the environment. Actual exposure time of the algae to the particulates would be a major factor governing the actual amount of utilizable particulate P in the environment. Suspended solids in streams generally contain large amounts of silt-size particles of clay and silt aggregates, which tend to settle more readily than smaller sized particles, thus perhaps further reducing realized availability of particulate P.

Caution should be exercised in using Williams et al.'s availability percentages because their data may have been biased by sample handling. They froze their sediment samples prior to assaying, which may have substantially altered their characteristics and behavior in the test system.

Logan [19], Logan et al. [29], Verhoff et al. [21] and Verhoff and Heffner [22] undertook a study of the biological availability of particulate phosphorus entering Lake Erie as measured by chemical fractionation, and the rate of availability (uptake) of P in river waters. Logan et al. [20] reported on both aspects of the study, while the other three works cited above presented the results of parts of this study.

Logan [19] and Logan et al. [20] analyzed 66 samples from 36 tributaries in the U.S. portion of the Lake Erie watershed in an attempt to determine the short-term and longer-range P availability potential. The solids fraction that was considered available in the short-term was the NaOH extractable form; the difference between the citrate-dithionite-bicarbonate (CDB) extractable and the NaOH extractable P was considered to represent the portion which potentially can be released over long periods of time during anoxia. The HCl-P_i was considered unavailable. No work was done, however, to substantiate the validity of these assumptions. They found that when averaged on a geographical regional basis, the NaOH extractable P represented about 30-40% of the total inorganic P in the particulate samples, except for the New York area in which this fraction was 14% of the total inorganic P. Data presented indicated that the NaOH-P was 6-15% of the total particulate P concentration (PP_T). The fraction potentially available over a long term (which includes short term availability), the (NaOH + CDB) - P_i fraction, represented on the average 17-27% of the total sediment P content for all four geographical areas. These percentage estimates may be somewhat in error due to the fact that the measured and calculated total P values differed.

Logan [19] and Logan et al. [20] pointed to the importance of assessing the individual character of the systems being considered in using chemically derived P bioavailability estimates. According to those authors, the key factors are the rate at which P becomes available, the duration of the exposure period, and the potential for reexposure of algae to particulate P.

Verhoff et al. [21], Verhoff and Heffner [22] and Logal et al. [20] attempted to assess the rate of conversion of total P to available forms in three Lake Erie tributaries. In interpreting the fraction of unavailable P that may become available in terms of natural water systems, the conversion rate must be considered in light of the characteristics of the receiving water, such as the retention time of the particulates in the waterbody or in the photic zone. Verhoff and Heffner collected a water sample from each of the four Lake Erie tributaries during major storm events. They prepared triplicate 12- to 14-liter subsamples of each river water, added AAP nutrients (without P), and incubated for 100 days or more, periodically harvesting the supernatant for determination of algal P uptake (assumed to be the same as insoluble P). The centrifugate was returned to the assay tank. For the Sandusky River, 0.087% of the total P in the sample was removed by algae per day over the 80-day incubation period. This rate was 0.268% and 0.191% per day for the Honey Creek and Broken Sword samples, respectively. According to Logan et al. [20], there was insufficient growth in the Cattaraugus Creek (NY) sample test to make a harvest. Verhoff and Heffner indicated that the Cattaraugus Creek contained a substantially greater apatite (nonavailable) fraction of total P than the others, and consequently should stimulate much less growth. Verhoff and Heffner indicated that this assessment corresponded to the results of Cowen and Lee [8] for western New York samples. However, examination of the Cowen and Lee study showed that when their tributary samples were autoclaved prior to assay, considerably greater P availability in the samples was found. As discussed previously, Cowen and Lee attributed the lower growth in the New York samples to competition for P by native algae and bacteria, rather than the presence of "unavailable" (apatite) P as suggested by Verhoff and Heffner.

Verhoff and Heffner found that their conversion rates "corresponded vavorably" with a number of literature values. However, the literature values cited were for conversion of P in the dark, indicating a rate of mineralization of organic P to dissolved inorganic P, rather than the removal of particulate inorganic forms of P by microbes for assimilation. Verhoff and Heffner's conversion rates are considerably less than rates reported for conversion of various extractable forms of P to algal P reported by Sagher et al. [11] and Sagher [10]. The rates may have been limited by the apparent lack of mixing of the settled sediments and overlying water. Further, since indigenous populations were used, this represents net microbial uptake rate, rather than algal uptake rate; what may be available to zooplankton may not be the same as that available to algae. As pointed out by Logan et al. [20] with reference to this work, there may have been preferential removal of the fine particles as a result of the harvest method; also, at the end of incubation, substantial amounts of algae were present in the sediment which would not have been accounted for in the microbial removal of P. Further, Verhoff and Heffner reported problems with algae growing on the sides of the flasks, and crusts forming on the surface of the sediments. This may have affected the results of their work. If their conversion rates are appropriate, on the order of 30-50% of the total P in these river waters would become available in a year's time. By then substantial amounts of the P would be removed from the lake or photic layer by settling and could result in minimal effect on the lake's productivity.

Logan et al. [20] showed that during the incubation period of about 100 days, generally about 405 of the NaOH extractable P in the sediment was removed, leaving several hundred $\mu g/g$ NaOH extractable P in the sediment. The Cattaraugus sample showed only 6% removal of NaOH-extractable P and was the only one to show a decrease in the HCl-extractable fraction (after NaOH extraction). The other three samples all showed increases in this fraction ranging from 4 to 50%. This was generally not observed by other investigators. Based on the work of other investigators, the 40% removal of NaOH-P_i is low. Because of reported problems with loss of sample during the procedure, the results of the fractionation portion of this experiment may be unreliable.

Logan et al. [20] commented on the significance of these findings relative to Lake Erie. Sediments from the more urbanized tributary areas and high clay areas should have the highest bioavailable P load, based on the fact that their character is reportedly representative of the native soil P levels. Since these areas of the Lake Erie Basin drain to the shallower western basin, their impact on algal production may be more significant. This is all under the assumption that the chemical analysis of a soil or river sediment accurately predicts ultimate P availability. However, controlling the sediment load to the Lake will cause proportionately less available P to be removed, since the available P is supposedly more prevalent in the fine-grained materials; the suspended solids control programs tend to be better at removing the larger-size fractions. Logan et al. suggested that since the availability of sediment P to algae is kinetically controlled, dynamic P lake models may be more appropriate for predicting future change in the individual basins than static P-loading models. Based on what is known about the applicability of the P loading models of Vollenweider [23] and Rast and Lee [24], this statement lacks technical foundation. The applicability of various P-loading models, as influenced by the availability of the P load, will be discussed in a subsequent section.

Armstrong et al. [25] investigated the amounts and fractionation of P forms in particulates from several locations (generally near the mouth) of five Great Lakes tributaries draining mostly rural, agricultural and forested areas, inlcuding the Genesee (above Rochester, New York), Grand, Maumee, Menomonee (draining mainly urban areas) and Nemadji Rivers. Composite water samples were collected during particular flow periods at various times over a year. Armstrong et al. processed the data by averaging the data for all sampling sites in each tributary. The averaged flow and suspended solids levels were fairly representative of historical data according to those authors. Samples of Lakes Michigan and Erie recessional shoreline material were also examined. The maximum amount of "available" P in the river particulate matter was assumed in this study to be equal to the NaOH-extractable portion, although Armstrong et al. did not evaluate the appropriateness of this assumption. The fraction of the total particulate P (PP_T) which was taken up by an anion exchange resin over an 18-hr period was considered to be the fraction most readily available for use by algae. The percent NaOH extractable P of PP_T for the five tributaries ranged from 14 to 37%, a range which compares well with that of Cowen and Lee [8] and Cowen [9], of 22-27% NaOH-P for urban runoff, and 11-28% for Genesee River samples. The resin extractable fraction was 7-17% of the PP_T, representing 43-50% of the NaOH-P fraction as averaged for each tributary. The resin P fraction also compared well with that found by Cowen and Lee [8] and Cowen [9]. The organic P fraction of the particulates was not measured by Armstrong et al., since they indicated this fraction is mineralized very slowly, yielding only a small fraction of available P. They did not, however, present any data to support this position.

Armstrong et al. suggested that the mechanism of the uptake of NaOH-P was desorption due to low solution phosphorus concentration, which is maintained by P-deficient algae. Cowen and Lee [8] also found that the amount of extractable P increased when dissolved P levels in the water were lower.

Examination of P fractions associated with various particle size fractions led Armstrong et al. to conclude that the particle size fractions all had similar percentages of the various P fractions measured. Recessional shoreline samples showed only 1-4% of the particulate P was extractable with NaOH. Greater than 68% of these materials were HCl extractable. "Available" P loads (based on dissolved P and NaOH-P being available) were calculated for the five tributaries investigated by Armstrong et al. [25]. Between 23 and 58% of the total P loads (mean of 44%) were calculated to be available. Using the approach developed by Cowen and Lee [7,8] and Cowen [9], Lee and Jones [26] determined the percent available P load to Lake Ontario from the four major U.S. tributaries. As shown in Table I, between 30 and 46% of the

Table I. Distribution of Total and Soluble Orthophosphate Tributary Load from Major U.S. Tributaries to Lake Ontario[a]

Tributary River	Mean Total P Load (metric tons/day)	Mean Soluble Ortho P Load (metric tons/day)	Soluble Ortho P/Total P (%)
Niagara	20.8	2.4	12
Genesee	1.8	0.34	20
Oswego	2.9	0.92	32
Black	0.5	0.08	16

Tributary River	Potentially Available P/ Total P (%)	Percent of U.S. Tributary Total P Load	Percent of U.S. Tributary Soluble Ortho P Load
Niagara	30	80	64
Genesee	36	7	9
Oswego	46	11	25
Black	32	2	2

[a] Adapted from Casey et al. [28] after Lee and Jones [26].

total P loads from these rivers is potentially available. This compares well with the estimates of Armstrong et al. [25] for a variety of tributaries to each of the Great Lakes.

Monteith and Sonzogni [27] reported on the "available" P in over 160 U.S. Great Lakes shoreline soils representing 49 soil profiles. Available P was reported as the P fraction extracted in 2 hr with 0.05 N HCl. As discussed by Monteith and Sonzogni, although this technique may not be the most appropriate for assessing biological availability, those data were the only ones provided which may even roughly approximate the amount of available P in the particulate shoreline soils. The literature cited by those authors which discuss the use of mild acid extraction for estimating available forms indicates that it very likely provides an overestimate of both short-term and long-term contaminant availability. For the soils evaluated in that study, an average of 43% of the total soil P was extracted as orthophosphate by dilute HCl. From the data presented, they indicated that there appeared to have been more 0.05 N HCl-extractable P associated with the clayey fraction than with the sandy or loamy soil fractions.

Thomas [29] summarized data derived during Task D studies of PLUARG. Water samples were collected from ten Great Lakes tributaries (Bronte Creek, Humber, Credit, Welland, Niagara, Grand, Cedar, Thames, Saugeen and Nottawasaga Rivers) and analyzed, generally monthly, for bulk mineralogical properties and forms of P. The concentrations of NAIP (nonapatite inorganic P), organic P and apatite P in each river were highly variable over the annual cycle, but the annual mean values (when expressed as a percentage of the total sediment P) were fairly constant. What was termed "available P" (NAIP) was on the order of 20-40% of the tributary particulate total P, which is on the same order as that to be potentially available by Cowen and Lee [7,8] and Cowen [9]. According to Thomas, available sediment (particulate) P as measured by these chemical techniques represents 0.9, 3.6 and 1.2% of the total P loadings to Lakes Huron, Erie, and Ontario, respectively. These estimates cannot be compared directly to those of Lee and Jones [26] or Armstrong et al. [25], since Thomas' estimates of available P do not include the soluble P in the tributaries. Thomas also concluded from the data that there was little variation in the texture and mineralogy of particulates in the streams, indicating consistency in sediment sources, mixing and transport, and that the particulates reflect the basin soils. Further, he indicated that based on these data, available P cannot be predicted by land use or form but must be measured for individual lakes or watersheds.

It is important to note that this study did not include any assessment of available P using bioassay techniques. As is discussed in other parts of this review, the authors have reservations about the ability of NaOH extraction of particulate matter to assess the biologically available P associated with

Great Lakes tributary particulates in some cases, especially those with elevated organic P content. It has recently been learned, however, that Thomas and co-workers have subsequently conducted bioassay studies on these tributary waters and that these data will be published in the near future. Such data will potentially allow a much better assessment of how well the NaOH extraction procedure applied to particulate matter plus the soluble orthophosphate approximates the amount of algal-available P in a variety of tributary waters.

Atmospheric Inputs

A number of investigators have attempted to assess the atmospheric contributions of phosphorus to waterbodies. While most of these investigators determined various chemical fractions of P in their rainfall and dustfall samples, only Cowen [9] and Cowen and Lee [8] did studies to determine the availability of the P from this source to algae. Cowen [9] and Cowen and Lee [8] conducted bioassays with *Selenastrum* on particulates from three samples of Madison, Wisconsin snow collected in April 1973. In general, the total soluble P concentrations were the same as the dissolved reactive P level. Less than 25% of the particulate P in these snow samples was available to *Selenastrum* in 18 days. One sample did not show any detectable growth. Cowen and Lee collected 13 precipitation samples from various locations within New York State. These samples contained small amounts of dissolved reactive P, compared to the total soluble P. Bioassays with *Selenastrum* on the total sample showed that in only three samples was the algal available P more than 10% of the total P concentration. In 12 of the samples, the percent algal available P was less than the percent dissolved reactive P.

Brezonik [30] reviewed the literature on atmospheric contributions of a variety of chemicals and found that ortho P was typically on the order of 0-0.005 mg P/l, and total P on the order of 0.02-0.15 mg P/l. He also reported on the concentrations of P in rainwater collected August 31-September 1, 1969, at nine stations near Gainesville, Florida. Total P ranged from 0.02 to 0.65 mg P/l and ortho P from 0.004 to 0.043 mg P/l; soluble ortho-P ranged from 7 to 66% of the total P. Phosphorus was also measured every 5-10 minutes during two storm events in the summer of 1969 at Gainesville, Florida. Both total and ortho-P decreased over time after start of the rainfall. The ortho-P fraction was 50-100% of the total P in the samples evaluated.

Brezonik [30] also presented literature values for P in snowfall. For the two studies cited for snowfall, ortho-P levels were 3.5 and 30 μg P/l, the former being 58% of the reported total P. A summary by Brezonik of literature estimates of the percent of the total P load contributed by rainfall shows that it ranges from a few percent to as much as 74%.

Murphy [31] evaluated the phosphorus in 89 samples of rainwater collected from atop a 10-m high building in a densely populated urban Chicago area. Air particulates were also collected using a high volume air sampling pump. The average concentrations of total P in rainfall were 0.034 mg P/l, and soluble orthophosphate (52 samples) was 0.012 mg P/l. Concentrations were found to be higher in lower volume rainfall events and lower in higher volume events. Concentration in eight snowfall samples ranged from 0.016 to 0.054 mg P/l, and 0.006 to 0.196 mg P/l for "orthophosphate" (assumed to be determined on unfiltered sample), with most concentrations in the 0.02-0.05 mg P/l range.

Peters [32] collected eight rainwater samples, six samples of rainwater plus dustfall, and one snow sample from a pasture-forest area above the eastern shore of Lake Memphremagog, near Montreal, Canada. Samples were analyzed for total P (TP), soluble P (SP) and soluble reactive P (SRP). Although these samples were frozen prior to analysis, the author indicated that this type of preserving did not affect the integrity of Lake Memphremagog samples evaluated. He did not, however, evaluate the reliability of this preservation technique for their atmospheric samples. Using radiotracer techniques, Peters concluded that on the order of 38% of the atmospheric P was exchangeable, which he equated with being available for algal growth.

A total of 188 precipitation samples was collected from six locations around Lake Michigan and analyzed for total P by Murphy and Doskey [33]; 131 of these were also analyzed for dissolved reactive phosphate. Weighted (unclearly defined) average total P concentrations for the six stations ranged from 0.016 to 0.036 mg P/l; weighted average dissolved reactive P ranged from 0.006 to 0.014 mg P/l. The higher concentrations were typically found at the stations at the south end of the lake. On a stationwide average, soluble ortho-P ranged from 30 to 56% of the total P concentration. A series of 23 rainfall samples was chemically fractionated further to determine the amount of hydrolyzable, organic and total reactive P in the samples. The weighted average dissolved reactive P was 0.01 mg P/l; hydrolyzable (less dissolved reactive P), 0.0044 mg P/l; organic P, 0.009 mg P/l; and total reactive, 4.68 mg P/l.

Twenty-two snow samples from four of the above cited locations were also analyzed by Murphy and Doskey [33]. Total P ranged from 0.007 to 0.058 mg P/l (mean 0.026 mg P/l) and dissolved reactive P from 0.002 to 0.024 mg P/l (mean 0.007 mg P/l). On a per sample basis, soluble reactive P was 39% of the total P. In a number of samples, the total reactive P concentration was considerably greater than the dissolved reactive P. Murphy and Diskey [33] concluded that 50% of the total P present in precipitation is a reasonable estimate of the amount of P that will ultimately become available, although they did not conduct any bioassays to substantiate this conclusion.

Delumyea and Petel [34,35] studied atmospheric inputs of phosphorus along the shore of and on Lake Huron between April and October 1975. It is important to note in reviewing their results that what they have termed "available" P is operationally defined as the soluble ortho-P plus that leached at pH 2 in H_2SO_4. It does not, however, have any demonstrated relationship to that fraction which is available to stimulate algal growth in the lake water. Concentrations of acid-leached P in six, monthly, integrated samples of rainfall and dry fallout from each of 11 shoreline stations showed concentrations to be highly variable from month to month, ranging from a few μg/l to 0.7 mg/l. Although they reported some problems due to sample loss through evaporation, especially in the summer, average concentrations appeared to be highest in June and September and lowest in August.

Total soluble phosphorus concentrations in 21 event rain samples collected from eight shoreline stations between June and October 1975 were also highly variable, ranging from a few tenths of a μg P/l to 36 μg/l. Concentrations appeared to be highest in mid-July, decreasing through October.

Availability of Soluble Orthophosphate

A review of the literature on the availability of soluble orthophosphate, generally defined as molybdate reactive phosphate that will pass through a 0.45-μ pore size membrane filter, leads to the conclusion that this form of phosphorus is usually readily available to support algal growth. However, under certain conditions, the soluble molybdate reactive P may not be readily available. A number of investigators, notably Rigler [36,37] and Lean [38,39], have reported that substantial parts of the soluble molybdate-reactive phosphorus is not available to support algal growth. On the other hand, Walton and Lee [40] have found, using algal assay procedures and a variety of waters, that there was a good correlation between the expected algal growth (based on the soluble orthophosphate content of water samples) and the actual growth achieved under standardized culture conditions. Examination of the conditions governing the results of these studies shows that the two groups of investigators were working in markedly different concentration ranges. Rigler and Lean worked with water samples having soluble ortho-P concentrations in the μg P/l range, whereas Walton and Lee [40] worked with waters having concentrations of 10 μg P/l. It is well known that the molybdate method is not necessarily specific for soluble orthophosphate, but includes other forms of P and other materials, such as arsenate, usually present in trace amounts in aquatic systems. At higher P levels, the soluble ortho-P dominates the other P forms and many interfering substances. It is therefore not surprising to find at low concentrations of phosphorus, at or near those that are generally believed to limit algal growth, that there would be a greater discrepancy between molybdate reactive P and that found available through bioassay.

From a phosphorus management strategy point of view, the fact that the molybdate test does not accurately measure algal available phosphorus at a few μg P/l concentrations levels or less is of no major consequence since P control programs must be directed toward those sources which have concentrations well above that level.

There is another condition under which the molybdate-reactive phosphate at greater than 10 μg P/l levels will not correspond well to algal available P in the standard assay procedure. As discussed by Cowen and Lee [8] this may occur for those samples which contain large amounts of biodegradable organic carbon. Under these conditions, large bacterial populations may develop in the algal cultures which will absorb the phosphorus as part of respiration, making it unavailable for algal growth. Aquatic bacteria, because of their generally higher growth rates compared to algae, can out-compete algae for P, provided that there is a sufficient concentration of organic carbon to support bacterial growth. While this has not been investigated, it is likely that bacterial absorption of P would only temporarily make the P unavailable for algal growth. Upon bacterial death and lysis, substantial parts of the P in bacterial cells would be released and made available for algal growth. The algal assays on high oxygen demand—high BOD waters would tend to show lower P availability than would be realized in the water over extended periods of time. Therefore, caution should be exercised in interpreting results of algal assays for available P.

Summary of Studies on Algal Availability of P from Diffuse Sources

Cowen and Lee [7,8] and Cowen [9] have conducted essentially the only large-scale study to examine the relationships between biological availability of P as measured by bioassay-determined cell growth and a variety of chemical extraction methods. Other studies, such as those of Sagher [10] and Sagher et al. [11], have demonstrated a relationship between an amount of P removed from various soil P fractions and converted to algal P uptake, and the percentage of the total particulate P represented by the soil P fractions. The amount of particulate P available to algae, according to the studies of Cowen and Lee, is approximately equal to an amount between that extracted by an anion exchange resin and by NaOH, the latter giving a long-term maximum. This amount is often approximately equal to 10-30% of the particulate P concentration. Sagher's [10] study indicated that the NaOH-P removed by the algae represented on the order of 30-40% of the initial total inorganic P concentration of the soils. If the sediment organic matter were taken into account in his computations, a range of percent total particulate P available on the order of that found by Cowen and Lee would likely be approached.

The results of NaOH extractions on Great Lakes area soils and sediments performed by investigators whose work is discussed herein, have typically shown that 10-30% of PP_T is extractable. This is in close agreement with the finding of Cowen [9] and Cowen and Lee [7,8], which showed available P from urban and rural drainage was approximately equal to soluble reactive P + 0.2 times the particulate P. A number of these investigators have generally agreed that this range would represent a maximum, long-term availability based on ideal laboratory conditions; the realized availability of particulate would likely be somewhat less than the NaOH-extractable amount. However, if it is assumed, based on the work of Cowen and Lee, and Sagher and others, that the NaOH extractable P is a crude estimate of the algal available part of the particulate P in a Great Lakes tributary water sample, then it may be concluded that the relationship developed by Cowen and Lee is a suitable approach for generally estimating the amount of algal available P entering a waterbody. There has been sufficient work done to substantiate this relationship in the Great Lakes basin so that it can be used for development of P management strategies for the control of phosphorus from "typical" land and urban runoff. There are situations where this procedure should not be used without verification. Some of these systems are discussed below and in subsequent sections of this paper.

Although several investigators have reviewed either experimentally or through the literature the question of the availability of organic P fractions, none have focused on this issue. They have assessed that in the time frame considered (generally several weeks' exposure), mineralization of organic P to readily available forms is negligible. For the most part, however, these investigators have not dealt with high organic P systems. This issue should receive attention, especially since (as discussed in another section of this chapter) as erosion control practices are put into effect, the proportions of sediment P fractions entering tributaries and lakes may change. An example of this is the "no till" agricultural practice designed to limit soil erosion. It has been observed that while this practice may reduce the amount of NaOH-P or other sediment inorganic P fractions entering a water, it may significantly increase the organic P load. The significance of this change relative to eutrophication-related water quality is not known at this time. It does indicate, however, since the organic P fractions are not extracted by NaOH or anion exchange resins, that before such extraction precedures are adopted as standard techniques for determining "biologically available" particulate P, the availability of the organic P fraction in higher organic P systems should be evaluated. It also points to the importance of assessing through bioassays the suitability of chemical extraction techniques for estimating availability of particulate phosphorus for a wider variety of systems.

It should be noted that, in reviewing the literature on this topic it has been found that many investigators have not provided sufficient detail and clarity of presentation in their reports to enable a reviewer to ascertain what was actually done in many cases. All papers and reports devoted to this and related topics should contain a detailed, clearly presented experimental procedures description to enable other investigators to duplicate their work should they choose to do so. While this is applicable to all scientific and engineering studies, it is especially important for studies of available forms of contaminants since the answers obtained in these studies are highly dependent upon how the experiments were conducted.

IMPACT OF P LOADS ON WATER QUALITY

The eutrophication program of the Organization for Economic Cooperation and Development (OECD) has demonstrated for the U.S. [24,41], as well as nationally (unpublished), that strong correlations exist between the P loads to waterbodies (lakes, impoundments and estuaries) normalized by mean depth and hydraulic residence time, and the phytoplankton biomass, as measured by planktonic algal chlorophyll and Secchi depth. It is evident that P is the primary controlling factor in the overall trophic state of many lakes and impoundments. Further, it is possible to quantitatively predict a waterbody's eutrophication-related water quality based on the normalized phosphorus load using the OECD eutrophication modeling approach. Lee and Jone [26] and Lee et al. [42] have demonstrated that the OECD modeling approach is applicable to the Great Lakes.

While it is possible to relate P load to eutrophication-related water quality in lakes and impoundments, no such relationships exist at this time for rivers and streams. There are no meaningful P load or concentration limits that can be used to estimate the impact of a particular P source on eutrophication-related water quality in a stream or river. It is evident, however, that streams and rivers can have much higher concentrations of available P without producing significant deterioration of water quality than can lakes, impoundments, estuaries and marine waters.

The discharge of P to a stream or river may have a markedly different effect on a downstream lake or impoundment compared to the effect that would result if P were discharged directly to the lake or impoundment. These differences result from the fact that the algal availability of the phosphorus can change between the point where it enters the stream or river, and where it enters the lake or impoundment. Although it is possible that some unavailable P forms may be converted to available forms in transit downstream to a waterbody, the predominent reaction likely causes available forms to be converted to unavailable forms. This would be of particular significance for wastewater treatment plants located some distance inland from the waterbody of concern, since for some river systems appreciable parts of the available forms of phosphorus would be converted to unavailable forms during river transport to the lake or impoundment. This is the result of sorption reactions, and in calcareous areas, precipitation of phosphorus, and the incorporation of P into aquatic plant biomass occurring within the stream, which results in part of the P becoming refractory and therefore unavailable to support downstream aquatic plant growth. In a similar vein, the OECD studies [24,43] have shown that arms or bays of a larger lake can be effective nutrient traps, in which available P entering the area is converted to a substantial extent to unavailable forms through phytoplankton growth and subsequent death. Such situations can result in the main body of a waterbody having much better water quality than would be predicted based on the total P load to the entire waterbody (including the arm or bay).

From a P management strategy point of view, it is certainly inappropriate to require the same degree of P removal from wastewaters discharged directly to a waterbody as from wastewaters discharged to a tributary of the waterbody, especially if the discharge is a considerable distance upstream from the waterbody. There is need for studies to define the aqueous environmental chemistry of P in river and stream systems. They should be focused on determining the amount of P available at the point of entry into the river or stream systems which remains available at the lake/tributary interface. It certainly behooves those municipalities located considerable distances inland from the waterbody of concern to support studies of this type. Such studies could prove to be highly cost-effective in reducing the treatment costs associated with P removal. It is important to emphasize that the factors governing conversion of available P to unavailable P are reach-of-stream specific. Therefore, studies of this type conducted on one stream are likely to have limited applicability to another stream or to a different reach of the same stream.

While the above discussion has focused on wastewater treatment plant effluent, it is equally applicable to other P sources such as agricultural runoff. It is apparent for the Great Lakes basin, as well as other areas, that the amount of algal-available P that reaches the Great Lakes, per acre of crop under identical froming and other conditions, is less for those farms located considerable distances inland compared to those on the lake's shore. Greatest attention in developing P control measures for diffuse sources should be given to control of the available forms of P from those sources adjacent to the lakes.

AVAILABLE PHOSPHORUS IN DOMESTIC
WASTEWATER TREATMENT PLANT EFFLUENTS

Current Great Lakes Practice

Domestic wastewaters represent a significant source of phosphorus for the Great Lakes. It is generally assumed that the P present in domestic wastewaters is available to support algal growth. However, this has not been demonstrated and it is reasonable to expect that at least some part of the total P (especially colloidal P and the nonreadily settleable organic P fraction) in domestic wastewater treatment plant effluents is not available to support algal growth. This unavailable portion would likely constitute even a greater percentage of the total P in an effluent that has been treated for phosphorus removal because, as discussed by Lee [44], some of the phosphorus would be tied up with iron or aluminum as particulate P. Studies on the direct addition of iron and aluminum salts to lakes to reduce watercolumn P concentration have shown that the P incorporated in hydrous oxide floc is not readily returned to the watercolumn over extended periods of time and is essentially made unavailable for algal growth. There is a need to conduct studies on the amount of P that is available for algal growth in secondarily treated wastewater, as well as wastewater treated for P removal. The characteristics of such studies are discussed in a subsequent section.

It should be noted that concomitant with P removal by alum or iron precipitation techniques, a number of toxic chemicals, such as heavy metals, organic materials, etc., are also removed. Further, there is generally a reduction in the classical pollutant load (BOD, suspended solids, etc.) to the receiving water. With increased emphasis being placed on improving the quality of municipal effluents over that typically attained with secondary treatment, it is possible that the focal point of increased removal of contaminants from wastewater effluents should shift from phosphorus to other chemicals. It is important to conduct studies which properly define the costs and the benefits associated with removal of toxic chemicals by iron and alum precipitation techniques to develop management strategies for those contaminants from municipal wastewater sources. In developing these strategies, the availability of the contaminants to aquatic life must be assessed properly. It is well known [45,46] that iron and aluminum hydrous oxides are efficient scavengers of many toxic materials and can convert some of them to forms unavailable to aquatic life. Lee [45], in his review of the role of hydrous oxides in the control of contaminants in aquatic systems, pointed out that there are often marked differences in the abilities of freshly precipitated (in situ precipitated) and aged hydrous metal oxides to remove contaminants from aquatic systems. Malhotra et al. [47] found that the removal of P from a wastewater effluent by Fe and Al precipitation was much more effective when the hydrous oxide floc was formed in situ in the presence of P, compared to the addition of preformed floc.

Proposed Great Lakes Practice

Domestic wastewater treatment plants throughout the Great Lakes' Basin, and in other areas as well, are practicing phosphorus removal by the addition of alum, lime or iron. As discussed above, with these techniques a 1 mg P/l effluent concentration is readily attainable at low cost. The 1978 Great Lakes Water Quality Agreement [1] calls for additional phosphorus removal from domestic wastewaters to at least 0.5 mg P/l for municipalities situated in the drainage basin of the lower Great Lakes to achieve the proposed phosphorus loads in the Agreement. There are several aspects of this objective that must be considered, one of the most important of which is the expected improvement in water quality which would result from achieving the 0.5-mg P/l level. This topic has been reviewed in detail by Lee and Jones [26] and Lee et al. [42]. They have found, based on total P load and using the OECD eutrophication modeling approach, that the reduction in effluent concentration to 0.5 mg P/l will result in a nonperceptible change in water quality in the open waters of the Great Lakes (Table II), compared to the 1 mg P/l effluent limitations of the 1972 Great Lakes Water Quality Agreement.

As discussed by Lee and Jones [26], the potential changes in nearshore water quality must be evaluated on a case-by-case basis. In general, however, it is doubtful that there would be many locations in the nearshore waters where a significant additional improvement in water quality would result from wastewater treatment plants in a region further reducing effluent P concentrations from 1 to 0.5 mg P/l. Therefore, considering the significant increase in cost associated with reducing domestic wastewater treatment plant effluent P concentrations from 1 to 0.5 mg P/l, and a possible lack of improvement occurring in Great Lakes water quality as a result of this further reduction, it does not appear that adoption of a general requirement that all domestic wastewater treatment plants in the lower Great Lakes Basin must achieve 0.5 mg P/l effluent concentration is technically valid or cost-effective. However, for those waterbodies in which it appears that a significant impact on water quality would result from decreasing effluent total P concentrations from 1 to 0.5 mg P/l, studies should be conducted to determine the amount of available P that will be removed by this action, since much of the total P in a 1-mg P/l effluent which has been treated for P removal may be tied up with hydrous iron or aluminum oxides and therefore unavailable for algal growth in the receiving waters.

Table II. Summary of Predicted Changes in Water Quality
in the Great Lakes Under the Base Loads, Reduction Scenarios
and 1978 Water Quality Agreement Target Loads[a]

Lake	Base Load	Detergent Phosphate Reduction Scenario	1.0 mg/l Effluent Limitation Scenario	0.5 mg/l Effluent Limitation Scenario	Target Load
Chlorophyll-a (μg/l)					
Superior	1.0	0.9	0.9	0.9	0.9
Michigan	2.0	2.0	1.8	1.8	1.8
Huron	1.2	1.2	1.2	1.2	1.1
Erie	5.0	4.7	4.1	3.8	3.3
Ontario	2.3	2.3	2.1	1.9	1.7
Secchi Depth (m)					
Superior	9.1	9.2	9.3	9.3	9.3
Michigan	6.0	6.2	6.4	6.4	6.5
Huron	7.9	7.9	8.0	8.0	8.3
Erie	3.8	3.9	4.1	4.3	4.7
Ontario	5.7	5.7	6.0	6.2	6.7
Hypolimnetic Oxygen Depletion ($g\ O_2/m^2/day$)					
Erie (Central Basin)	0.43-0.75	0.40-0.67	0.38-0.61	0.36-0.58	0.32-0.52

[a]See Lee and Jones [26] or Lee et al. [42] for the basis of these estimations.

Although for the Great Lakes, little additional improvement in water quality will likely be achieved by reducing domestic wastewater treatment plant effluent P concentration to below 1 mg P/l, there may be situations in which removal of P below that readily attainable by iron, alum or lime treatment (i.e., below 1 mg P/l) will result in a significant improvement in water quality in the receiving waters. There are parts of the U.S., such as near Lake Dillon near Vail, Colorado, where municipal wastewater treatment plants are required to achieve 0.2 mg P/l in their effluents. The cost-effectiveness of and water quality benefits to be realized by achieving this concentration in domestic wastewater treatment plant effluents being discharged to Lake Dillon is under investigation [48]. The appropriateness of having domestic wastewater treatment plant effluent limitations of less than 1 mg P/l in other areas must be evaluated on a case-by-case basis, with both short-term and long-term bioassays being used to assess the availability of phosphorus in the treatment plant effluent.

The Detroit, Michigan situation, in which the municipality's wastewater discharges are considerably above the 1 mg P/l effluent standard (apparently because of a hydraulic overloading of the plant) represents a special case in which work should be done to determine how much of the so-called excess total phosphorus load is available to support algal growth in Lake Erie. The Detroit treatment plant, like a number of others in major metropolitan areas, receives large amounts of industrial wastes which contain appreciable quantities of iron. Therefore it is possible that most of the phosphorus currently discharged from the plant is not in a form that is available to support algal growth. Therefore, further reduction in the effluent P concentration may have little or no impact on eutrophication-related water quality in Lake Erie. It is conceivable for treatment plants that are hydraulically overloaded and have excess P in their effluents, that the addition of alum to the effluent could be a short-term, cost-effective way of reducing the available P content of the effluent. The P would be tied up in the aluminum hydrous oxide floc, and would be incorporated into the sediments as the floc settles in the receiving waters. This approach represents a modified version of direct lake treatment with alum or iron for phosphorus control. Recent work by the authors has shown significant potential for the use of water treatment plant alum sludge as a means of removing P from wastewaters and aquatic systems. If further work substantiates these initial findings, this approach could significantly reduce the cost of P removal under certain situations.

Approach for Assessing Available P in Domestic Wastewaters

The approach for assessing available P in wastewaters must be different from that used to assess P availability in soils, river suspended sediments and urban drainage. The studies on the amounts of available P in domestic wastewater treatment plant effluents must involve long term bioassays. Chemical extraction tests such as NaOH extraction will likely not be reliable in estimating availability of P in the effluents because such tests do not properly assess the availability of organic P. There is also some question about the algal availability of some forms of iron and aluminum phosphate present in domestic wastewater treatment plant effluents and in aquatic systems which are solubilized by NaOH. It should also be noted that U-tube culture systems which employ a membrane to separate the P source from the culture may also prove to be highly unreliable in estimating available P in domestic wastewater effluents because of bacterial and other growths on the membrane during long-term tests, which could interfere with the transfer of phosphorus from the leaching-mineralization part of the tube to the algal culture.

Some investigators have assumed that disappearance of soluble ortho-phosphate from an algal culture can be directly equated with algal uptake of phosphorus and, therefore, its availability to support algal growth. There are situations where this assumption is erroneous. The only proper way to assess availability is to measure growth (i.e, increase in biomass). What may appear to be uptake of soluble ortho-P may actually be precipitation of phosphorus with calcium, iron or aluminum species. Further, P could be sorbed onto detrital organic remains of algae and/or bacteria, and yet be counted as "available" P. There is a substantial marine literature which shows that P tends to strongly sorb onto organic aggregates arising from the polymerization of apparently dissolved, most probably colloidal organic and inorganic molecules present in aqueous systems. Further, as discussed in the section of this chapter devoted to condensed phosphates and their ability to support algal growth, there are biochemical transformations of phosphorus compounds which produce soluble forms not measured by the molybdate test without acid hydrolysis, and which are not immediately available to support algal growth. The long term availability of these compounds needs to be determined.

All investigators should be extremely cautious about trying to short-cut bioassays involving measurement of algal growth as a measure of phosphorus availability. This is especially true for domestic wastewater effluents. It is emphasized that growth experiments should not be based on growth rates, but rather on actual biomass yield under established plateau conditions.

It is evident from the above discussion that a considerable research effort should be undertaken on the amounts of P in domestic wastewater treatment plant effluents that are or can become available to support algal growth in aquatic systems. Although it probably is appropriate to assume that chemical precipitation techniques removing P to 1 mg/l levels in domestic wastewater treatment plant effluents are cost-effective in removing available P, the efficacy of P removal to less than 1 mg/l in the effluents should be based on the amounts of algal-available P being removed, which has not yet been defined. As with other sources of phosphorus, consideration must be given not only to P availability in the effluents, but the availability of P over extended periods of time under the conditions tha will prevail in the receiving waters. Particular attention must be given to the iron-phosphate systems because reducing conditions such as occur in anoxic hypolimnia tend to promote the release of iron-bound P from sediments. It should be noted, however, that even under extended periods of anoxia, appreciable quantities of P in sediment are not released to overlying waters under conditions that would promote release. Therefore, even particulate iron hydrous oxide phosphate that enters an anoxic hypolimnion would likely only partially become available to support algal growth. Small amounts of money spent on research in defining algal available P in domestic wastewater treatment plants and in receiving waters could save large amounts of money in the control of P from domestic wastewater.

Studies on the availability of P in domestic wastewater treatment plant effluents must be conducted to include a substantial number of samples collected over an annual cycle from several different types of treatment processes, such as iron or alum addition to the primary settling tank, iron or alum addition to the activated sludge system and the tertiary use of iron, alum or lime in the secondary settling tank or final sedimentation tank. It is possible that significant seasonal differences in algal availability of P will be found among treatment plants using the same removal processes, as well as among different processes. A few grab samples from a limited number of treatment plants could readily result in development of information not representative of the typical situation.

Condensed Phosphate

Today detergent phosphate represents about 20-35% of the total P concentration in domestic wastewaters in areas where no detergent P limitations are in effect. Detergent P is typically in the form of pyro- and tripolyphosphates (condensed phosphates). It is generally believed that these forms of phosphorus are readily available to support algal growth, based on the supposition that there is rapid and complete hydrolysis of the condensed phosphate to soluble orthophosphate during secondary wastewater treatment. While hydrolysis appears to be rapid during conventional secondary treatment, examination of the literature raises questions about the extent of this hydrolysis to soluble orthophosphate. In a study of hydrolysis of condensed phosphates in algal culture systems [49], it was found that the bacteria present in the algal culture and/or extracellular phosphatases excreted by the algae caused a rapid hydrolysis of pyro and tripolyphosphate. However, this hydrolysis did not result in the formation of soluble orthophosphate, or in the same number of algal cells as found in standard cultures containing equivalent amounts of soluble orthophosphate. Typically, for the same phosphorus concentration, condensed phosphate produced about half as many cells as soluble orthophosphate. Based on the studies by Clesceri and Lee [49], it appears that as part of the enzymatic hydrolysis of the condensed phosphate, part of the P is converted to a form which is not measured in the standard molybdate test for soluble orthophosphate, and which is also not available to support algal growth in the several-week incubation period used in these tests. It is not known at this time whether extended periods of incubation would cause conversion of this unidentified form of P to soluble orthophosphate available to support algal growth. Obviously, this is an area that needs attention to evaluate the significance of phosphorus used as a builder in detergents, in stimulating algal growth in the aquatic environment. It is possible that up to 50% of the P present in detergents in the form of condensed phosphates may be converted to forms not available to algae as part of enzymatic hydrolysis.

OTHER FACTORS THAT NEED CONSIDERATION

The review of the literature in an earlier section showed that the amount of algal-available phosphorus in a sample of Great Lakes tributary water can be estimated by using the sum of the soluble orthophosphate concentration and about 0.2 times the particulate phosphorus content. It was further shown that for many systems, the anion exchange resin-extractable P and hydroxide-extractable P provide reasonable estimates of the amount of phosphorus that will become available to support algal growth in about a two-to-three week culture period. There are several important limitations and factors that must be considered in the use of this information on the development of a phosphorus mangement strategy for a waterbody. Some of these factors are discussed in this section of this paper.

Organic Phosphorus Sources

The studies that have been conducted thus far on algal available phosphorus have considered predominantly inorganic systems or were conducted for sufficiently short periods of time so as to preclude the mineralization of organic P to orthophosphate. There are some situations in which particulate or soluble organic phosphorus represents significant parts of the total P content of an aquatic system. In these types of systems, the hydroxide extraction or anion exchange procedures for estimating available P are not reliable and should not be used. Further, the standard algal assay procedures only assess the amount of organic P that becomes available during the course of the bioassay test, typically 2-3 weeks. It is likely that the rates of mineralization of most organic phosphorus compounds will be such that a much longer incubation period would be needed to assess their long-range availability.

As noted previously, this situation could be of particular significance in agronomic practices in that in shifting from normal to minimum to no-till farming practices, the availability of P in the particulates derived from the land may be drastically altered. Under normal tillage practices, it is expected that 10-30% of the P associated with the erosional particulate matter would become available to support algal growth. Under no-till conditions, where erosion is minimal, the character of the particulate matter will be substantially different from that of normal tillage; it will be predominantly organic, rather than inorganic in nature. The availability of P from the organic particulate matter from no-till systems is unknown at this time. It is possible that, although switching to a no-till method of farming would reduce erosion, it might increase the overall net flux of available phosphorus from a unit area of land. This is an area that needs immediate attention in order to develop meaningful, cost-effective programs for the control of phosphorus from agricultural sources.

Another farming practice which has been shown to result in a particularly significant increase in the amount of P derived from agricultural lands is the spreading of manure on frozen soil. In some years when the spring thaw occurs while the soil is still frozen, the P associated with the manure is carried to a nearby watercourse, rather than being incorporated into the soil. Further discussion of the significance of manure spread on frozen land as a cause of water quality deterioration has been presented by Sonzogni and Lee [50].

Other nonpoint sources of P which have been found to have large amounts of organic phosphorus present are wetlands and marshes. Lee et al. [51] found that marsh drainage from several Wisconsin marshes contained relatively large amounts of apparently soluble organic phosphorus compounds. Those investigators did not, however, determine the rate at which these compounds would be converted to orthophosphate through mineralization processes. Wetlands would also be expected to contribute large amounts of particulate organic phosphorus to downstream waterbodies. In the upper Midwest during the late winter-early spring high flow periods, the organic phosphorus present in detrital plant remains is transported from marshes to nearby watercourses [51]. Lee et al. [51] and Jones and Lee [52] reported that over the annual cycle the amount of phosphorus entering many wetlands is expected to be approximately equal to the amount transported from the wetlands during high flow. As Jones and Lee [52] pointed out, although total P_{in} may equal total P_{out} for many wetlands, the presence of wetlands could have a significant impact on the algal availability of phosphorus from various sources. This is of importance with respect to not only conventional wetlands systems, but also those used for treatment of domestic wastewaters. With increasing attention being given to this method of wastewater disposal, it is important that the efficacy of such systems be judged in terms of their ability to remove algal available P over the annual cycle.

From a Great Lakes phosphorus management strategy point of view, the effect of wetlands in changing the forms of P that enter the Great Lakes is likely to be small because of the relatively small amount of wetland area in the Great Lakes watershed. The above-mentioned phenomenon could, however, be important in localized areas of the Great Lakes or for other lakes. There is need to determine both short-term and longer term availability of phosphorus present in wetland drainage. With this type of information, it would be possible to assess the significance of wetlands in affecting the phosphorus management strategy for a waterbody.

Mineralization of Phosphorus

In assessing the longer term availability of phosphorus, consideration must be given to the regeneration or mineralization of the phosphorus present in algal cells. There is considerable controversy in the literature about the extent to which algae are degraded. From their studies on axenic algal cultures, they determined that bacteria and microfauna increase the rate and extent of aerobic algal decompositon. Mixtures of autotrophic and heterotrophic organisms had a refractory portion of 12-86% that was not degraded within a year under aerobic conditons. Older cultures exhibited a narrower range of 30-70%, while the mean of 75 different cultures was 40% refractory material.

Foree et al. [54 studied nutrient regeneration of algae in laboratory tests. According to them, the two major factors controlling this regeneration are: (1) the extent of organic decomposition; and (2) the initial nutrient content of the algal cells. With respect to the latter, Foree et al. pointed out that in algae which have taken up "luxury" phosphorus, the extra P is loosely bound in the cells and is therefore more readily released than P more tightly bound within the cell. They suggested that the critical initial P content of algae, above which excess P regeneration occurs, is about 0.3% by weight. They concluded that after 0.5-1 year of decomposition under aerobic conditions, an average of 50% (range of 0-94%) of the initial P remained refractory (undecomposed), whereas under anaerobic conditions 40% of the P remained refractory. According to Foree et al., this difference between aerobic and anaerobic regeneration is not significant. Two additional comments were made concerning these results. First, regeneration values determined in that study related to the maximum rate of decomposition, and second, other means of essentially permanently removing regenerated or other phosphate from the water (e.g., precipitation, sorption) were not considered. It is likely, therefore, that Foree et al.'s estimate of the amount of P regeneration from algae is an overestimate of what would actually be expected to be available in the environment. The results of their studies were consistent with the conclusions of the American Water Works Association Committee [55] review, that 30-70% of algae and associated organics are readily degraded. The remainder is resistant to degradation, being consumed at a rate of a few percent per year.

Golterman [5] concluded that the phosphate not returned from algae into the biochemical cycle was about 1-5% per cycle through algae. He pointed out, however, that this could represent a significant amount over a 200-day growing season. With growth cycles of 5-10 days, this could represent a loss (i.e., phosphate made unavailable) of 20-100% of the phosphate taken up. The product formed in this P-cycle is detrital phosphorus which is resistant to bacterial attack.

DePinto and Verhoff [56] investigated nutrient regeneration from algal cultures by growing axenic cultures of *Chlorella* and *Selenastrum* for about seven days, then placing them in the dark, inoculating some with bacteria. They indicated that the rate and extent of algal regeneration was dependent on initial concentrations of available P, as well as the amount of bacteria present. For initial concentrations of 186 μg P/l in the assay medium, soluble ortho-P regeneration in the dark ranged from about 80 to 95% over 1-2 months. For initial concentrations of 80 μg P/l, regeneration of soluble ortho-P in the dark was about 30-50%. Assuming that there is a real dependence of regeneration on the initial concentration, the upper ranges of percent regeneration found by DePinto and Verhoff would not be expected in the field because very few P-limited waterbodies would be expected to contain 0.2 mg/l available P during the summer when temperatures are at the levels evaluated by DePinto and Verhoff. Because of their experimental design, it would therefore be expected that their estimates of regeneration would be high. Their regeneration estimates also represent a one growth cycle regeneration; it may be expected that each time the regenerated phosphorus is reused, 50-70% of it will become refractory. The amount of regeneration found by DePinto and Verhoff is similar to that found by Jewell and McCarty [53] and Foree et al. [54] for regeneration over a year's time. Their rate data, however, would not be expected to be applicable to natural water systems because of the nature of their experimental design. One of the common problems encountered by investigators in this area is designing a test system which can properly simulate natural water conditions.

While the range of values for the unmineralizable P fraction of dead algae found in the literature is wide, it appears that on the order of half of the P content of dead algal cells remains unavailable. The amount of mineralization of algal P to available P depends to a large extent on the specific physical, chemical and biological characteristics of the area of concern. These results have important implications in the development of a P management strategy. For example, the development of refractory organic compounds is an important mechanism in some river systems by which P is converted from available to unavailable forms. The same situation exists within arms or bays of waterbodies.

Other Forms of Phosphorus

There are a number of commerically used organic P compounds which may potentially stimulate algal growth in receiving waters. These compounds include P-C, P-O-C, and P-O-P ring and chain bonded materials, such as those used in treatment of cooling waters for industrial cooling towers. Some of these compounds would likely enter aquatic systems as blow-down to storm or municipal sewers. Generally, the polymeric inorganic compounds are hydrolyzed in aquatic systems. While some of the organic P compounds are resistant to hydrolysis, they would be expected to eventually hydrolyze at least in part to soluble orthophosphate. It is possible that these compounds would show a phenomenon similar to that found by Clesceri and Lee [49] for condensed phosphate of only partially hydrolyzing to soluble orthophosphate. While the authors are not aware of the amount of these compounds used, they speculate that this use is small, compared to the total P load to the Great Lakes and therefore represents an insignificant factor in developing a P management strategy for the Great Lakes. While likely unimportant in the Great Lakes system, P used in cooling towers, as well as in industrial processes, could be significant in influencing water quality in other waterbodies. This would have to be investigated on case-by-case basis for each particular waterbody. When it appears that the total P load from these sources is potentially important, then studies should be done to ascertain what part of that total P load enters the waterbody in a form available to support algal growth.

IN-LAKE AVAILABILITY AND LOAD-RESPONSE MODELING

Thus far, this discussion has focused on the availability of P in samples of water or particulate matter from various sources, such as agricultural runoff, domestic wastewaters, etc. Also, consideration has been given to the availability of P in samples of river water obtained at the lake-tributary interface. What is needed, however, is an assessment of the amount of P contributed to the lake or waterbody of concern which is available to support algal growth within the body of water. Numerous investigators, such as Cowan and Lee [8], have pointed out that what may be available to an algal culture in a sample of water from a particular source over a two-week incubation period will not likely be the same as what becomes available in the lake. There is a wide variety of physical, chemical, and biological processes which influence the algal availability of P in aquatic systems.. Unfortunately, in the early 1970s the U.S. EPA and other agencies greatly curtailed funding for research on the aqueous environmental chemistry of phosphorus. We know very little more about the actual behavior of P in aquatic systems than was known in the 1960s when funds were available. It is certainly likely that the attempts to simulate environmental systems in laboratory beakers or jugs, for such purposes as measuring the rate and extent of mineralization of organic forms of P, the slow release of P from apatite species, and similar kinds of reactions, are likely to have limited applicability to real world situations. For example, it is generally stated in the literature that apatite phosphorus is not available to support algal growth. This statement is typically based on relatively short-term (i.e., a week or two) laboratory experiments. However, longer-term experiments such as those of Gerhold and Thompson [57] have shown that P in at least some forms of apatite is available to support algal growth. This is confirmed by the work of Golterman [5] and Golterman et al. [4].

This raises an interesting question concerning the erosion of Lake Erie and Lake Ontario bluffs. It is generally stated that this erosion does not contribute to the fertilization of these lakes [6]. However, examination of the results of algal culture experiments [6] on the erosional material derived from the bluff erosional material where the predominant form of P is apatite. When one considers the total amount of erosion that takes place (estimated to contribute 40% of the total P load to Lake Erie), and the fact that the previously conducted growth experiments only incorporated a few weeks of incubation, there could be appreciable algal growth in these lakes resulting from this source. To determine the actual significance of apatite erosion from Lake Erie bluffs or other sources, a much better understanding of the behavior of apatite P in the lake system of concern must be achieved. The same is true of P from other sources, such as erosional material from land runoff, particulate P in wastewater effluents or atmospheric dustfall.

Because it is unlikely that funding will become available to provide the types of information needed, it is reasonable to ask whether a P management strategy can be developed based on available P inputs. There are those who advocate that because of the many unknowns about the behavior of P in aquatic systems, the only appropriate approach is to base a P management strategy on the total phosphorus input, without any attempt to assess what part of the total will be unavailable. This isessentially the policy which has been followed until now. However, in the future such an approach could be tremendously wasteful of the financial resources available for pollution control with little actual improvement in eutorphicatin-related water quality in the lakes. Therefore, there is substantial agreement that P management strategies should be based on the best information available on the amounts and sources of available forms of P that enter the lakes.

Development of P Management Strategies

The first step in the development of a P management strategy is to determine the amounts of P derived from each principal source. Next, assuming that all P from each source is 100% available, an estimation must be made of the potential improvement in water quality that will result from controlling each source to the extent possible. The estimate should be based on the OECD eutrophication modeling approach described below. If it appears that the controllable P from a given source will have a significant impact on water quality, then consideration should be given to the amount of total P from that source likely to become available to support algal growth. It may be possible to use some of the generalized relationships discussed earlier in this paper (such as 20% of the particulate P from diffuse sources being available for algal growth). It may be necessary, however, to conduct laboratory and field studies to evaluate for the particular aquatic system in question, the amount of algal available P contributed from each potentially significant source. It has been recommended that there is a need to standardize bioassay procedures being used to assess the availability of P. Such an approach, however, will not necessarily improve the quality of the data generated. While general guidelines on how to conduct such a test should be formulated, a "cook book" approach which provides step-by-step instructions should be avoided. All bioassays conducted to assess P availability must be tailored to some extent to fit the particular system under investigation. All users of such bioassays must be thoroughly familiar with their appropriate use and limitations.

Standardization of methods and approaches has carried through to the design of treatment plants for P and other contaminant removel. The approach typically used today in the design of wastewater treatment works and the development of control programs for diffuse sources is based on what is readily achievable in conventional treatment (i.e., 90% removal of some contaminants from domestic wastewaters), rather than the characteristics of the receiving waters. However, in the 1980s, in the U.S., waste load allocation under the provisions of best available treatment (BAT) will require that contaminant control programs be based on the characteristics of the receiving waters. This approach requires a case-by-case evaluation of the costs and water quality benefits that will be derived from a particular control program. Much greater emphasis will have to be placed on field and laboratory studies designed to measure the impact of a contaminant from a particular source on receiving water quality. Because of the fact that many contaminants such as P exist in unavailable form in sources and aquatic systems, the heart of the field evaluation program will have to be devoted to defining the art of the contaminant load that is available to affect water quality in the waterbody in question. As discussed earlier for P removal in wastewater treatment plants to 0.5 mg P/l in the effluent, the expenditure of funds for defining, quantifying and assessing the significance of available forms of contaminants could be highly cost-effective in helping to design contaminant control programs that minimize cost, yet provide maximum environmental protection.

The key to the above-mentioned process is the availability of a reliable load-response model that can predict for any given contaminant control program, the improvement in water quality that will arise from program implementation. The OECD eutrophication study provides a technically valid approach for assessing load-response relationships for P-based eutrophication control programs. The characteristics of this modeling approach and its ability to handle available forms of P are discussed below.

OECD Eutrophication Modeling Approach

The Organization for Economic Cooperation and Development (OECD) eutrophication program conducted an evaluation of P load-eutrophication response relationships for 200 waterbodies in 22 counties based on the work of Vollenweider [58,59,23]. The bulk of the results of the overall OECD eutrophication study has not yet been published, because the project was just being completed at the time of preparation of this chapter. However, the results of the U.S. part of the study have been published [24]. Rast and Lee [24], using Vollenweider's [58,59,23] framework and the data from the 40 U.S. OECD waterbodies, developed statistical relationships between P load to a waterbody, as normalized by mean depth and hydraulic residence time, and planktonic algal chlorophyll, Secchi depth (water clarity), and hypolimnetic oxygen depletion rate. The authors have had the opportunity to review the results of the total OECD eutrophication study and have found that the load-response relationships for the waterbodies outside the U.S. are essentially the same as those found by Rast and Lee [24]. Further, the authors of this chapter have, subsequent to the completion of the U.S. OECD studies, studied approximately 60 waterbodies principally located in the U.S. and including all of the Great Lakes, and found that they fit the same P load-eutrophication response relationships. Thus, there are over 300 waterbodies across the world that fit the P load eutrophication response relationships developed by Vollenweider [58,59] and Rast and Lee [24].

It is now possible to predict for waterbodies in which the maximum planktonic algal growth is limited by phosphorus, the mean and the maximum [60] planktonic algal chlorophyll, Secchi depth due to phytoplankton light scattering, hypolimnetic oxygen depletion rate and overall fish yield [61], based on mean depth, hydraulic residence time and P load. Figure 1

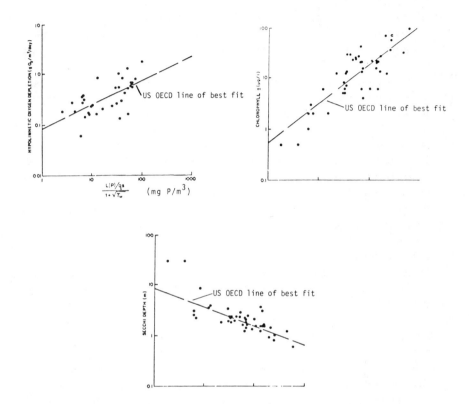

Figure 1. U.S. OECD data applied to phosphorus load, mean chlorophyll *a*, mean Secchi depth, and hypolimnetic oxygen depletion rate relationships [4].

presents a summary of the relationships developed by Rast and Lee [24] based on the U.S. OECD waterbodies. It should be noted that although the Vollenweider-OECD relationships are based on total P load, they implicitly consider the available P load. This is because many of the waterbodies included within the OECD study have about the same ratio of point to nonpoint source P loads. Waterbodies with high percentages of domestic wastewater point source P loads tend to become limited by nitrogen or some other factor, and therefore would not be expected to fit the Vollenweider-OECD load-response relationships. In essence, it appears that there is a relatively constant error relating total P load to available P load for many of the U.S. OECD waterbodies. The remarkable insensitivity of lakes' eutrophication responses to small changes in P loads, such as would be associated with considering total P load vs available P load for a particular waterbody effectively overshadows the significance of this error in formulating load-response relationships for a waterbody. Although water quality is generally insensitive to minor changes in P load (such as reducing domestic wastewater treatment plant effluents to 0.5 mg P/l in the Great Lakes basin), such changes may be very expensive to achieve.

In an effort to assess whether the availability of the P load to a waterbody is accounted for in the OECD modeling approach, or causes a bias in the position of a waterbody in the P load-response relationships, the U.S. OECD waterbody data sets and relationships were reexamined. Those waterbodies (or portions thereof) to which the point source total P load (mostly readily available to algae) was less than 50% of the total load (56% of the 45 points) were identified on the U.S. OECD normalized P load-chlorophyll diagram (Figure 1) to determine if they were positioned differently, as a group, the whole set of OECD waterbodies. It was found that there was no discernible demarcation between the two sets of data points. Those waterbodies with less than 50% of their P loads from point sources were essentially equally distributed among those waterbodies having greater than 50% of their P loads in readily available forms. This is an indication that the Vollenweider-OECD modeling approach handles implicitly in its normalization components, the availability of P load to waterbodies.

Rast and Lee [24] and Lee et al. [4] indicated that altering the P load to a waterbody should result in the load-response point for the given waterbody moving parallel to the U.S. OECD lines of best fit shown in Figure 1. Actually this is in error to the extent that the ratio of available P to unavailable P in the P load being removed from the waterbody deviates from the average ratio for the waterbodies from which the regression lines were derived. If the P control program is directed primarily toward diffuse sources, in which most of the P is in an unavailable form, then it would be expected that there would be less improvement in water quality per unit P removed than would be predicted based on the U.S. OECD lines of best fit. The converse would be true for those control programs directed toward the available forms of phosphorus.

In addition to providing an important engineering tool upon which to develop eutrophication control programs for lakes, impoundments and estuaries, these studies have provided valuable insights into the behavior of various forms of P in waterbodies. First and foremost, it is clear from these and other studies, such as Lee et al. [16], that the sediments of waterbodies are very important sinks for P. This is supported by the P residence time studies of Vollenweider [62] and Sonzogni et al. [63]. Many waterbodies have been found to have P residence times of a year or less. This means that the P which enters the waterbody each year is either deposited into the sediments or is carried out of the waterbody in its outflow. Although there is release of P from sediments during certain times of the year, which can have near-term effects on water quality, the driving force for eutrophication of many waterbodies is the external P load. Therefore, the long-term, slow release of P from sediments is apparently insignificant in determining overall trophic state for many waterbodies. It is concluded by the authors that available P incubation studies which incorporated incubation periods of a few weeks to a month or so are probably adequate to describe the potentially significant amounts of available P entering waterbodies. It is important to note that the rates of P becoming available in the field vs laboratory are likely to be markedly different. Also, the P concentrations in the water will likely affect the extent to which P becomes available from particulate forms. Ultra-low P levels in laboratory study system would tend to cause P to become available which would normally never become available in an actual waterbody. This situation again points to the importance of gaining a much better understanding of P cycling within lakes in order to understand the extent to which particulate P becomes available to support algal growth within the waterbody.

The OECD eutrophication studies have demonstrated that the approach of Miller et al. [64] of using results of algal assays for available forms of P in a particular sample is invalid to estimate the benefits that would be derived by controlling the available P load to a waterbody. One cannot use, contrary to what was implied by Miller et al., a direct ratio of available load to response for estimating benefits (i.e., a 10% reduction in available load will not necessarily result in a 10% reduction in algal biomass). Consideration must be given not only to the P load, but also the mean depth and hydraulic residence time of the waterbody, and to the fact that the load-response correlation from the OECD studies is a log-log relationship.

Techniques such as measurement of phosphatases, acetylene reduction, etc., developed by Fitzgerald [65,66], provide useful information for assessing the amount of available P in a particular sample. These techniques have application to determining whether the algae present in a sample have been grown under a P-limited or surplus condition. They have little or no applicability, however, to determining the amount of available P in a particular sample.

Other Considerations

Studies on the aqueous environmental chemistry of P must focus attention on nearshore waters, since it is in this region that the greatest uptake and release from particulate matter will occur. Work in this region of a waterbody is especially difficult because of the lack of understanding of the hydrodynamics of these regions, which affects the mixing of nearshore and offshore waters, and the mixing of tributary and nearshore waters, as well as the sedimentation dynamics of particles. The hydrodynamic information coupled with appropriately designed laboratory bioassay tests will provide the kinds of information needed to determine how much of the P load to a waterbody will become available to support algal growth.

In conducting in-lake studies of P availability, particular attention must be given to the role of hydrous iron oxides in controlling P availability. The studies of Jones and Lee [46] showed that hydrous iron oxide plays a dominant role in controlling the amount of algal-available P released from waterway sediments. Failure of U.S. pollution control agencies in the Great Lakes to understand the role of hydrous iron oxides in controlling the release of contaminants from sediments and the use of bulk sediment criteria has resulted in expenditures of over $200,000,000 by the American public for alternative, more expensive methods of dredged sediment disposal involving on-land, rather than open water disposal. As discussed by Jones and Lee [46], the so-called confined disposal could be more detrimental to Great Lakes water quality than the previously used, less expensive open water disposal techniques. Similarly, Plumb and Lee [67] and Lee [45] have demonstrated through the use of algal assays that the approximately 50,000 tons of taconite tailings discharged daily to Lake Superior by the Reserve Mining Company did not represent a source of available P to the lake as a result of the formation of hydrous iron oxide on the surface of the tailings particles, which sorbed any P release from the tailings and made the P unavailable to support algal growth under the oxic conditions that prevail in Lake Superior.

The dynamic modeling approach for estimating load-response relationships attempts to describe the hydrodynamics and aquatic chemistry of phosphorus with particular emphasis on tracking the soluble orthophosphate-available P within aquatic systems. One of the reasons why the dynamic modeling approach has limited predictive capabilities is the fact that the sedimentation and sediment/water exchange processes are not properly modeled. Such modeling is not likely to be accomplished until appropriate studies have been conducted on the behavior of P in aquatic systems. The dynamic models, in considering the availability of the various forms of phosphorus, do not properly reflect the natural water system. There is no justification for claiming that the dynamic modeling approach is a more valid approach for estimating nutrient load-eutrophication response relationships than the statistical approach developed by Vollenweider. Evidence points to the opposite conclusion.

CONCLUSIONS AND RECOMMENDATIONS

Phosphorus management strategies for the Great Lakes and other water-bodies should be based on the control of the algal available P load to the waterbody, rather than the total P load. The available forms of phosphorus from a particular source should be determined by algal bioassays in which the increase in algal biomass is measured. Other bioassay or chemical extraction techniques should only be used after they have been verified for the particular source in question by comparison to the results of concurrently run algal bioassays. It appears that NaOH-extractable and anion exchangeable phosphorus correlate fairly well with bioassay-available P for urban drainage and for phosphorus derived from normal tillage agricultural runoff. Neither of these two techniques as normally used would properly assess the availability of organically bound phosphorus. There is a need for P availability studies on domestic wastewater treatment plant effluents, atmospheric dust-fall and precipitation, high organic P sources, such as no-till or minimum-till farming, and wetland drainage.

Research needs to be conducted on the aqueous environmental chemistry of phosphorus in rivers and in lakes with particular reference to defining the amounts of phosphorus that remain or become available within the system, or are converted to unavailable forms. This research should be specifically focused on obtaining information needed for development of phosphorus management strategies. At this time, it is recommended that the available P load from urban stormwater drainage and normal-tillage agricultural runoff that enters the lake directly, or has a limited distance of tributary travel between source and lake, be estimated based on the relationship:

$$\text{available P} = \text{soluble ortho P} + 0.2 \text{ (particulate P)}$$

Before a P management plan which includes a reduction in the P from domestic wastewater treatment plant effluents below 1 mg P/l is implemented, definitive research must be done to evaluate the amount of available P removed by the treatment proposed. Also, before implementation of minimum or no-till farming for the purpose of controlling P input to a waterbody, studies should be conducted to define the relative amounts of available P derived from normal tillage vs other farming practices.

Also, studies should be conducted on the amount of condensed P present in domestic wastewater treatment plant influents and in combined sewer overflow, which becomes available to support algal growth in the treatment plant and in the Great Lakes.

ACKNOWLEDGMENTS

This paper was supported by the Colorado State University Experiment Station, Department of Civil Engineering at Colorado State University, Fort Collins, CO, and EnviroQual Consultants and Laboratories, Inc., Fort Collins, CO. In addition, background materials for this chapter were provided by staff of the International Joint Commission Great Lakes Regional Office, Windsor, Ontario. We wish to acknowledge the assistance of Twyla Smith in typing the manuscript.

REFERENCES

1. International Joint Commission. "Great Lakes Water Quality Agreement of 1978," Agreement with Annexes and Terms of Reference, Between the United States of America and Canada, signed at Ottawa, November 22, 1978. (International Joint Comission, Great Lakes Regional Office, Windsor, Ontario) 1978.
2. U.S. Environmental Protection Agency. "Algal Assay Procedure Bottle Test," U.S. EPA, Corvallis, OR (1971).
3. U.S. Environmental Protection Agency. "Marine Algal Assay Procedure Bottle Test," U.S. EPA, Corvallis, OR (1974).
4. Walton, C. P., and G. F. Lee. "A Biological Evaluation of the Molybdenum Blue Method for Orthophosphate Analysis," *Verh. Internat. Verein. Limnol.* 18:676-684 (1972).
5. Golterman, H. L., C. C. Bakles and J. Jakobs-Mogelin. "Availability of Mud Phosphates for the Growth of Algae," *Verh. Internat. Verein. Limnol.* 17:467-479 (1969).
6. Pollution From Land Use Activities Reference Group (PLUARG). "Annual Progress Report from the International Reference Group on Great Lakes Pollution from Land Use Activities (PLUARG)," Windsor, Ontario, July (1977).
7. Cowen, W. F., and G. F. Lee. "Phosphorus Availability in Particulate Materials Transported by Urban Runoff," *J. Water Poll. Control Fed.* 48:580-591 (1976a).
8. Cowen, W. F., and G. F. Lee. "Algal Nutrient Availability and Limitation in Lake Ontario during IFYGL, Part I," U.S. EPA-Report No. 600/3-76-094a, U.S. EPA, Duluth, MN (1976b).
9. Cowen, W. F., "Algal Nutrient Availability and Limitation in Lake Ontario during IFYGL," Ph.D Thesis, University of Wisconsion-Madison (1974).
10. Sagher, A. "Availability of Soil Runoff Phosphorus to Algae," Ph.D. Thesis, University of Wisconsin-Madison (1976).
11. Sagher, A., R. F. Harris and D. E. Armstrong. "Availability of Sediment Phosphorus to Microorganisms," Technical Report WIS WRC 75-01, Water Resources Center, University of Wisconsin, Madison, WI (1975).

12. Huettl, P. J., R. C. Wendt and R. B. Corey. "Prediction of Algal-Available Phosphorus in Runoff Suspensions," *J. Environ. Qual.* 8:130-132 (1979).
13. Chiou, C. J., and C. E. Boyd. "The Utilization of Phosphorus from Muds by the Phytoplankter, *Scenedesmus dimorphus* and the Significance of These Findings to the Practice of Pond Fertilization," *Hydrobiologia* 45:345-355 (1974).
14. Porcella, D. B., J. S. Kumagai and E. J. Middlebrooks. "Biological Effects on Sediment-Water Nutrient Interchange," *J. San. Eng. Div. Am. Soc. Civil Eng.* SA4:911-926 (1970).
15. Stauffer, R. E., and G. F. Lee. "The Role of Thermocline Migration in Regulating Algal Blooms," in *Modeling the Eutrophication Process*, E. J. Middlebrooks, D. H. Falkenborg and T. D. Maloney, Eds. (Ann Arbor, MI: Ann Arbor Science Publishers, Inc., 1975), pp. 73-82.
16. Lee, G. F., W. C. Sonzogni and R. D. Spear. "Significance of Oxic Versus Anoxic Conditions for Lake Mendota Sediment Phosphorus Release," in *Proceedings of International Symposium on Interactions Between Sediments and Fresh Water, 1976,* (The Hague: W. Junk, Purdoc, 1977), pp. 294-306.
17. Li, W., D. E., Armstrong and R. F. Harris. "Biological Availability of Sediment Phosphorus to Macrophytes," Technical Report WIS WRC 74-09, Water Resources Center, University of Wisconsin, WI (1974).
18. Williams, J. D. H., H. Shear, and R. L. Thomas. "Availability to *Scenedesmus* of Different Forms of Phosphorus in Sedimentary Materials from the Great Lakes," Draft Report, Canada Centre for Inland Waters, Burlington, Ontario, Canada.
19. Logan, T. J. "Chemical Extraction as an Index of Bioavailability of Phosphate in Lake Erie Basin Suspended Sediments," Final Project Report, Lake Erie Wastewater Management Study, U.S. Army Corps of Engineers, Buffalo, NY (1978).
20. Logan, T. J., F. H. Verhoff and J. V. DePinto. "Biological Availability of Total Phosphorus," Technical Series Report, Lake Erie Wastewater Management Study, U.S. Army Corps of Engineers, Buffalo, NY (1979).
21. Verhoff, F. H., M. R. Heffner and W. A. Sack. "Measurement of Availability Rate for Total Phosphorus from River Waters," Final Report, Lake Erie Wastewater Management Study, U.S. Army Corps of Engineers, Buffalo, NY (1978).
22. Verhoff, F. H., and M. R. Heffner. "Rate of Availability of Total Phosphorus in River Waters," *Environ. Sci. Technol.* 13:844-849 (1979).
23. Vollenweider, R. A. "Advances in Defining Critical Loading Levels for Phosphorus in Lake Eutrophication," *Mem. Ist. Ital. Idrobiol.* 33:53-83 (1976).
24. Rast, W., and G. F. Lee. "Summary Analysis of the North American (U.S. Portion) OECD Eutrophication Project: Nutrient Loading-Lake Response Relationships and Trophic State Indices." U.S. EPA Report No. EPA-600/3-78-008, U.S. EPA, Corvallis, Oregon (1979).

25. Armstrong, D. E. , J. R. Perry and D. Flatness. "Availability of Pollutants Associated with Suspended or Settled River Sediments Which Gain Access to the Great Lakes," Draft Final Report on EPA Contract No. 68-01-4479, U.S. EPA, Chicago, IL (1979).

26. Lee, G. F., and R. A. Jones. "Water Quality Characteristics of the U.S. Waters of Lake Ontario during the IFYGL and Modeling Contaminant Load-Water Quality Response Relationships in the Nearshore Waters of the Great Lakes," Report to National Oceanic and Atmospheric Administration, Ann Arbor, MI (1979).

27. Monteith, T. J. and W. C. Sonzogni. "U.S. Great Lakes Shoreline Erosion Loadings," PLUARG report, International Joint Commission, Windsor, Ontario (1976).

28. Casey, D. J., P. A. Clark and J. Sandwick. "Comprehensive IFYGL Materials Balance for Lake Ontario," U.S. EPA Report No. 902/9-77-001, U.S. EPA, Rochester, NY (1977).

29. Thomas, R. L. "Forms of Phosphorus in Particulate Materials from Shoreline Erosion and Rivers Tributary to the Canadian Shoreline of the Great Lakes," Draft Report, Canada Centre for Inland Waters, Burlington, Ontario, Canada.

30. Brezonik, P. L. "Nutrients and Other Biologically Active Substances in Atmospheric Precipitation," Proc. 1st Special Symposium on Atmospheric Contribution to the Chemistry of Lake Waters, Int. Assoc. Great Lakes Res. (1975).

31. Murphy, T. J., "Sources of Phosphorus Inputs from the Atmosphere and Their Significance to Oligotrophic Lakes," WRC Research Report No. 92, University of Illinois Water Resources Center, Urbana, IL (1974).

32. Peters, R. H. "Availability of Atmospheric Orthophosphate," *J. Fish. Res. Bd. Can.* 34:918-924 (1977).

33. Murphy, T. J., and P. V. Doskey. "Inputs of Phosphorus from Precipitation to Lake Michigan," U.S. EPA Report No. EPA-600/3075-005, U.S. EPA, Duluth, MN (1975).

34. Delumyea, R. G., and R. L. Petel. "Atmospheric Inputs of Phosphorus to Southern Lake Huron, April-October 1975". U.S. EPA Report No. 600/3-77-038, U.S. EPA, Duluth, MN (1977).

35. Delumyea, R. G., and R. L. Petel. "Wet and Dry Deposition of Phosphorus into Lake Huron," *Water, Air Soil Poll.* 10:187-189 (1978).

36. Rigler, F. H. " Radiobiological Analysis of Inorganic Phosphorus in Lake Water," *Verh. Internat. Verein. Limnol.* 16:465-470 (1966).

37. Rigler, F. H. "Further Observations Inconsistent with the Hypothesis that the Molybdenum Blue Method Measures Orthophosphate in Lake Water," *Limnol. Oceanog.* 13:7-13 (1976).

38. Lean, D. R. S. " Phosphorus Dynamics in Lake Water," *Science* 179: 678-679 (1973).

39. Lean, D. R. S. "Movements of Phosphorus between Its Biologically Important Forms in Lakewater," *J. Fish. Res. Bd. Can.* 30:1525-1536 (1973).

40. Walton, C. P., and G. F. Lee. "A Biological Evaluation of the Molyb-denum Blue Method for Orthophosphate Analysis," *Verh. Internat. Verein. Limnol.* 18:676-684 (1972).
41. Lee, G. F., W. Rast and R. A. Jones. "Eutrophication of Waterbodies. Insights for an Age-old Problem," *Environ. Sci. Technol.* 12:900-908 (1978).
42. Lee, G. F., W. Rast and R. A. Jones. "Use of the OECD Eutrophication Modeling Approach for Assessing Great Lakes' Water Quality," Environmental Engineering, Colorado State University, Fort Collins, Occasional Paper No. 42 (1979).
43. Lee, G. F., M. Abdul-Rahman and E. Meckel. "A Study of Eutrophica-tion, Lake Ray Hubbard, Dallas, Texas," Environmental Engineering, Colorado State University, Fort Collins, Occasional Paper No. 15 (1977).
44. Lee, G. F. "Review of the Potential Water Quality Benefits from a Phosphate-Built Detergent Ban in the State of Michigan," Paper presented at State of Michigan Hearings on Proposed Phosphate Deter-gent Ban, Lansing, MI, December (1976).
45. Lee, G. F. "Role of Hydrous Metal Oxides in the Transport of Heavy Metals in the Environment," *Prog. Water Technol.* 17:137-147.
46. Jones, R. A. and G. F. Lee. "Evaluation of the Elutriate Test as a Method of Predicting Contaminant Release during Open Water Disposal of Dredged Sediment and Environment Impact of Open Water Dredged Material Disposal, Vol. I: Discussion," Technical Report D-78-45, U.S. Army Corps of Engineers, Waterways Experiment Station, Vicksburg, MS (1978).
47. Malhotra, S. K., G. F. Lee and G. A. Rohlich. "Nutrient Removal from Secondary Effluent by Alum Flocculation and Lime Precipitation," *Air Water Poll.* 8:487-500 (1964).
48. Horstman, H. K., and G. F. Lee. "Studies on the Eutrophication of Lake Dillon, Colorado," Environmental Engineering, Colorado State University, Fort Collins, Occasional Paper (1979).
49. Clesceri, N. L., and G. F. Lee. "Hydrolysis of Condensed Phosphates I: Non-sterile Environment," *Air Water Poll.* 9:723-742 (1965).
50. Sonzogni, W. C. and G. F. Lee. "Nutrient Sources for Lake Mendota—1972," *Trans. Wisc. Acad. Sci., Arts Lett.* LXII:133-164 (1974).
51. Lee, G. F., E. Bentley and R. Amundson. "Effect of Marshes on Water Quality," In: *Ecological Studies 10, Coupling of Land and Water Systems,* (New York: Springer-Verlag, New York, Inc., 1975), pp. 105-127.
52. Jones, R. A. and G. F. Lee. "An Approach for the Evaluation of the Efficacy of Wetlands-Based Phosphorus Control Programs on Down-stream Water Quality," in *Environmental Quality through Wetlands Utilization,* Sym. Proc., Coordinating Council on the Restoration of the Kissimeee River Valley and Taylor Creek-Nubbin Slough Basin, (Tallahassee, FL: 1978), pp. 237-243.

53. Jewell, W. J., and P. L. McCarty, "Aerobic Decomposition of Algae," *Environ. Sci. Technol.* 5:1023-1031 (1971).
54. Foree, E. G., W. J. Jewell and P. L. McCarty. "The Extent of Nitrogen and Phosphorus Regeneration from Decomposing Algae," Proceedings of Water Pollution Research Conference, 1970 (Elmsford, NY: Pergamon Press, Inc., 1971).
55. American Water Works Association (AWAA) Water Quality Division Committee, "Chemistry of Nitrogen and Phosphorus in Water," *J. Am. Water Works Assoc.* 62:127-140 (1970).
56. DePinto, J. V., and F. H. Verhoff. "Nutrient Regeneration from Aerobic Decomposition of Green Algae," *Environ. Sci. Technol.* 11:371-377 (1977).
57. Gerhold, R. M., and J. R. Thompson. "Calcium Hydroxyapatite as an Algal Nutrient Source," Paper presented at American Chemical Society Meeting, New York, NY, September 7-12, 1969.
58. Vollenweider, R. A. "Scientific Fundamentals of the Eutrophication of Lakes and Flowing Waters with Particular Reference to Nitrogen and Phosphorus as Factors in Eutrophication," Technical Report DAS/CSI/68, OECD, Paris (1968).
59. Vollenweider, R. A. "Input-Output Models, With Special Reference to the Phosphorus Loading Concept in Limnology," *Schweiz. Z. Hydrol.* 37:53-84 (1975).
60. Jones, R. A., Rast, and G. F. Lee. "Relationship between Mean and Maximum Chlorophyll *a* Concentrations in Lakes," *Environ. Sci. Technol.* 13:869-870 (1979).
61. Lee, G. F., B. W. Newbry, and R. A. Jones. "Relationship between Phosphorus Load and Fish Yield in Waterbodies," (submitted).
62. Vollenweider, R. A. "Possibilities and Limits of Elementary Models Concerning the Budget of Substances in Lakes", *Arch. Hydrobiol.* 66:1-36 (1969).
63. Sonzogni, W. C., P. C. Uttormark and G. F. Lee. "A Phosphorus Residence Time Model: Theory and Application," *Water Res.* 10:429-435 (1976).
64. Miller, W. E., J. C. Greene and T. Shiroyama. "The *Selenastrum capricornutum* (Printz) Algal Assay Bottle Test," U.S. EPA Report No. EPA-600/9-78-018, U.S. EPA, Corvallis, OR (1979).
65. Stewart, W. D. P., G. P. Fitzgerald, and R. H. Burris. "Acetylene Reduction Assay for Determination of Phosphorus Availability in Wisconsin Lakes," *Proc. Nat. Acad. Sci.* 66:1104-1111 (1970).
66. Fitzgerald, G. P. "Bioassay Analysis of Nutrient Availability," in *Nutrients in Natural Waters,* (New York: John Wiley & Sons, Inc., 1972), pp. 147-169.
67. Plumb, R. H., and G. F. Lee. "Phosphate, Algae and Taconite Tailings in the Western Arm of Lake Superior," Proc., 17th Conf., International Assoc. Great Lakes Res., pp. 823-836 (1974).

SECTION IV

POINT AND NONPOINT SOURCE PHOSPHORUS MANAGEMENT STRATEGIES

CHAPTER 12

IMPLEMENTATION INCENTIVES IN PHOSPHORUS MANAGEMENT STRATEGIES FOR THE GREAT LAKES

B. T. Bower

Resources for the Future
Washington, DC

INTRODUCTION

Four premises compose the basis for the comments in this chapter. First, the sources of phosphorus inputs into the Great Lakes are reasonably well known, recognizing that the proportions from different sources vary significantly among the lakes. Second, various physical measures to reduce discharges of phosphorus by various degrees from sources—other than the atmosphere—and the associated costs can be reasonably estimated. Third, the desired ambient water quality (AWQ) has been defined for each lake in terms of specific, quantitative water quality indicators, such as concentrations of dissolved oxygen, pesticides and heavy metals, and biomass of fish of particular species. Fourth, the linkages between the inputs of phosphorus from various sources and the resulting AWQ can be estimated. That is, natural systems models exist which can translate the time and spatial patterns of phosphorus discharges from various sources into the resulting time and spatial pattern of AWQ.

The extent to which these premises are valid has essentially been indicated in various reports of the International Joint Commission (IJC) and in previous chapters in this book. The most difficult of the premises is the fourth, the next most difficult is the first, next the second, and the least difficult, the third. That the third is characterized as the least difficult only means that some desired levels of AWQ in the lakes have been established, in quantitative

311

terms. It does not mean that these necessarily are the best or the most rational standards. It simply means that over the years of operation of the IJC and related governmental agencies and private groups of both sides of the border, some degree of consensus has been achieved that improvement in, and maintenance of, AWQ is desirable. The knowledge represented by these premises must exist before the development and evaluation of management strategies are possible.

DEFINITIONS

Given these premises, what next is required are some definitions. These are presented not because they will necessarily be accepted, but rather so that the meanings of the terms used in the remainder of the paper are clear. They are operational definitions which have been found to be useful in discussions of environmental quality management in a variety of contexts.

Physical measures are those physical actions by the generators of phosphorus, which measures reduce the discharge of phosphorus into the environment. They include, for example: reducing the quantity of phosphates contained in a given amount of phosphate fertilizer applied on each acre of agricultural lands and golf courses; constructing terraces on agricultural lands to reduce the amount of phosphorus discharged from such lands with eroded soil; installing facilities at municipal sewage treatment plants to remove phosphates.

Implementation incentives are the inducements—positive/negative, direct/indirect, prescriptive/proscriptive—which stimulate the adoption and continuous operation and maintenance of the physical measures at least at their design levels, by the various dischargers of phosphorus. (For a more extensive discussion of implementation incentives see Bower et al. [1].) Examples of implementation incentives are: grants for the construction of terraces of agricultural lands; grants for construction of phosphorus removal facilities at municipal sewage treatment plants; grants for operation and maintenance of phosphorus removal facilities at municipal sewage treatment plants; a limit on the kilograms of phosphorus equivalent discharged per day into a water body; charge on each kilogram of phosphorus equivalent discharged into surface water bodies from point sources—and perhaps from nonurban, nonpoint sources—surcharge on each kilogram of phosphorus equivalent in fertilizer applied to agricultural land and to golf courses; ordinance limiting phosphate content of cleansing agents and other products, percent by weight. As should be clear, implementation incentives can be imposed on inputs, production processes, product outputs, residuals modification facilities, and residuals discharges.

An institutional arrangement for phosphorus management is the set of governmental agencies which have the authorities and responsibilities to impose the implementations incentives on the various phosphorus discharging activities, and to carry out other tasks of water quality management. A phosphorus management strategy for the Great Lakes then consists of three components:

1. the physical measures for reducing discharges of phosphorus into the lakes;
2. the implementation incentives which induce the installation and operation and maintenance of those physical measures over time; and
3. the institutional arrangement which ties the first two components together.

All three must be considered simultaneously in developing management strategies as illustrated in Table I.

Two points with respect to the foregoing merit emphasis. One, considerable confusion has resulted from the use of the semantic dichotomy, "structural-nonstructural". For example, various reports from EPA and from the Water Resources Council contain the statement that a nonstructural measure is a substitute—or an alternative—for a structural measure. An example cited is that an effluent charge is a substitute for a sewage treatment plant. It would be difficult to conceive of a worse misconception. An effluent charge is that which *induces* the activity, if the charge is high enough, to reduce its discharge. It may do this by building a sewage treatment plant, changing a production process, changing raw materials, changing product specifications, reducing output or by some combination of these. The charge is the stimulus—the incentive—to action; the sewage treatment plant or process change is the physical measure adopted in response to the charge One is not a substitute for the other; one induces the other.

Two, the specification of an effluent charge, a discharge standard, a soil loss limit, is only the first step in devising an implementation incentive. In reality what is involved is an implementation incentive system—a set of procedures, rules, activities and/or sanctions relating to the imposition of the implementation incentive. For example, what must be done with respect to a discharge standard is:

1. define what is meant by noncompliance with the standard, e.g., excess discharge of 1 kg or 100, 1 day in 10 or 1 day in 100;
2. devise a monitoring, sampling, analysis and inspection system, including frequency of sampling and inspection and quality control of analyses;
3. devise a reporting and evaluation procedure; and

Table I. Illustration of Physical Measure-Implementation Incentive-Institutional Arrangement Combinations [2]

Residual(s) Affected	Physical Measure	Applicable Activity	Implementation Incentive	Institutional Arrangement (Locus (Loci) of Authority)
Windblown particulates	Plant/maintain windbreak trees	Agricultural operation	Provision of free tree stock	State Department of Environmental Protection, State Department of Agriculture
All from municipal sewage treatment plant	Better operation of conventional primary and secondary sewage treatment plant	Municipal sewage treatment plant	Training of operators by state agency (cost borne by state)	State Department of Environmental Protection
All from municipal sewage treatment plant	Better operation of conventional primary and secondary sewage treatment plant	Municipal sewage treatment plant	Salary increment for licensed operators	Municipality
All from municipal sewage treatment plant	Better operation of conventional primary and secondary sewage treatment plant	Municipal sewage treatment plant	Require operators to have state license (after training and certification exam)	State Department of Environmental Protection
Chromium	Recovery system	Manufacturing operation	Provide detailed information on design/cost of alternative recovery systems	Environmental Protection Agency (federal)

Phosphates	Limit phosphate content in detergents	Detergent production	Ordinance	Municipality, state, federal government
Suspended solids fertilizers	Terraces	Agricultural operation	Regulation requiring terraces on all crop land with slopes $> x$ %; with state bearing 50% of construction cost	State Department of Environmental Protection
Glass containers	Standardize shapes and sizes, i.e., wide mouth, straight sides	Container producers	National legislation	Federal, based on National Bureau of Standards design specifications
Particulates	Various, e.g., precipitators, baghouses, change quality of raw material input	Activities generating particulates, e.g., manufacturing, energy generation, incinerator	Effluent charge: Z cents per kg discharged, 2 Z cents per kg discharged above discharge standard	State Department of Environmental Protection

4. establish the sanctions and the mechanisms for applying the sanctions when noncompliance occurs.

If no sanctions are imposed for nonperformance, no implementation incentive will be effective. All too often these aspects of implementation—including their costs—are overlooked, ignored or inadequately considered in developing environmental quality management strategies. The implementation system is shown in Figure 1.

APPLICATION TO THE GREAT LAKES

The foregoing is prologue. How can the considerations set forth above be related to phosphorus management strategies for the Great Lakes? The objective of a strategy is to reduce the day-to-day use of the environment—in terms of reduction in discharges of phosphorus—to within some specified limits to achieve and maintain desired levels of AWQ. What are some of the implementation incentives which might be utilized in this context and what are some of the problems associated with them? To answer this question the specific sources of phosphorus inputs into the lakes must be considered.

SOURCES OF PHOSPHORUS LOADINGS

The IJC 1978 environmental strategy report [3] provides the basic data on sources of phosphorus loadings. Based on those data, Table II shows the percentages of the estimated 1978 phosphorus loads into each of the five lakes contributed by various sources, subdivided as follows:

- direct discharges, e.g., discharges from municipal treatment plants, discharges from industrial activities, urban nonpoint discharges or atmospheric discharges.
- tributary discharges, e.g., point discharges from municipal treatment plants, point discharges from industrial activities, diffuse urban nonpoint, agricultural nonpoint, and forest and other nonpoint discharges.
- load from upstream lake

The result of that exercise is the listing of the relative importance of the sources for the five lakes, shown in Table III.

Several comments and questions are stimulated by these data. First, the aggregate data—that is, taking each lake as a whole—mask sublake, subarea or local water quality management problems. Second, there apparently is negligible direct nonurban discharge into the lakes. Third, given the fact that the lake surfaces comprise about 30% of the total area of the Great Lakes basin, it is not surprising that *direct* "discharges" from the atmosphere

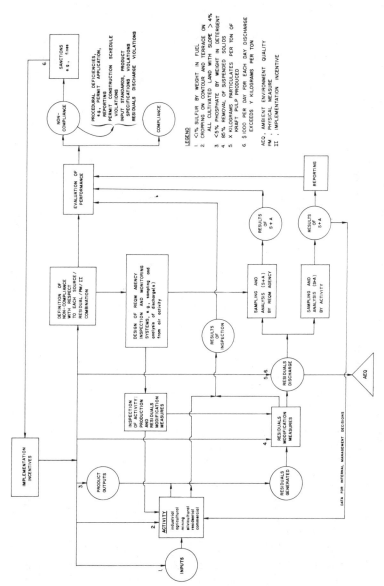

Figure 1. Flow diagram of execution of an REQM strategy [2] by a REQM agency or agent of such agency.

Table II. Estimated 1976 Phosphorus Loads into Great Lakes by Source [3].

	Lake Superior	Lake Michigan	Lake Huron	Lake Erie	Lake Ontario
Direct					
Discharges from municipal treatment plants	2	16	3	32	17
Discharges from industrial activities	2	<1	<1	2	<1
Urban nonpoint discharges	<1	a	<1	<1	3
Atmospheric	37	26	23	4	4
Total	<42	43	26	38	24
Tributary b					
Discharges from municipal treatment plants	5	23	8	7	7
Discharges from industrial activities	<1	4	2	<1	<1
Diffuse:					
Urban nonpoint discharges	4	4	6	10	5
Agricultural nonpoint discharges	4	21	24	32	19
Forest and other nonpoint discharges	45	5	10	6	4
(diffuse subtotal)	53	30	50	48	28
Total	<59	57	60	56	35
Load from upstream lake			14	6	41
Total	~100	~100	~100	~100	~100

aIncluded with tributary diffuse levels.
bSome portion of total atmospheric load falls on land areas, and reaches the lake via diffuse tributary discharges.

Table III. Relative Importance of Sources of 1976 Estimated Phosphorus Loadings into Great Lakes

	Lake Superior	Lake Michigan	Lake Huron	Lake Erie	Lake Ontario
Most Important	Forest and other nonpoint	Municipal sewage treatment plants	Agricultural nonpoint	Municipal sewage treatment plants	Load from upstream
	Direct atmospheric	Direct atmospheric	Direct atmospheric	Agricultural nonpoint	Municipal sewage treatment plants
	Municipal sewage treatment plants	Agricultural nonpoint	Load from upstream	Urban nonpoint	Agricultural nonpoint
Least Important			Municipal sewage treatment plants		Urban nonpoint

contribute such significant portions of total loadings into the three upper lakes: Superior, Michigan and Huron. Some portion of the loadings from tributary diffuse sources represents deposition from the atmosphere on land surfaces, with subsequent runoff to the lakes [3]. However, sufficient data are not available to estimate relative proportions. Fourth, to what extent do the direct (and indirect) atmospheric inputs of phosphorus to the Great Lakes represent inflows from outside the basin in contrast to material entrained from within the basin and transported downwind as "gifts from upstream," analogous to the source, "load from upstream lake?" For Lakes Superior and Michigan, the predominant wind patterns would imply that the direct atmospheric loads to those two lakes—representing about 37% and 26% of total phosphorus loadings to the respective lakes—are primarily from sources external to the Great Lakes basin. Developing and executing implementation incentive systems is much more difficult in such cases where interregional transfers of residuals are involved. Fifth, the different distributions of sources of phosphorus loadings among the five lakes suggest that—if efficient and effective management is to be achieved—management strategies should be tailored to the different situations in the lakes. Sixth, for Lake Ontario, it seems patently clear that any management strategy—including its implementation incentive system—which is not directed at reducing the load from upstream, estimated to be about 40% of the total load, will not enable the desired AWQ to be achieved for that lake.

IMPLICATIONS FOR IMPLEMENTATION INCENTIVES

Assuming that the data compiled under the aegis of the IJC are sufficiently accurate to comprise a basis for developing management strategies, what are the implications of the relative importance of sources with respect to implementation incentives?

Agricultural nonpoint discharges of phosphorus are critical to the development of efficient and effective management strategies, particularly where agricultural production occurs on hydrologically active areas. It is necessary to look at the various factors which influence residuals generation in agricultural activities to be able to assess the utility of various possible implementation incentives which might be imposed on agricultural activities. Residuals generation in agricultural activities is a function of: (1) prices of factor inputs, including energy; (2) production activity restrictions; and (3) off-farm employment opportunities. (The time pattern of generation is, of course, directly related to weather conditions.*)

---

*Variability in generation and discharge within any given year and from year to year adds substantially to the difficulty of devising effective implementation incentives. This problem is not addressed in this chapter.

Production activity restrictions refer to restrictions on the number of acres which can be planted to a particular type of crop in an agricultural operation, e.g., cotton, tobacco, wheat acreage limitations. The objective of a limitation is to reduce total output of the specified crop. However, a common response to such restrictions has been the intensification of production activity on the remaining acres, e.g., more fertilizer per acre, so that crop production often has decreases substantially less than intended and predicted, and more residuals have often been generated per acre and in total.

Off-farm employment opportunities affect the combination of factor inputs and production processes utilized on agricultural activities under direct management of the farm operator, e.g., family farm or tenant-operated. The extent to which a farmer works off the farm affects the levels and types of inputs he uses, both those directly related to his labor inputs and indirectly in relation to money made available by his off-farm employment for purchase of inputs.

The above facts of life must be considered in analyzing residuals generation by agricultural activities in both the short and long run. Necessary for that analysis is an understanding of the interrelationship among soil loss, factor inputs, long-run productivity and net farm income. The "soil tolerance level" represents the annual maximum per acre soil loss consistent with maintaining the productive capacity of the soil indefinitely. How long can soil losses *above* the soil tolerance level for a given soil type be sustained before productivity begins to decline, given that the length of time is a function of cumulative soil loss and that productivity is a function of both the basic soil type, e.g., depth and fertility, and management inputs, e.g., fertilizer? Up to some level of cumulative soil loss, the addition of fertilizer—and perhaps other inputs—can maintain productivity. However, beyond a certain level these inputs can no longer compensate for soil loss. Further, even during the period when "external" inputs prevent product output from decreasing, net farm income can decrease because of the additional inputs needed to sustain yield. For example, one University of Missouri study found that the net income of farms on moderately eroded and severely eroded fields decreased $18 and $33 per acre, respectively, because of reduced corn yields [4]. With respect to implementation incentives for water quality management, the question is, to what extent do the farm operator's practices to maximize net income coincide with, or operate in the same direction as, physical measures to reduce soil erosion.

Reduction in soil loss from a given agricultural activity can be achieved by some combination of:

1.   alternate crop sequences, such as corn-soybeans, corn-oats-meadow, continuous meadow as alternative to continuous corn;

2. alternative livestock operations, such as forage- or range-fed stock instead of primarily grain-fed stock, closed instead of open feedlots, hog farrow-to-finish, cow-calf production, cattle feeding;

3. alternative crop-livestock combinations;

4. tillage practices, such as minimum tillage in contrast to conventional tillage, spring and/or fall tillage with residues, e.g., tillage in conjunction with leaving 1-1.5 tons of crop residues on the surface of the soil at planting time; and

5. conservation practices such as terracing and contour plowing/planting/harvesting.

Each combination has a corresponding set of labor, equipment, energy, fertilizers, pesticides and capital inputs and a resulting time pattern of residuals generation under assumed meteorological conditions. Depending on the resources available to the agricultural operation—labor, quantity and quality of land, equipment—one or some combination of these physical measures will yield the least cost response to whatever constraints are imposed on residuals discharges.

Two factors are particularly important in estimating responses to incentives by managers of agricultural operations: (1) debt capital requirement; and (2) net taxable income/after-tax cash income. The former has been defined operationally to consist of: (1) operating capital; (2) intermediate capital, i.e., seven-year loans; and (3) investment capital, e.g., for terraces. Net taxable income is total revenue minus total fixed and variable costs; after-tax cash income is cash revenue minus cash expenses, including income taxes. These definitions are used because: (1) *most* decision-makers are interested in maximizing after-tax cash income; and (2) they enable explicit inclusion of alternative implementation incentives relating to depreciation, interest and other tax provisions such as treatment of expenses. The tripartite classification of debt capital also simplifies analysis of likely responses to various implementation incentives.

Given the possible physical measures relevant to a particular agricultural operation, the factors affecting the choice of physical measures to reduce soil loss from the operation are:

1. the mix of soil types of which the agricultural operation consists;

2. the resources available to the activity, including returns from off-farm equipment if any;

3. tenure status;

4. time horizon of the agricultural operation's decision-maker;

   5. factor input prices and product output values*;

   6. types and magnitudes of production incentives; and

   7. types and magnitudes of implementation incentives imposed on the activity to reduce residuals discharges from the activity.

Relevant implementation incentives include:

   1. limit on annual soil loss for each acre of each type of soil[+];

   2. limit on mean annual soil loss per acre for the entire agricultural activity;

   3. charge on each ton of the total tons of soil lost from the area of an agricultural activity; e.g., suspended sediment (residual) discharged, with no limitation on source, e.g., specific soil types;

   4. required tillage/residue practices and/or required installation of terraces/grassed waterways;

   5. public subsidy—state and/or federal—for terrace/grassed waterway construction[§];

   6. permitted duction of "soil and water conservation expenditures" from taxable income [5];

   7. rapid depreciation of investment in soil loss prevention facilities, such as terraces and grassed waterways;

   8. surcharges on prices of fertilizers containing phosphorus;

   9. technical assistance;

  10. restricting use of certain soils; and

  11. various combinations of the foregoing.

The first seven assume that reduction in soil loss will result in reduced discharges of phosphorus. (If phosphorus discharges could be measured directly,

---

*The variables are interrelated. For example, land values are affected by the response; debt servicing capacity is a function of net income and affects the range of possible responses. What is involved is a complex interrelationship among tenure, implementation incentives to induce reduction in soil losses, farm land values, debt servicing capacity and factor prices.

[+]For example, sections 4670 and 467A.42-467A.53 of the Iowa Conservancy Law extend the authority of soil conversation district commissioners to allow them to classify land on the basis of eerdibility, establish soil loss limits for each land class, and require landowners to undertake the necessary measures to enable meeting the limits. See Code of Iowa, 1975, *Iowa Conservancy Law,* and Acts of the Second Regular Session, 65th General Assembly of Iowa.

[§]Current (1978) laws in Iowa provide for the state's paying 50% of the costs of terraces and for voluntary terracing and 75% of the costs for mandatory terracing.

then limits and/or charges on phosphorus discharges could be imposed.) Presumably, aerial photographs and the universal soil loss equation would be used in the application of the first three, with some type of sanction, e.g., a fine, imposed for failure to meet the limit of the first two. The sanction with respect to the fourth might be achieved by making any crop support payments contingent on the installation and utilization of the specified practices.

It is important to emphasize that the implementation incentive or incentives chosen can affect long-run productivity of the agricultural operation. For example, given a typical agricultural activity with a mix of soil types of different degrees of erodibility, the imposition of a charge on each ton of soil loss—regardless of source on the agricultural activity—will likely lead to soil losses exceeding the soil tolerance level for one or several of the soil types, because the costs of reducing soil losses differ significantly among soil types. In contrast, if a limit on soil loss were established for each acre at a level approximately equal to the tolerance level for each acre, long-run productivity would be maintained, provided that the soil loss limit could be enforced.

The effectiveness of capital subsidies for such physical measures as terracing depends primarily on (1) the time horizon of the farm operator; (2) the net cash and the marginal tax rate of the agricultural activity; and (3) national agricultural production objectives and associated incentives. National programs to encourage agricultural production have often been contradictory to national soil conservation programs [4]. At various times the stimuli to increase production have resulted in "plowing up" thousands of acres of soil conservation practices; the incentives to increase production were greater than the incentives to reduce soil loss. In order to ensure that discharges of phosphorus will remain below some acceptable level, the incentives—including sanctions—for reducing discharges must be greater than the incentives to high productivity eschewing soil erosion reduction measures. The effectiveness of capital subsidies for terracing, etc., may also be limited by the reduction in effectiveness of such measures over time in the absence of adequate maintenance, for which there is no subsidy. The failure of many municipal sewage treatment plants to operate at design levels over time, subsequent to a grant for construction, has been well publicized.

In addition, the continued trend in design and manufacture of modern farm equipment toward bigger and heavier equipment has discouraged the maintenance of existing, and the installation of additional, physical measures to reduce soil losses. The importance of production technology merits emphasis. The blandishments of the equipment manufacturers have induced many farmers to purchase and use heavier equipment. The results tend to be more soil compaction and hence more soil erosion, more energy use, and less flexibility in field operations and hence more difficulty in maintaining soil erosion reduction measures, not to mention the need to produce more in order to pay for the more expensive equipment. This shift toward heavier equipment may be less of a problem in the Great Lakes basin than it is in the Great Plains and in the Pacific Northwest wheat country.

It should also be pointed out that, if an implementation incentive system were effective in inducing an agricultural operator to shift to minimum tillage to reduce soil erosion for crops where such a practice were feasible, the shift might or might not result in a net environmental gain. In some case maintaining crop yields with minimum tillage involves the use of substantially increased amounts of herbicides. Decreased discharge of phosphorus may yield an increased discharge of herbicide.

Finally, provision of technical assistance to agricultural operators—financed by federal and state governments—is an important incentive. Such information transmission should not be limited to technical information; it should also include information on the various types of financial incentives and assistance available to the operator. Some, perhaps many, farmers do not know about investment tax credits, rapid depreciation or expensing provisions. Even where they know these exist, they, like the rest of us, might well have difficulty in understanding and filling out the requisite forms.

Discharges for municipal sewage treatment plants are major sources of phosphorus inputs to the Great Lakes, via both direct discharges and via tributary discharges. The 1978 Canadian-U.S. agreement [6] shows: (1) the 1976 "established" phosphorus loads; and (2) the loads which are estimated would be achieved if all municipal plants of over 1 mgd capacity achieved an effluent concentration of 1 mg/l. But it is not clear from the various reports whether the "established" loads for 1976 are actual loads based on actual performances of plants throughout the year, or are what the loads would have been if design removals and effluent standards had been achieved every day. Unless the municipal sewage treatment plants in the U.S. portion of the Great Lakes basin are—for some reason—much better performers than the average of those in the U.S., as found in various studies by the Government Accounting Office and others, it is not likely that the "established" numbers were actually achieved.

What might induce the achievement of the specified 1 mg/l phosphorus "standard" by the municipal sewage treatment plants? Various implementation incentives are possible, directed at different parts of the problem. Examples include:

1. municipal or state/provincial ordinances reducing the percentage of phosphorus by weight, or banning it, in household and commercial cleansing agents;
2. federal grants for construction of phosphorus removal facilities at municipal sewage treatment plants;

3. federal/state grants for operation and maintenance of facilities;
4. state/provincial programs for training and licensing of plant operators;
5. publication every day in the daily paper of the performance of the municipal plant on the previous day, analogous to reporting the weather or air quality;
6. limit on total quantity of phosphorus discharge per day for each plant of over 1 mgd capacity; and
7. charge of 30-40¢/kg for every kilogram of phosphorus discharged above the daily limit.

Associated with the last two—discharge standard and effluent charge with discharge standard—would be the requirement that each plant of larger than 1 mgd capacity would have to measure discharge continuously and sample proportionately to determine the actual daily discharge. Random, unannounced inspections and sampling/analysis by state/provincial personnel would be made to check the accuracy of measurements, sampling and analysis of samples.

If no effluent charges are imposed, and only the daily discharge limit were to exist, then the state/provincial agency must define noncompliance and the sanctions for noncompliance. Thus far the U.S. at least has been notoriously unsuccessful in imposing sanctions on municipalities where the municipal sewage treatment plant(s) fail to meet the discharge standards. A discharge standard is not likely to be taken very seriously if it is clear that there is little likelihood that sanctions will be imposed. The mayor, the city council, even the treatment plant manager are not likely to be fined and/or jailed when plant performance does not meet the standard. Nor is the plant likely to be closed down. It may well be that effluent charges may be the most effective incentive to induce adequate operation—along with operator training and improved salaries—provided that the charges are paid directly or are subtracted from general revenue grants to the municipality.

Urban nonpoint discharges are another major source of phosphorus inputs to the Great Lakes. However, it is difficult to develop efficient and effective implementation incentives for such discharges unless more is known about the specific sources within any given urban area. For example, if fertilizing gardens, parks and golf courses comprised the major source within an urban area, surcharges could be levied on all phosphorus fertilizers sold in the municipality/state/province for nonagricultural use.

Evidence presented at this conference indicated that eroded soil is often a major source of phosphorus, because of the in situ content. This suggests three sources of phosphorus: (1) construction sites; (2) stream channels and stream beds within urban areas; and (3) hilly residential areas from which most native vegetation was removed during construction. Various local jurisdictions have sediment ordinances which specify certain required practices where construction is underway, such as debris basins, baled hay along the perimeter of the site, no more than 10% of ground uncovered at any time. But the major problems are having sufficient personnel to inspect sites and imposing sufficiently stringent fines for noncompliance to act as deterrents. The potential spreading of "Proposition 13 fever" is likely to exacerbate this problem, unless permit fees required by the local jurisdiction for land development were sufficiently high to cover personnel costs which, in fact, would be the logical procedure.

Erosion from stream banks and stream beds may well be a major source of phosphorus loads in already developed urban areas, especially in rolling and hilly terrain. The increase in impermeable area means increased runoff for any given storm intensity and decreased time of concentration, resulting in higher carrying capacity and more stream channel and stream bed erosion. Once an urban area has been developed, there are no incentives which are likely to induce much action, because there are relatively few, possible, physical measures to reduce runoff and phosphorus discharges from in situ eroded soils.

Pending more specific information on sources of phosphorus in nonpoint runoff within an urban area, recourse must be to implementation incentives imposed on the municipality at the discharge point of the storm water pipe. Possibilities are:

1. limit on phosphorus discharge for any given size storm;
2. grant for construction of detention basins;
3. charge on each kilogram of phosphorus discharged in storm water, or each kilogram above some limit; and
4. federal/state/provincial technical assistance to determine sources within the urban area.

Land use zoning has often been cited an an implementation incentive which would reduce discharges by either restricting uses on certain types of erosion-prone land or by specifying a certain pattern of development, such as cluster development, or both. Zoning is the responsibility of local governmental jurisdictions. Generally these have the legal authority as well. Thus far there do not seem to have been any effective incentives developed at the state level to induce local jurisdictions to use zoning more effectively for the objective of reducing discharges. In addition, developmental pressures are very difficult to resist at the local level, especially when the economy is not doing very well. Local officials are likely to draw back from any actions which could be construed as being antidevelopment.

CONCLUSIONS

1. Management strategies must include all three components: physical measures, implementation incentive systems and institutional arrangements.

2. No institutional arrangements currently exist to handle crossregional boundary flows of phosphorus residuals into the Great Lakes via the atmosphere.

3. Cost-effectiveness of strategies, which is measured by dividing costs of physical measures to reduce discharges by the estimated amount of discharge reduction, is not adequate. Unless the physical measures can be put in place and operated continuously over time, the imputed reductions will not be achieved, and therefore the posited strategies will not be effective. All three companies and their associated costs must be simultaneously considered.

4. The choice of implementation incentives to be included in management strategies is a function of such factors as political structure, cost of administering the incentive, flexibility, and consistency with existing legal powers. Although difficult, at least an attempt should be made to assess these factors in relation to each implementation incentive, to aid in the selection of a mangement strategy.

REFERENCES

1. Bower, B. T., C. N. Ehler and A. V. Knesse, "Incentives for Managing the Environment," *Environ. Sci. Technol.* 11(3):250-254 (1977).
2. *Handbook for Analysis for REQM.* (Washington, DC: Resources for the Future, 1979).
3. *Environmental Strategy for The Great Lakes System,* Final Report to the International Joint Commission from the International Reference Group On Great Lakes Pollution from Land Use Activities (PLUARG), IJC, Windsor, Ontario, (1978).
4. Risser, J. "Soil Erosion Creates a Crisis Down On the Farm," *Conserv. Found. Lett.* (December, 1978).
5. *Farmer's Tax Guide,* Publication 225, Department of Treasury, Washington, DC (1978).
6. Great Lakes Water Quality Agreement of 1978, p. 26.

CHAPTER 13

EXPERIENCE WITH PHOSPHORUS REMOVAL AT EXISTING ONTARIO MUNICIPAL WASTEWATER TREATMENT PLANTS

S. A. Black

Wastewater Treatment Section
Pollution Control Branch
Ontario Ministry of the Environment
Toronto, Ontario

INTRODUCTION

The Ontario Ministry of the Environment (Ontario Water Resources Commission until 1973) initiated research into processes of phosphorus removal from municipal wastewaters in 1963. This research followed the completion of a review of then current literature and laboratory eutrophication studies [1] which demonstrated phosphorus to be the most potentially limiting nutrient to algal growth in Ontario's lakes.

Initially, effort was directed toward the utilization and subsequent removal of phosphorus by activated sludge bacteria and by algae in tertiary polishing lagoon systems. Because of incomplete and inconsistent removals by such methods, however, emphasis was soon placed on chemical and combined/biological processes.

To effectively introduce phosphorus removal to wastewater treatment in Ontario, processes had to be selected which could be used at existing primary and secondary biological treatment facilities with a minimum of capital expenditure.

Laboratory investigations, involving the use of aluminum sulfate, ferric chloride and lime, indicated lime to be an economical chemical for possible widespread application of phosphorus removal throughout Ontario. This was predicted upon its apparent effectiveness in precipitating phosphorus from wastewater and the widespread availability of large quantities of low-cost lime in Ontario. Aluminum sulfate and ferric chloride, however, were found to be equally effective.

Results of extensive model studies led to the ministry's first plant-scale investigation of phosphorus removal at Richmond Hill in 1968 [2]. Following completion of these studies, the province's first full-scale permanent phosphorus removal facility was installed in 1970 at the ministry-owned and operated Newmarket/East Gwillimbury Water Pollution Control Plant (WPCP) [3]. This facility employed lime treatment of raw sewage. Hydrated lime was received in bulk, stored as a slurry and fed to a flash-mix tank installed ahead of the primary clarifiers of the 2.0 million imperial gallons per day (migd) conventional activated sludge plant.

Concurrently, as a result of a 1969 International Joint Commission Report [4] recommending that phosphorus discharges from all sources in the lower Great Lakes be reduced to the lowest practical level, the Province of Ontario announced a policy requiring the installation of phosphorus removal facilities at municipal and institutional wastewater treatment plants in both the lower Great Lakes areas and in inland recreational waters.

Initially, the policy required a minimum removal of 80% of the phosphorus from wastewater treatment plant influents with the need for higher levels of removal to be determined by further studies of the receiving waters. This criterion was subsequently superseded in the lower Great Lakes by the signing, in April 1972, of the Canada-United States International Agreement on Great Lakes Water Quality [5] which called for an effluent objective of 1 mg/l total phosphorus.

Under this agreement, permanent phosphorus removal facilities were to be operational by December 31, 1973 in the most critically affected areas of the province; by December 31, 1975 in areas with less critical receiving waters, and three years after notification in all other areas of the province where problems were found to exist. Figure 1 outlines the scheduled phosphorus removal compliance dates for the southern section of the Province of Ontario.

To implement and accelerate pollution control programs, the governments of Canada and Ontario in August 1971 signed an agreement [6] that secured funding for a $250 million capital works program aimed at upgrading sewage collection systems, treatment works, including the installation of phosphorus removal equipment. An additional $6 million over a 5 yr term of the agreement was provided for related research studies. This was later extended for an additional 2-yr period and provided an additional $1 million in research funding.

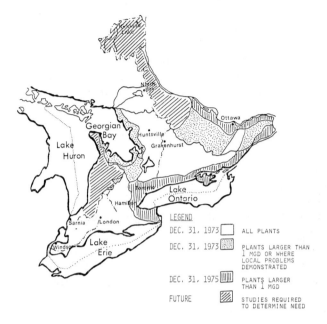

Figure 1. Province of Ontario—southern section phosphorus removal program scheduled compliance dates [16].

Under this funding program, the effectiveness of aluminum sulfate, ferric chloride and lime as prime coagulants for use in existing sewage treatment facilities was further evaluated. A predictive methodology involving jar testing techniques [7] was developed whereby selection could be made of the prime coagulant best suited for phosphorus removal at a particular plant. Temporary full-scale treatability studies were then carried out at selected plants to confirm the jar testing precedures and determine design and operational parameters.

Once the methodology of the treatability studies was established, municipalities requiring phosphorus removal were encouraged to conduct their own treatability studies. Municipalities in the lower Great Lakes basin were reimbursed money spent on these studies, while those municipalities not eligible for reimbursement were provided with equipment and/or technical expertise whenever possible.

By December 31, 1975 there was a total of 168 sewage treatment plants in Ontario with permanent phosphorus removal facilities in operation. By December 31, 1977 this number had increased to 211, representing about 85% of the provincial total sewage hydraulic capacity.

This chapter relates the experience and knowledge gained from the phosphorus removal implementation program. It outlines the methodology developed for determining the most appropriate chemical, the approximate dosage and point of application for use at individual facilities. It discusses problem areas and outlines design criteria gained from eight years of operation of full-scale phosphorus removal facilities in the Province of Ontario. Some basic cost data of treatability studies and full-scale implementation are presented.

OCCURRENCE OF PHOSPHORUS IN SEWAGE

Phosphorus is an essential element in the metabolosm of organic matter. Its presence in a treatment plant is essential to the proper functioning of biological waste treatment processes.

Phosphorus may occur in the form of organic phosphorus found in organic matter and cell protoplasm, as complex inorganic phosphates (polyphosphates) such as those used in detergents, and as soluble inorganic orthophosphate. Orthophosphate is the final breakdown product in the phosphorus cycle, and is the form in which phosphorus is most readily available for biological use.

In raw sewage there are substantial amounts of all three principal phosphorus forms. During biological treatment, organic compounds are decomposed and their phosphorus content converted to soluble orthophosphate. Similarly, in the biological treatment process polyphosphates are hydrolized to orthophosphate. Some orthophosphate is utilized in developing the biological mass, but the end result in a well-stabilized effluent is that a large percentage of the phosphorus is present as orthophosphate ion. Orthophosphates are readily removed by chemical precipitation processes.

Figure 2 shows the average phosphorus concentrations in Ontario raw sewages for the years 1967 to 1976. Major reductions occurred in 1970 and 1973. In August 1970, the Canadian Government passed control regulations limiting the allowable phosphorus content in household laundry detergents to 20% P_2O_5. This reduced the phosphorus concentration of the raw sewage from above 9.0 mg/l in 1969 to about 6.7 mg/l by 1972.

On January 1, 1973, additional control regulations were introduced which further limited the phosphorus level in laundry detergents to 5% P_2O_5. This further reduced the phosphorus content of raw sewage to about 5.2 mg/l in 1974. Since 1974 there has been a gradual rise to about 6.4 mg/l in 1977.

BIOLOGICAL REMOVAL OF PHOSPHORUS

In biological systems, phosphorus removal is limited to the nutritional requirements of the microorganisms. As microorganisms contain relatively

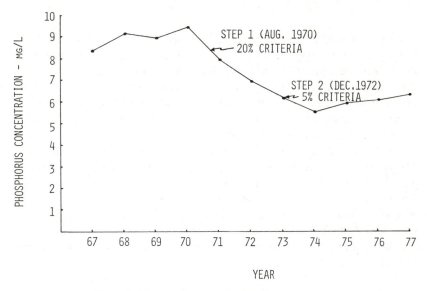

Figure 2. Phosphorus concentrations in raw sewage.

consistent quantities of phosphorus, nitrogen and organic carbon, biological uptake of phosphorus can be related to the quantity of biosolids produced. Biosolids grown under abundant nutrients may contain up to 12% nitrogen and 2% phosphorus, while those grown under nutrient-limiting conditions may contain as low as 5% nitrogen and 1% phosphorus. Assuming a solids yield of 0.5 g volatile suspended solids (VSS)/g BOD_5 removed, approximately 0.5 to 1 g of phosphorus will be removed per 100 g of five-day biochemical oxygen demand (BOD_5). For normal municipal wastewater, 90-95% BOD_5 removal will mean a 20 to 40% removal of phosphorus.

An earlier review of literature on phosphorus removal within the activated sludge process [2] indicated that, although higher biological uptake of phosphorus through plant and process modifications was possible, state-of-the art at that time was insufficient to meet the high removals required. Therefore, effort was directed toward the more positive chemical precipitation processes.

Recently, there has been a renewed interest in biological and combined biological/chemical phosphorus removal methods. Processes such as the Bardenpho [8] and Phostrip [9] processes are receiving considerable attention. A recent paper [10] reviewed the many variations of the above processes and threw more light onto the mechanisms involved and process controls. It would now appear that further research is well warranted in the area of biological uptake, perhaps followed by chemical treatment of individual subprocess flows, in view of the added phosphorus removal requirements under the Great Lakes Water Quality Agreement of 1978 [11].

Union Carbide's Phostrip system, investigated in pilot plant studies in 1976 [9], takes advantage of the biological phenomena of luxury uptake and anaerobiosis for the release of phosphorus. In the aeration tank, microorganisms take up excess phosphorus due to an environmental stress provided by a previous period of anaerobiosis. These microorganisms are settled in a secondary clarifer and part of the phosphorus-rich sludge is recycled back to the aeration basin, part is wasted, and part is sent to a stripping tank where the microorganisms release phosphorus under anaerobic conditions. The anaerobic sludge at the bottom of the stripper tank, which contains phosphorus-deficient microorganisms, is returned to the aeration basin where the phosphorus uptake cycle is repeated. The phosphorus-rich supernatant from the stripper, representing 10-15% of the influent flow, is transferred to a lime mix tank. Phosphorus precipitate settles out when the lime-phosphorus mixture is sent back to the primary clarifier. A schematic of the Phostrip system is given in Figure 3.

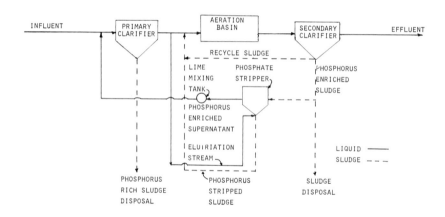

Figure 3. Phostrip system schematic.

Although there is still no full-scale operating facility of the Phostrip system, a recent study, undertaken by McNamee, Porter and Seeley Consulting Engineers, Ann Arbor, Michigan, for the city of Adrian, Michigan, showed that the Phostrip system would produce an effluent of less than 1 mg/l total phosphorus at a cost 34-37% lower than systems employing ferric chloride or alum [12].

PHOSPHORUS REMOVAL BY PRECIPITATION

The ionic forms of aluminum, iron and calcium have been found to be effective precipitants for phosphorus. Essentially, these metallic ions react with the orthophosphates to form insoluble phosphate precipitants but some polyphosphates and organic phosphorus compounds are also removed by a combination of complex reactions and sorption onto floc particles.

Aluminum

There is considerable flexibility in the location of alum addition points in a sewage treatment process: before the primary settling tank, in the aeration tank or following aeration before final sedimentation.

Alum addition before the primary settling tank provides additional suspended solids and organic removals concomitantly with phosphorus precipitation. These removals, however, occur as a result of reactions which may compete with soluble phosphorus for available alum, and the dosage requirements to reach a given phosphorus removal level are therefore increased by these reactions. In addition, polyphosphates present in raw sewage are not directly precipitated, with their removal being dependent upon adsorption and coagulation mechanisms.

Alum may be added directly to the aeration tank, but generally the best point of addition for alum in the activated sludge plant is in the effluent channel of the aeration basin which carries the mixed liquor to the final clarifier. Generally there is sufficient turbulence within this channel to provide adequate mixing.

The addition of alum following biological treatment capitalizes upon the phosphorus reactions within the aeration tank which convert organic phosphorus and polyphosphates to the readily precipitated orthophosphates. Optimum utilization of the alum for phosphorus removal is therefore provided.

The addition of aluminum sulfate (the major source of aluminum for phosphorus removal) will lower the pH of the wastewater, although most wastewaters contain sufficient alkalinity such that even large dosages of alum will not significantly depress the pH. In exceptional cases of low alkalinity wastewater, some pH adjustment following alum addition may be required.

Iron

Both ferrous (Fe^{2+}) and ferric (Fe^{3+}) ions can be used in the precipitation of phosphorus from wastewaters, although our studies have shown [13] that iron must be in the ferric form before significant phosphorus reduction is achieved. If the ferrous salts, however, are oxidized to the ferric state, the chemical reaction between the iron and the soluble reactive orthophosphates can efficiently take place. Thus, ferrous salts have been found ineffective in primary plants, whereas in secondary facilities, provided the aeration facility can provide adequate oxygen and retention time to convert the ferrous to ferric, ferrous salts have worked well. Approximately 0.15 g O/g Fe^{2+} is required, with a reaction time of 2-3 hr.

A number of iron salts such as ferrous sulfate, ferric sulfate, ferric chloride and waste pickle liquor may be used in phosphorus precipitation.

As with alum, ferric salt addition will also tend to depress the pH of the wastewater. Again, in cases of low alkalinity wastewater, pH adjustment may be required.

Lime

Calcium ions react with phosphate ions in the presence of hydroxyl ions to form hydroxyapatite, the reaction again being pH-dependent. However, in lime treatment for phosphorus removal from wastewater, the operating pH may be established by the ability to obtain good suspended solids removal rather than by phosphorus precipitation itself. Because of the high pH (> 9) resulting from lime addition, the lime treatment process is generally limited to primary treatment or post-secondary facilities.

The chief variable that affects phosphorus removal by lime is wastewater alkalinity. Unless a high pH is used with waters of low alkalinity (150 mg/l or less), a poorly settleable floc is formed. With high alkalinity wastewater, a pH of 9.5-10 can result in excellent phosphorus removal.

There are various lime precipitation schemes for the removal of phosphorus. Lime may be added before the primary clarifier in an activated sludge plant. However, because an excessively high pH would interfere with the biological process, lime addition is limited to a pH of about 9.5. As a result, 80 to 85% phosphorus removal in this manner is about the best that can be expected. Additional phosphorus removal could be achieved by adding small amounts of aluminum or iron to the secondary process. The use of lime in primary treatment has the added benefit of greatly improving thd organic and suspended solids removal efficiencies in the primary clarifier, thereby significantly decreasing the load on the aeration system.

A second alternative is lime treatment following biological treatment involving tertiary rapid mix, flocculation and sedimentation basins. This mode of chemical addition is generally considered when very low effluent phosphorus concentrations are required. Improved efficiencies of phosphorus removal have been obtained [3] by recirculating some of the lime sludge back to the rapid mix tank.

Polymers

Although polymers may be used to increase the flocculation of the phosphate precipitates, the polymer itself does little in improving phosphorus capture. In Ontario, polymers are rarely used because of their high costs and instead emphasis is placed upon providing adequate sedimentation facilities for the metal-phosphate precipitates.

A recent study [14] funded under the Canada/Ontario Agreement evaluated design and performance criteria for primary clarifiers for the removal of physical-chemical flocs in primary plants.

The study concluded that under similar design conditions, the addition of ferric chloride increased suspended solids removal to 70% as compared to 40-50% without chemical addition. The addition of ferric chloride plus polymer achieved a suspended solids removal efficiency of 85%. This study demonstrated the effectiveness of polymers in improving clarifier operation of primary plants. In specific instances, the use of a polymer may delay the need for plant expansion and the use of a chemical plus a polymer may well negate the need for secondary treatment in particular watershed areas.

PHOSPHORUS REMOVAL TREATABILITY SYSTEMS

Phosphorus removal treatability studies are an essential aspect in the implementation of a phosphorus removal program. They are used as a means of selection of the chemical most suitable for a particular sewage, for determining the approximate chemical dosage, the optimum point of application and chemical compatability with the particular sewage treatment plant process. Treatability studies for Ontario's phosphorus removal implementation program were carried out in two phases. Initial jar testing studies were followed by fairly long-term, full-scale temporary studies. As a result of the extensive treatability studies carried out under the Canada/Ontario Agreement, however, it is now felt that the full-scale temporary studies need only be carried out in specific instances where some particular sewage characteristic or plant design and operating conditions prevail.

This chapter will only briefly outline the purpose, procedure and results of the treatability study program. Detailed guidelines for conducting treatability studies are available in a Ministry of the Environment publication [7].

Jar Testing Procedures

The jar testing procedure established for water treatment plants development [15] was used with slight modification to evaluate chemical addition to sewage for purposes of phosphorus removal. The procedure uses an established timing sequence of rapid mixing, slow mixing and quiescent settling to ensure adequate mixing and reaction time for sufficient flocculation and sedimentation. The procedure does not attempt to duplicate the hydraulic and mixing conditions within a particular plant, but was developed as a standard technique for yielding predictive results.

The preliminary testing is done on batch raw sewage and final effluent grab samples using the three primary coagulants aluminum sulfate, ferric chloride and lime at a wide range of dosages. Additional jar tests are subsequently conducted to narrow down the range of coagulant dosage needed to effect the required degree of phosphorus removal, expressed either as a percentage removal or a residual phosphorus concentration.

Finally, a number of jar tests are conducted using dosages that attempt to bracket the range of removal efficiency desired, i.e., dosages yielding removals lower, equal to and higher than the objective. It is essential that these jar tests be conducted over an extended period of time in order that a representative variety of sewage characteristics are encountered. Grab samples for jar testing should be taken at different times of the day and on various days of the week. Once the range of chemical dosages has been determined, it is essential not to alter these dosages used in the jar testing in order to avoid deliberately skewing the results. In order to have sufficient data to draw meaningful conclusions, it was arbitrarily considered that a minimum of ten data points should be obtained for each chemical dosage on each waste stream being studied.

Proper analysis of the data obtained from jar testing is essential in obtaining meaningful information on the relative efficiencies of each chemical and point of addition. Rather than looking at the average total phosphorus removal for each set of data, plots of each set of data points are made on arithmetic probability paper. Such probability curves have two significant aspects: (1) the relative vertical position of the point indicates the degree of effectiveness of the coagulant used at that specific dosage; and (2) the slope of the line indicates the expected reliability of phosphorus removal using that particular coagulant.

A typical set of data using alum as the coagulant for raw sewage is shown in Figure 4. This figure indicates that in order to achieve a residual phosphorus concentration of 1 mg/l, 50% of the time, a dosage of 150 mg/l is probably adequate. A dosage of 200 mg/l, however, produces considerably more consistent results, while a dosage of 250 mg/l represents a higher than required application.

Using similar curves for each coagulant, the dosage required to yield a given degree of phosphorus removal (or residual phosphorus value) for a specified frequency may be determined and results compared.

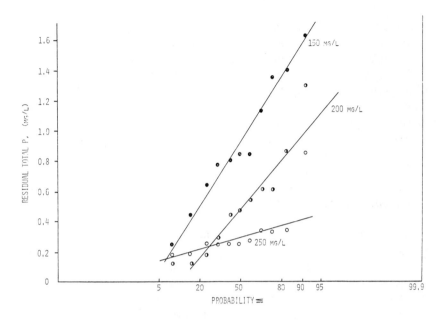

Figure 4. Jar test results, alum on raw sewage [16].

The relative economics of the chemicals and dosage points for phosphorus removal can then be determined on the basis of optimum dosage, delivered cost and availability of chemical.

Full-Scale Temporary Studies

The results of the preliminary jar tests are effective in determining the optimum chemical to be used for phosphorus removal at a particular WPCP as well as the generalized point of its application. Full-scale temporary studies are then conducted using the optimum chemical to achieve the following: confirm the suitability of the poant design to phosphorus removal; confirm the chemical dosage; optimize the point of application; provide data on sludge production and characteristics; determine what final effluent quality can be expected; and provide preliminary operational cost data for phosphorus removal.

For primary sewage treatment plants, full-scale temporary studies should be approximately 6 weeks in duration. Eight-week full-scale studies are considered necessary for secondary treatment facilities. The above study durations assume that jar tests have been used to establish the optimum chemical and general point of application (i.e., to raw sewage or secondary effluent) and that the chemical selected and its point of application yield satisfactory results. If not, the study duration must be expanded accordingly. Additionally, the presence of a significant industrial waste load may necessitate more extensive studies in both the jar testing and full-scale work.

The primary objective of the full-scale temporary phosphorus removal study is to demonstrate the feasibility of achieving a specific phosphorus removal objective at a particular sewage treatment plant without requiring extensive plant modifications or additions.

Further details of the treatability studies carried out in Ontario are presented in a report by Boyko and Rupke [16].

PROCESS OPERATION AND DESIGN CONSIDERATIONS FOR P REMOVAL FACILITIES

Based upon the experiences of the Ministry of the Environment through the design, installation and operation of its own ministry-operated facilities and through providing assistance to municipally operated facilities, the following comments and observations are presented. In introducing this chapter it may be stated that phosphorus removal can be achieved at any existing wastewater treatment facility with the addition of aluminum, iron salts, or lime. Although generally, any one of these three chemicals can adequately remove phosphorus at a given plant, jar testing and possibly full-scale testing must be carried out in order to optimize chemical dosage, point of addition, etc., and to confirm the suitability of the existing plant design for phosphorus removal.

Sewage Composition

No particular relationship could be established between required chemical dosage and raw sewage characteristics, as even influent phosphorus concentration could not be used to predict the required chemical dosage. The studies did show, however, that the presence of industrial wastes in appreciable quantities generally had a rather severe detrimental effect on the ability of the prime coagulant to remove phosphorus. This was particularly true when phosphorus removal was attempted across the primary clarifier and frequently dosage had to be considerably increased.

Chemical Source

Although alum, ferric chloride and lime have most commonly been used as phosphorus removal chemicals, any waste product containing appreciable quantities of aluminum, iron or calcium cations may have a potential use for phosphorus removal. Such products include waste pickle liquors, spent carbide limes, and aluminum finishing wastes. It should be noted that iron is only effective in precipitating phosphorus when in the ferric form, and therefore, the use of ferrous iron is restricted.

Materials such as ferrous sulfate must be oxidized to ferric form before being added, or are effective only a secondary aerated system. Waste carbide limes may contain high concentrations of inert materials which may prove excessively abrasive to pumps and piping.

The source of the waste material must be considered. Frequently, waste products contain high concentrations of heavy metals which may affect plant processes or the end use of the sewage sludge. Another thing to consider is the variability of the waste product. If the material is delivered directly from the source, each load may vary considerably in strength. An intermediate holding and mixing facility is recommended for reducing this variability. Therefore, for each situation, suitability must be judged on the basis of content of reactive chemical, continuous availability of supply, uniformity of shipments, and the presence of excessive undesirable contaminants. A recent report by Scott [17] evaluates the various sources of iron waste by-products as phosphorus removal materials in Ontario.

Dosage

As previously mentioned, lime dosage requirements are closely related to the alkalinity of the wastewater. Dosages of aluminum and iron may be approximated by the stoichiometry of their reactions with phosphorus, although an accurate estimate may only be determined through jar testing procedures. Even then, the dosage required to achieve a desired level of phosphorus removal in an operating facility is generally slightly lower than that predicted by jar testing. This is likely due to the effect of recycling accumulated precipitate, thereby increasing chemical utilization.

Prested et al. [18] compared jar test dosages of alum or ferric chloride to the mixed liquor of secondary treatment facilities to full-scale dosages to achieve an average of 1 mg/l phosphorus in the effluent. It was found that the overall jar test to full-scale dosage ratio was 1.15:1 indicating the jar test to slightly overestimate full-scale requirements.

Dosage Control

An attempt to correlate chemical dosage to various chemical characteristics of the sewage [19] failed to determine any suitable means of predicting chemical dosage. Chemical dosage is generally controlled by flow-pacing the chemical feed pump to the raw sewage feed rate. This may be accomplished by automatic equipment, or in the case of small facilities, by manual adjustment.

Point of Addition

The point of addition of the various phosphorus removal chemicals has already been discussed. Addition point chosen will depend upon the facility design, chemical used and jar test results. It should be pointed out, however, that flexibility of point of addition should be designed into the chemical feeding system. Additionally, it may well be advisable when considering alum or ferric chloride to provide chemical storage and feeding facilities to handle either chemical. Sewage composition or the relative cost of chemicals may change with time justifying a change in point of addition or chemical used.

Mixing Requirements

Adequate mixing of the prime coagulant into the sewage stream is essential. In cases where initial results from full-scale plant studies did not approach the jar test predicted results, inadequate mixing or excessive hydraulic loadings were invariably the cause. Efficient mixing was obtained in full-scale facilities in the following ways:

1. injection of the chemical into the aeration side of the raw sewage pumps;
2. chemical addition into a preaeration tank or aerated grit chamber;
3. addition of chemical at a Parshall Flume or similar constriction in a flow channel; and
4. chemical addition into the channel or pipe between the aeration tank and the final clarifier providing additional mechanical or air mixing is available.

In several cases, the P removal efficiency was doubled by increasing the intensity of mixing at the point of chemical addition.

Clarifier Hydarulic Loading

In all activated sludge systems where chemical addition to the aeration tank discharge was used to precipitate phosphorus, the hydraulic load on the final clarifier had a marked effect on effluent quality. At hydraulic loads above 1.60 m/hr, a steady deterioration in effluent quality occurred (Figure 5). Final effluent suspended solids (SS) values of over 25 mg/l were consistently obtained when the hydraulic rate on the clarifiers increased to over 1.60 m/hr. Below 1.60 m/hr, there appears to be no further decrease in effluent SS values. The minimum expected SS value is 10 mg/l from these systems. The effluent biochemical oxygen demand (BOD) values were observed to follow the fluctuations of the effluent SS very closely. With

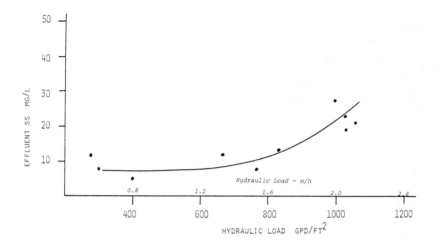

Figure 5. Effect of hydraulic load of secondary clarifier chemical addition to aeration section.

satisfactory solids separation, BOD values of 10-15 mg/l were consistently obtained in the studies.

It must be emphasized that phosphorus removal efficiency is very closely tied in with solids removal efficiency. No matter how much chemical precipitant is added, unless the effluent suspended solids level can be reduced to below about 15 mg/l, it is impossible to achieve an effluent total phosphorus concentration of less than 1 mg/l, even though the soluble phosphorus may be as low as 0.1 mg/l. Extremely low levels of effluent total phosphorus (< 0.3 mg/l) may only be achieved through the use of tertiary filtration processes.

Sludge Production

Relatively consistent results in sludge production and characteristics from phosphorus removal facilities have been found to date depending upon the use of metal (iron or aluminum) salts or lime as the prime coagulant.

Metal Salts

In secondary facilities, the addition of metal salts for phosphorus removal generally increases the weight of dry solids produced per given volume of sewage treated by 5-25%, varying from 275 to 325 g/m^3 sewage treated. In addition to this increase in sludge mass, there is also a decrease in the solids concentration of the raw sludge, reducing the total solids content from the normal 4.0-5.5% to 3.5-5.0%. The combination of these two effects, tending to increase the wet sludge volume by approximately 35%, has resulted in some digester operating problems which will be discussed later.

In primary plants, the sludge production increases more dramatically. A 100% increase in dry solids production can be expected, the majority of which can be attributed to increased solids capture from the sewage. Depending upon the raw waste characteristics, the solids production may range between 250 and 300 g/m^3 of sewage treated with phosphorus removal from the normal 140 g/m^3 without. Additionally, a 20% reduction in sludge solids concentration may be expected.

Lime

The use of lime for phosphorus removal in secondary plants has an even more dramatic effect on sludge production than does the use of metal salts. With the addition of 200 mg/l lime at the Newmarket WPCP [20], sludge production increased from 147 g/m^3 to 374 g/m^3, while the solids content of the sludge increased from a normal 4-5% to 9% with values approaching 30% at times. The net effect was a 25% increase in sludge volume to be handled.

The use of lime in a primary plant can be expected to increase sludge solids production about threefold, while at the same time increasing solids content of the sludge from the normal 5-7% to 12-17%; the net effect being a 50% increase in total sludge volume.

Sludge Digestion

In general, sludges produced from phosphorus removal facilities have been found to be readily digestible in both existing anaerobic and aerobic digesters. Some initial operating problems, however, have been experienced, but these relate more to the increased sludge production than to the nature of the sludge itself. The increased digester hydraulic loading has, on occasion, resulted in inadequate heat exchanger capacity with resultant operational problems due to loss of digester temperature. Additionally, because of increased volatile solids loading on the digestion system, some problems have been experienced from inadequate gas-liquid separation resulting in foaming within the primary digester.

In no instance has an inhibitive effect of the accumulated metal salts been experienced in digester operation, although in the Newmarket facility [20] during start-up, digester operation was completely disrupted because of erratic lime dosing. This resulted in periodic massive doses of high pH sludge being pumped to the digester until the buffering capacity of the digester was exceeded.

Later, under continuous operation, the digester was found to operate very effectively if a sludge blanket of 0.3-0.5 m was maintained in the clarifier. This partially neutralized the raw sludge to a pH of 8.5-9.

Phosphorus resolubilization within the digester has been found to be insignificant in relation to the total plant loading regardless of the chemical used.

Sludge Utilization and Disposal

The most common method of sewage sludge disposal in Ontario is land application of liquid sludge. Stabilized sludge is transported by tank truck and spread thinly on agricultural land for utilization as a source of nitrogen and phosphorus for crop growth. Other methods of sludge disposal used in Ontario include incineration and ash landfilling, and dumping into sanitary landfulls and other designated disposal areas (Table I).

Table I. Sludge Disposal Methods Practiced in Ontario [22]

Method of Disposal	No. of Plants	Percent of Total Plants	Wt of Sludge (dry tons/yr)	Percent of Total Wt
Application to Agricultural Lands	98	63.2	52,900	34.0
Incineration	3	2.0	62,000	39.8
Landfill Application	17	11.0	35,000	22.5
Dumpsite	14	9.0		
Storage Lagoon	7	4.5	5,800	3.7
Drying Beds	16	10.3		
TOTAL	155	100	155,700	100

Although research is continuing on all aspects of sludge disposal, enough studies have been conducted to date [21] to indicate that the same disposal practices which were used prior to phosphorus removal continue to apply. With respect to land utilization of sludges, chemical addition to primary treatment facilities for phosphorus removal will increase the precipitation of heavy metals from the sewage. This may adversely affect the utilization of the sludge. Little effect is noted with secondary treatment plant sludges, as secondary facilities are generally very effective by themselves in precipitating out heavy metals.

PHOSPHORUS REMOVAL FROM LAGOONS

The requirement of phosphorus removal from lagoons in Ontario is dependent upon location and/or size of facility (Figure 1). High levels of phosphorus removal may be achieved at all types of lagoon but the approach taken depends considerably upon the type.

Aerated Lagoons

For aerated lagoons, alum or ferric chloride may be fed into the influent to the aerated cell. Frequently, this is done at the raw sewage pumping station to ensure a high degree of mixing prior to reaching the lagoon. Again, the full-scale design should be based upon a wide range of jar testing, but excluding lime, carried out on the raw sewage as previously outlined. Lime has not been found to be effective [23,24]. Full-scale, temporary studies should not be necessary. Several aerated lagoons in Ontario are now effectively achieving phosphorus removal in this manner [25].

Conventional Stabilization Ponds

Phosphorus removal from conventional stabilization ponds may similarly be achieved by the continuous addition of alum or ferric chloride to the pond influent [23]. Again, the full range of jar testing, excluding lime [23], should be conducted to determine optimum chemical and approximate dosage, but as with aerated lagoons, full-scale testing should not be required unless there is some concern over mixing. Chemical addition for phosphorus removal will have little to no effect on other effluent parameters such as SS and BOD.

Seasonal Retention Lagoons

Although phosphorus removal may be effective in seasonal retention lagoons by the continuous addition of alum or ferric chloride to the influent raw sewage, it is generally achieved in Ontario by batch treatment of the lagoon contents with chemical prior to discharge [26]. When considering batch chemical treatment of lagoons, jar tests are conducted on the lagoon contents. Onsite testing is not considered to be essential, and limited sampling is adequate because of the long retention period within the retention lagoon, making rapid changes in pond characteristics unlikely. The dosage to be used for full-scale treatment should be the one which reduced the phosphorus concentration to approximately 0.5 mg/l. Again, lime has not been found to be satisfactory for batch treatment [20]. In the batch treatment of lagoons, the chemical, as a liquid, is contained in a 600- to 700-liter plastic tank mounted amidships in a 5-m, 70 hp outboard motorboat. Chemical dispersal throughout the lagoon is by means of a 50-mm siphon discharging into the prop wash. The main application objectives in lagoon batch treatment are rapid, even distribution of the chemical and good mixing.

Batch chemical treatment of seasonal retention lagoons using alum and ferric chloride has been highly successful. Effluent quality after treatment is usually less than 10 mg/l BOD and 20 mg/l SS, with total phosphorus generally less than 0.5 mg/l. Due to a slight deterioration in effluent quality after about two weeks, it is desirable to complete the discharge of the lagoon contents within 8-10 days after treatment [26].

PHOSPHORUS REMOVAL COSTS

Costs associated with phosphorus removal of existing wastewater treatment facilities are not only related to the capital and operating costs of full-scale operation but must also include the costs of jar testing and, if required, full-scale, temporary studies.

The following costs are derived from Ontario's experiences and are broken down into treatability studies and full-scale costs [27].

Treatability Study Costs

Jar Testing

Jar testing treatability costs have varied over a wide range depending somewhat upon the type of plant but more so on whether the studies were contracted out or performed by plant operations staff. Average jar testing costs are presented in Table II.

Table II. Jar Testing Treatability Costs

Type of Facility	Average Cost ($)	
	Plant Staff	Outside Contractor
Primary	964	5020
Secondary	817	5877
Lagoons	588	

Jar testing is considered essential to proper facilities design. However, if carried out by plant personnel under proper supervision, as seen from the above table, costs are quite moderate. These costs, however, do not include costs for all three chemicals, and in the case of secondary facilities, two points of application.

Full-scale Temporary Studies

Full-scale temporary studies are not necessarily required in every instance, but should be conducted where there is any concern over being able to achieve proper mixing of the chemical, or the suitability of other aspects of plant design. Again, the full-scale testing may be conducted by plant staff or by outside consulting firms. The costs in Table III do not differentiate between the two methods, instead they present maximum, minimum and average costs. The costs are for one chemical only. If two chemicals are tested, the costs would double.

Table III. Full-Scale Temporary Study Costs

Type of Facility	Maximum	Cost ($) Minimum	Average
Primary	62,000	7,000	19,100
Secondary	83,000	1,000	13,100
Lagoons	8,000	800	4,100

One reason for the wide range in costs presented in Table III is related to the cost of equipment. In some cases, ministry equipment was loaned to the municipality, whereas in other cases the cost of the equipment was included in the treatability study costs.

Generally, full-scale temporary studies are not conducted on lagoons. The costs indicated above resulted from specific studies on lagoons carried out by operating staff.

Full-Scale Phosphorus Removal Costs

The full-scale operating costs presented below are taken from a recent survey of 64 mechanical wastewater treatment plants which had been practicing phosphorus removal for 3-6 years prior to the survey [27].

Capital Costs

Capital costs for phosphorus removal may be broken down into the following categories with average unit costs for a 200-m^3/hr plant as shown:

Chemical Storage Tank	$ 7,000
Feed Pumps	2,500
Mechanical (piping, etc.)	1,000
Electrical	1,000
Structural (tank base, etc.)	1,000
Engineering Contingency	1,500
	$14,000

These costs assume that a 22.7-m³ fiberglass reinforced plastic tank is used for liquid chemical storage and is placed on a suitable gravel or concrete pad outside. The tank and exposed piping would require heat tracing for winter operation.

As the size of the plant increases, the capital cost per volume of plant design capacity also increases. Figure 6 shows the average capital costs of phosphorus removal related to capacity for the municipal wastewater treatment plants surveyed.

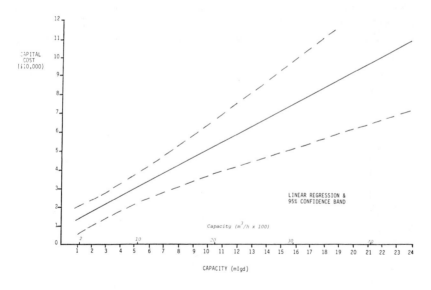

Figure 6. Phosphorus removal equipment capital costs.

Chemical Costs

Chemical costs are invariably the highest of the costs associated with phosphorus removal. Since the price of any one chemical varies considerably with the quantity of chemical required and the distance from the source of supply, actual chemical costs must be determined on an individual plant basis taking into account dosage required, size of plant, source of chemical, etc.

Delivered chemical costs during the summer of 1976, at the time of the survey [27], ranged as follows:

Alum	-	$70.95-83.74 (per ton dry alum)
Iron Salts	-	$ 0.16- 0.21 (per pound of Fe)
Lime	-	$34.50-44.35 (per ton of bulk hydrated lime)

Actual chemical costs for operating facilities ranged from $2.48 to $76.88 per 200 m^3 of sewage treated in the plants surveyed, with the average cost being $23.00 per 200 m^3 treated.

Other Operating Factors

Additional operating costs, other than that for the chemical, are also associated with phosphorus removal, and include such things as routine maintenance, electricity to run pumps and heat tanks during the winter, water makeup and manpower associated with the above. Figure 7 presents

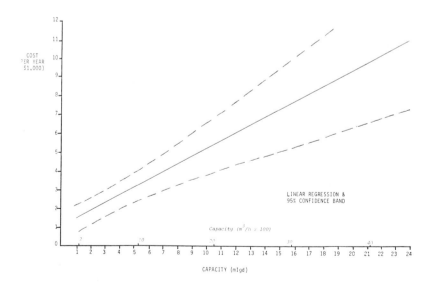

Figure 7. Phosphorus removal operational costs (excluding chemical costs).

the operating costs (excluding chemical costs) associated with phosphorus removal as a function of design capacity. Operating costs for a 200-m³/hr plant averaged out at approximately $1600 per year.

CONCLUSIONS

Based upon the experience of Ontario with its phosphorus removal program, it can be concluded that phosphorus removal can be implemented at any existing municipal wastewater treatment plant by chemical addition with little or no operational difficulty.

Treatability studies in the form of jar testing and possibly full-scale temporary studies are necessary to determine the optimum chemical, its dosage and point of addition, and to evaluate the suitability of the facility for phosphorus removal prior to full-scale permanent implementation.

Design and operational considerations which must be included in the implementation of phosphorus removal at a given sewage treatment facility include chemical source, dosage, feed control, point of addition and mixing, suitability of plant processes such as clarifier hydraulic loading, sludge handling and digester capacity and sludge utilization and disposal.

The major costs for phosphorus removal are related to the cost of the chemical itself which is dependent upon source, distance from plant, dosage and size of plant. Capital and other operating costs for a 200 m³/hr plant approximate $14,000 and $1,600 respectively.

REFERENCES

1. Black, S. A. "Lake Eutrophication—A Laboratory Investigation," Ontario Water Resources Commission Research Paper No. 4026 (1970).
2. Black, S. A. and W. Lewandowski. "Phosphorus Removal by Lime Addition to a Conventional Activated Sludge Plant," Ontario Water Resources Commission, Research Publication No. 36 (1969).
3. Black, S. A. "Lime Treatment for Phosphorus Removal at the Newmarket/East Gwillimbury WPCP—An Interim Report," Ministry of the Environment Research Paper No. W2032 (1972).
4. International Joint Commission. "Pollution of Lake Erie, Lake Ontario and the International Section of the St. Lawrence River," Volume I, Summary (1969).
5. "Canada-Ontario Agreement on the Lower Great Lakes Water Quality," (April 1972).
6. "Canada-Ontario Agreement on the Lower Great Lakes Water Quality," (August 1971).
7. Ministry of the Environment. "Guidelines for Conducting Treatability Studies for Phosphorus Removal at Wastewater Treatment Plants," (April 1972).

7. Ministry of the Environment. "Guidelines for Conducting Treatability Studies for Phosphorus Removal at Wastewater Treatment Plants," (April 1972).

8. Barnard, J. L. "A Review of Biological Phosphorus Removal in the Activated Sludge Process," *Water SA* 2:136 (1976)

9. Price, R. L. and R. W. Regan. "The Fate of Phosphorus in the *Phostrip* Process," presented at the 50th Annual WPCP Conference, Philadelphia, PA, October 3, 1977.

10. Jank, B. E. et al. "Nutrient Removal in Suspended Growth Systems Without Chemical Addition—Controlling Factors and Effects on Design," presented at the 51st Annual WPCF Conference, Anaheim, California, October 1978.

11. International Joint Commission. "Great Lakes Water Quality Agreement of 1978," (November 1978).

12. Union Carbide Environmental Systems Newsletter, 2(3) (undated).

13. Archer, J. "Phosphorus Removal—The Use of Ferrous Salts in Primary Wastewater Treatment Plants," Ministry of the Environment Paper (May 1976).

14. Heinke, G. W. et al. "Design and Performance Criteria for Settling Tanks for the Removal of Physical-Chemical Flocs, Volume II," Canada-Ontario Agreement Research Report No. 56 (1977).

15. American Water Works Association. "Simplified Procedures for Water Examination," AWWA Manual, M 12, p. 42 (1964).

16. Boyko, B. I. and J. W. G. Rupke. "Phosphorus Removal Within Existing Wastewater Treatment Facilities," Canada-Ontario Agreement Research Report No. 44, (1976).

17. Scott, D. S. "Use and Production of Iron Salts for Phosphorus Removal," Canada-Ontario Agreement Research Report No. 5, (1973).

18. Prested, B. P. et al. "Development of Prediction Models for Chemical Phosphorus Removal, Volume 1," Canada-Ontario Agreement Research Report No. 68 (1977).

19. Pollutech Pollution Advisory Services Ltd. "Chemical Dosage Control for Phosphorus Removal," Canada-Ontario Agreement Research Report No. 4, (1974).

20. Black, S. A. "Anaerobic Digestion of Lime Sewage Sludge," Canada-Ontario Agreement Research Report No. 50, (1976).

21. "Report of the Land Disposal of Sludge Sub-Committee Projects Conducted 1971-1978," Canada-Ontario Agreement Research Report No. 70, (1977).

22. Antonic, M. et al. "A Survey of Ontario Sludge Disposal Practices," Canada-Ontario Agreement Research Project No. 74-3-19 (in preparation).

23. Graham, H. J. and R. B. Hunsinger. "Phosphorus Reduction from Continuous Overflow Lagoons by Addition of Coagulants to Influent Sewage," Canada-Ontario Agreement Research Report No. 65, (1977).

24. Ahlberg, N. R. "Phosphorus Removal in a Facultative Aerated Lagoon—Grimsby WPCP," Ontario Ministry of the Environment, Research Branch, (July 1973).

25. Ontario Ministry of the Environment. "Operating Summary—Water Pollution Control Plants", (1977)).

26. Graham, H. J. and R. B. Hunsinger. "Phosphorus Removal in Seasonal Retention Lagoons by Batch Chemical Precipitation," Canada-Ontario Agreement Research Report No. 13 (1971).

27. Archer, J. "Summary Report on Phosphorus Removal," Canada-Ontario Agreement Report No. 83 (1978).

CHAPTER 14

OPERATIONAL EXPERIENCE IN PHOSPHORUS REMOVAL

J. W. G. Rupke

Rupke & Associates, Ltd.
Bradford, Ontario

PROFILE OF ONTARIO PHOSPHORUS REMOVAL FACILITIES

General

Since many of the newer pollution control plants have been constructed to serve the smaller urban towns and villages, a large number of small plants exist in Ontario. Of the total of 340 plants in Ontario, 39 provide primary treatment, 125 are lagoons and the remaining 176 are secondary plants. The medium size of the secondary plants is approximately 1.0 mgd with an average size of 4.7 mgd. The average is strongly influenced by the presence of several 50+ mgd plants servicing the larger cities.

A typical Ontario secondary plant has a capacity of 1.0 mgd, uses anaerobic digestion for sludge stabilization and employs land disposal for the digested sludge. Oxygen supply is equally split between diffused air systems and mechanical aerators.

Chemical Usage

The implementation of phosphorus removal in 1973 and 1975 resulted in an intensive government supported program [1] to demonstrate the most practical and cost-effective method of effecting phosphorus removal in existing plants. Without exception, chemical precipitation was used to remove phosphorus.

355

The typical plant installed a 5000-gal fiberglass reinforced plastic tank for liquid precipitants. This tank was insulated, heat treated and placed on an exterior concrete pad. Two parallel chemical feed pumps (one as standby) were installed within the buildings. Plastic chemical feed lines were installed to the feed and discharge ends of the aeration tanks as well as to the primary feed channels.

All feed equipment was to be compatible with all aluminum and iron salt solutions. The type of chemical used for phosphorus removal has changed significantly throughout the period of 1973 to the present. Table I shows the

Table I. Chemical Usage Distribution

Chemical	Initial (%)	1976 (%)	1979 (%)
Aluminum Sulfate	32	25	26
Iron Salts	60	64	70
Lime	5	6	2
None	3	5	2

initial chemical usage, 1976 usage and present usage. The gradual shift to iron salts at the expense of the lime and alum systems was due to the mechanical difficulty of the lime feed systems and the increasing costs of alum. More recently, alum has been regaining a portion of the market from the iron salts due to fears of heavy metal contamination of the digested sludges, and reported corrosion and staining due to iron salts.

In many cases waste iron salts or alum solutions are being used for at least a portion of the precipitant need. The relative abundance of free iron salts may also account for the trend of increasing iron usage.

Chemical Costs

Table II shows the average dosages of the commonly used chemicals and their approximate costs. Chemical costs vary widely due to transportation differences and sources of free material. However, in general, the values shown in Table II are representative of purchased material.

Free iron salts are frequently available in urban centers. However, difficulty with solids deposition blocking feed pumps, high toxic metal contamination, and load to load concentration variation have discouraged its use. The cost savings normally more than offset the increased operating costs, even if mechanical changes are required to allow the use of the waste material.

Table II. Average Doses of Commonly Used Chemicals and Their Costs

Chemical	Average Dosage (mg/l)	Chemical Costs	
		¢/lb	$/Mgal
Alum	65 mg/l as $Al_2 (SO_4)_3 \cdot 14 H_2O$	5.1	33.15
Iron Salt Fe^{2+}	11 mg/l as Fe	17	18.70
Fe^{3+}	11 mg/l as Fe	25	27.50

Sludge Production

Based on theoretical considerations, a measurable increase in sludge production can be expected upon the addition of a chemical precipitant for phosphorus removal. In secondary plants using iron and aluminum salts, the theoretical solids increase should be in the 25% to 40% volumetric increase range.

Actual sludge volumetric changes have not been measured in Ontario plants. The annual variation in digested sludge disposed of varies from 2000 to 4500 gal/Mgal of sewage treated. A plant-by-plant comparison of pre- and postphosphorus removal sludge data shows that the above noted values apply to both time periods. Many plants have shown marked increases in sludge production while others have had equally large decreases in sludge production. Similarly, no clear trend is evident in the sludge solids concentration data.

This lack of positive evidence is likely due to the data variations that develop due to changes in operation by the plant staff. Applied research data from full-scale plant operation reported in 1975 [2] shows that the phosphorus removal program has resulted in a 10-35% volumetric increase in sludge production. A 10% decrease in sludge concentration was also noted.

CASE HISTORY COSTS

Operation and Maintenance

The Regional Municipality of York operates 10 water pollution control plants ranging in size from 0.45 to 3.0 mgd. Within this size range the O&M manpower requirements are found to be approximately 100 man-hours per year. This includes a daily check of the feed equipment, pumps maintenance and analytical requirements for dosage control.

In all 10 plants, iron salts have proven to be the most cost-effective approach to achieve phosphorus removal. Two direct comparisons were done in an attempt to evaluate alternative chemicals. However, in both cases the cost of alum was significantly higher than the cost of ferrous chloride. The data are presented in Table III.

With the provision of a standby chemical feed pump, no significant downtime has been experienced in the four years of operation.

Capital Costs.

The capital costs of phosphorus removal by chemical precipitation has proven to be relatively low ranging from $15,000 to $25,000 for plants sized 4.0 mgd and less.

In most cases a 5000-gal FRP chemical storage tank has been installed outside the existing buildings. The tank should be insulated and heat traced to prevent solidification of the chemicals during the winter. Twin chemical feed pumps, plastic feed lines and some method of flow measurement complete the system.

In some cases mixing at the point of chemical addition may also be required.

Table IV shows the total costs associated with phosphorus removal. Manpower has been charged at $8.00/hr and capital at 10%/yr.

HIGHER QUALITY OBJECTIVE

The recently signed Canada-U.S. agreement on Great Lakes Water Quality calls for a 0.5-mg/l total phosphorus objective from municipal facilities in the near future. An analysis of existing data shows that this can readily be achieved in all facilities equipped with tertiary sand filtration for effluent polishing. A slight increase in precipitant dosage will reduce the total phosphorus level to 0.5 mg/l with a relatively slight increase in cost. It is estimated, based on limited existing data, that a 10 to 15% increase in chemical dosage would be required to reduce the phosphorus level from 1.0 mg/l to 0.5 mg/l. There are, however, only 15 tertiary filtration plants in Ontario.

A further analysis of the data does show 11 plants that achieve the 0.5 mg/l objective without tertiary sand filtration. In six of these plants, where sizing data were available, the secondary clarifier overflow rates were less than 300 gpd/ft^2.

A more detailed analysis of monthly operating data [3] from ten plants which periodically achieve the 0.5 total phosphorus effluent criteria shows that this effluent value is usually associated with secondary clarifier loadings of 500 gpd/ft^2 or less.

Table III. Comparison of Alum and Iron Salts for Phosphorus Removal, Region of York

Plant	Chemical	Dosage (mg/l)	Duration (days)	Residual P (mg/l)	Cost ($/Mgal)
Unionville	Ferrous Chloride	6.6 as Fe	51	0.5	11.22
	Alum	65 as $Al_2 (SO_4)_3 \cdot 14 H_2O$	35	0.88	32.50
Stouffville	Ferrous Chloride	16 as Fe	50	0.5	27.20
	Alum	60 as $Al_2 (SO_4)_3 \cdot 14 H_2O$	37	0.82	30.60

Table IV. Total Cost Based on 3.0 mgd Plant

		Annual Cost ($)
Chemicals	10 mg/l Fe^{2+} @ $0.17/lb	18,615
Manpower	100 hr @ $8.00/hr	800
Capital	10% of 15,000	1,500
Total		20,915
Unit Cost ($/Mgal)		19.10

The data are confusing, however, since chemical dosage information is not available at this time, and the effluent phosphorus level is also dependent on precipitant dosage level as well as clarifier loadings. All of these plants are aiming at the 1.0 mg/l total phosphorus effluent criteria. It is reasonable to assume that their dosage control would not deviate by more than 15% to 20% from that required to meet the objective. Based on this rather gross assumption, it can be concluded that the 0.5 mg/l objective can be met providing a small increase in chemical dosage is associated with a conservative clarifier loading of 500 gpd/ft^2 at average flow.

The use of polyelectrolytes to increase the floc settling rate may make it practical to achieve the 0.5 mg/l total phosphorus objective in most secondary plants, even with conventional clarifier loadings.

REFERENCES

1. Van Fleet, G. L. "Phosphorus Removal In Ontario," Phosphorus Removal Design Seminar (May 1973).
2. Boyko, B. I., and J. W. G. Rupke, "Technical Implementation of Ontario's Phosphorus Removal Program," Purdue Industrial Waste Conference, May 1973.
3. "Operating Summary Water Pollution Control Projects," Ontario Ministry of the Environment (1977).

CHAPTER 15

SLUDGE GENERATION, HANDLING AND DISPOSAL AT PHOSPHORUS CONTROL FACILITIES

N. W. Schmidtke

Wastewater Technology Centre
Environmental Protection Service
Environment Canada
Burlington, Ontario

INTRODUCTION

The problem of estimating sludge quantity, deciding how to handle the sludge and how to ultimately dispose of it have been with civilized men for some time, but have been mainly ignored. The concern regarding sludge quantities, handling and disposal is generally proportional to population density, and the degree and complexity of industrialization.

For example, consider for a moment that even when going back to early biblical records, nowhere does it mention that in the construction of the ark Noah considered or even anticipated the monumental sludge handling problem he would have to face once he had all his animals on board. His solution when faced once adrift with the problem, is left to your imagination.

It would appear that still too frequently design engineers today suffer from the "Noah Syndrome".

Early man considered sludge as a resource, something to be recycled. Many less industrialized nations still pursue this philosophy. Even highly industrialized nations not blessed with an abundance of resources have prescribed to a similar philosophy. In North America it took an energy crisis to redirect our thinking to the point where sludge is looked at from a utilization rather than disposal perspective.

It has been said that "history repeats itself." This then would also appear to be true when dealing with sludge.

This paper will focus on providing information on the effect of adding metal salts to existing wastewater treatment plants for phosphorus removal to 1 mg/l total as it impacts on sludge quantity, handling and disposal/utilization.

The information is based on data from 185 waste treatment plants surveyed in the Province of Ontario. The completeness of data varies considerably.

SLUDGE QUANTITIES

Waste treatment process design engineers are continually plagued by lack of information when it comes to designing sludge handling and disposal/utilization facilities. The length of the shortcut to data acquisition is pretty well proportional to the degree of confidence to be placed in the process capacity design.

Good data are hard to get and cost money. The original error made in sludge quantity estimation can be and is increased when attempting to estimate resulting sludge quantities due to chemical addition for phosphorus removal increases from a target of 1 mg/l to 0.1 mg/l, and can be attributed to increasingly greater deviations from stoichiometric relationships between influent phosphorus and effluent target phosphorus.
ships between influent phosphorus and effluent target phosphorus.

Sludge production is influenced by a number of variables. For chemical sludge production, the chemical used, wastewater characteristics, and point of chemical addition all play an important role. For biological sludge production, the type of process used in the conversion of substrate greatly affects the amount of biomass produced. The amount of sludge produced also varies with the nature of substrate oxidized. Higher sludge volumes result in winter than in summer because the autooxidation rate depends on temperature. The total volume of sludge produced from biological and physical/chemical systems, or any combination thereof is also influenced by clarifier performance, sludge recycle, and the degree of operator attention to the system.

The literature abounds with sludge production data. Figure 1 is just one example for municipal sludges and illustrates the degree of variability which can be in excess of 100%.

A most useful method of sludge quantity estimation consists of performing a mass balance around various treatment process components and coupling this with process efficiency assumptions [2].

Calculations to determine chemical sludge quantities based on stoichiometric relationships have been illustrated by Campbell [3].

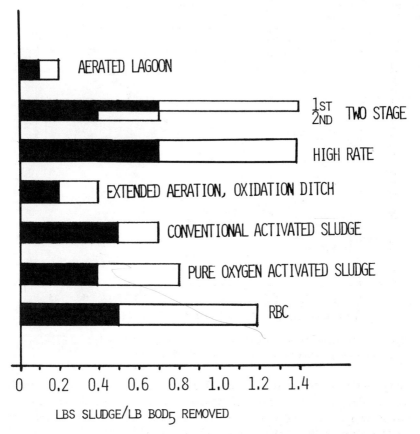

Figure 1. Biological sludge production [1].

Pilot Scale Activated Sludge Phosphorus and Nitrogen Removal Studies

A long-term pilot scale study for the removal of phosphorus and nitrogen [4] was conducted at the Wastewater Technology Centre (WTC). Sludge production was monitored over a 16-day consecutive period. Comparing the observed to calculated sludge production values for ferric iron addition based on stoichiometric relationships shows a 65% increase 50% of the time, as illustrated in Figure 2.

Full Scale Phosphorus Removal Studies—Primary Plant

Full scale phosphorus removal studies were conducted at the primary wastewater treatment plant at C. F. B. Borden [5]. The study lasted ten months and covered three phases of chemical addition for phosphorus removal

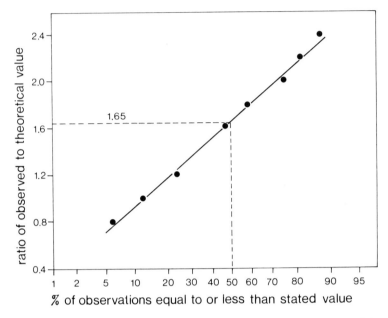

Figure 2. Observed to theoretical solids production ratio—ferric iron addition at a nitrogen and phosphorus removal activated sludge plant [4].

using lime, alum and ferric chloride. While the major objective of this study was to determine the optimum phosphorus removal precipitant and its dosage to achieve an effluent phosphorus objective of ≤ 1 mg/l, information on sludge production under various operational conditions was also collected. These data as summarized in Table I were compared to calculated sludge production values and are shown as a frequency distribution in Figure 3. In this case, the amount of sludge produced was overestimated by 28%, 50% of the time.

Ontario Treatment Plant Survey Data

Sludge production data were obtained in a 1975 survey of Ontario wastewater treatment plants where records prior to phosphorus removal were compared with plant records following installation of phosphorus precipitation systems [6].

Figure 4 summarizes the data from 15 conventional, primary plants (without precipitant addition) surveyed covering a range of hydraulic loadings from 0.26 to 11 mgd*. The data show that 50% of the time 1,995 gallons

---

*Imperial gallons used throughout this paper.

Table I. Camp Borden Sludge Production–Primary Plant [5]

P-Removal Precipitant		Sludge Mass Produced				Ratio of Measure to Calculated Value
Chemical	Dosage	Calculated		Measured		
–	mg/l	kg/m^3	lbs/10^6 gal	kg/m^3	lbs/10^6 gal	
Baseline	–	116	255	191	420	1.65
Lime	151[a]	1186	2609	1148	2526	0.97
	197	1389	3056	1400	3080	1.01
	275	3184	6426	1697	3425	0.53
	210	1876	3857	1399	2876	0.76
Alum	4.4[b]	201	520	332	859	1.65
	7.5	385	1020	283	750	0.74
	14.8	645	1730	253	679	0.39
	18.5	604	1601	294	779	0.49
Ferric Chloride	9.6[c]	290	760	304	796	1.05
	14.6	321	872	307	834	0.96
	19.0	502	1331	312	827	0.62
	26.6	515	1382	343	920	0.67

[a]As Ca(OH)$_2$.

[b]As Al^{3+}.

[c]As Fe^{3+}.

of sludge are produced for each million gallons of wastewater treated. This translates to 1,140 lbs dry solids for each million gallons treated (Figure 5). The total solids concentrations of the raw primary sludges varied from 3.5 to 8% with a mean of 5.7%.

The impact of chemical addition for phosphorus removal at primary plants is illustrated in Figure 6 for seven upgraded plants. In these plants, the average sludge solids concentration decreased from 6.0 to 5.3% after chemical addition. The sludge mass increased by 40%.

Sludge production data from 42 secondary plants using the conventional activated sludge process were analyzed. The plants have flow capacities ranging from 0.3 to 170 mgd. The raw sludge produced consists of both primary and waste activated sludge. In the case of conventional activated sludge plants, Figure 7 shows that 50% of the time at least 3,905 gallons of sludge are produced per million gallons treated. Solids concentrations varied from 2 to 7%, with a weighted average of 4.6%. Similarly, the dry

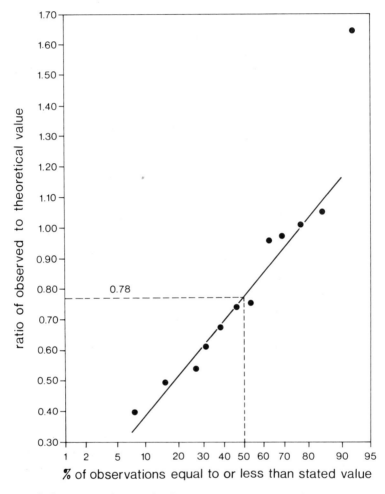

Figure 3. Observed to theoretical solids production ratio—precipitant addition to a full scale primary plant [5].

weight of solids produced at conventional activated sludge plants was equal to or less than 1,786 dry solids per million gallons of wastewater treated, 50% of the time (Figure 8).

Sludge production data for 15 upgraded secondary plants (primary and waste activated, chemical sludge) are illustrated in Figure 9. Fifty percent of the observations showed a solids production equal to or less than 1,725 lbs of dry solids per million gallons before chemical addition. This increased to 2,175 lbs of dry solids per million gallons after chemical addition and

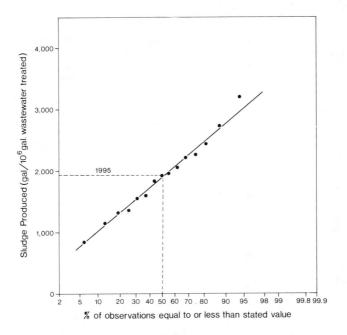

Figure 4. Probability distribution for sludge volume produced at conventional primary plants [6].

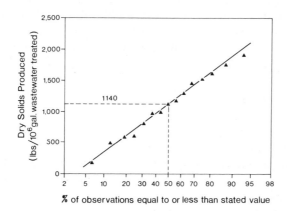

Figure 5. Sludge mass produced at conventional primary plants [6].

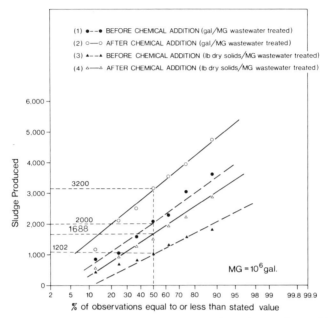

Figure 6. Probability distribution for sludge produced at primary plants with addition of metal salts [6].

represents a 26% increase in sludge mass. Following precipitant addition, the average total solids concentration decreased from 4.5 to 4.2%.

While metal salts are generally added to the aeration tanks, data analyzed from four installations where metal salts were added to the primary settling tank showed a decrease in solids produced. In this instance, the lower organic loading to the aeration tank due to additional organics removed in the primary, resulted in reduced biosynthesis.

Summary of Sludge Production

Sutton [4] underestimated sludge production by 65% when using stoichiometric relationships for a biological system with chemical addition. Stepko [5], however, overestimated sludge production resulting from chemical addition to a primary plant by 28%. These studies exemplify the problems associated with estimating sludge production from chemical stoichiometry.

From our experience, the best data base for sludge production exists in the Ontario survey of full scale treatment plants [6]. The data presented in earlier figures are summarized in Tables II and III for primary and activated sludge plants, respectively.

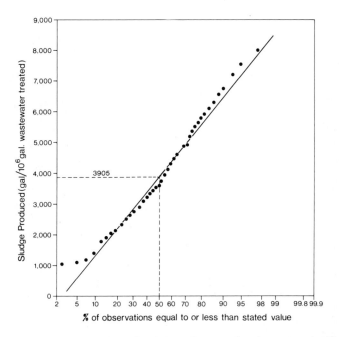

Figure 7. Probability distribution for sludge volume produced at conventional secondary (C.A.S.) plants [6].

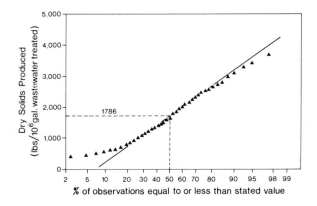

Figure 8. Sludge mass produced at conventional secondary (C.A.S.) plants [6].

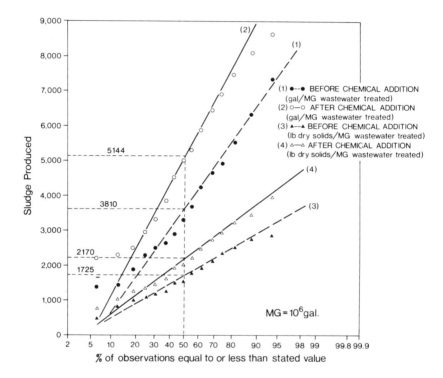

Figure 9. Probability distribution for sludge produced at secondary (C.A.S.) plants with addition of metal salts to aeration tank [6].

Table II. Primary Sludge Production Data [6]

Description	Units	Sludge Production		
		Prior to Chemical Addition	After Chemical Addition	Percent Change
Volume	gal/10^6 gal	2000	3200	+60
	gal/capita	0.29	0.46	–
	% of influent Q	0.20	0.32	–
Mass	lbs/10^6 gal	1202	1688	+40
	lbs/capita	0.17	0.24	–
Solids	percent	6.0	5.3	– 0.7
Number of Plants	–	7	7	–

Table III. Activated Sludge Production Data [6]

Description	Units	Sludge Production		
		Prior to Chemical Addition	After Chemical Addition	Percent Change
Volume	gal/10^6 gal	3810	5144	+35
	gal/capita	0.55	0.75	–
	% of influent Q	0.38	0.51	–
Mass	lbs/10^6 gal	1725	2175	+26
	lbs/capita	0.25	0.32	–
Solids	percent	4.5	4.2	– 0.3
Number of Plants	–	15	15	–

Based on the results of the Ontario survey [6], some generalizations concerning sludge production design are shown in Table 4.

The rule-of-thumb that sludge volume approaches 0.5% of the influent hydraulic load to a conventional plant is a good approximation. By using this estimate, the apparent margin of safety would allow upgrading of a

Table IV. Sludge Production—Suggested Design Data[a]

| System | lbs pcd | Sludge Quality | |
		Volume % of Influent	lbs d.s./10^6 gal
Conventional Primary	0.17	0.20	1200
Upgraded Primary	0.24	0.32	1700
Conventional A.S.[b]	0.25	0.38	1725
Upgraded A.S.[b]	0.32	0.51	2175

[a]Based on Q = 145 dpcd.   [b]Primary + waste activated.
d.s.   = dry solids.
pcd   = per capita/day.

conventional plant to include chemical phosphorus removal to 1.0 mg/l total phosphorus using metal salts without major expansion of sludge handling facilities.

Because few Ontario plants practice phosphorus removal using lime, no substantive data base for sludge quantity estimation exists. However, based on past experience at a number of pilot and full-scale facilities practicing phosphorus removal using lime, reasonable estimates of sludge production can be made.

The mass of sludge produced will depend largely on the wastewater alkalinity and the lime dosage required to attain a specific pH at which the target phosphorus effluent level is achieved.

Figure 10 illustrates that, having determined the pH at which the phosphorus effluent target will be achieved, the correlation indicates the lime/alkalinity ratio required. Knowledge of the wastewater alkalinity enables calculation of the required lime dosage [7]. Another correlation [9] for raw wastewaters from 20 Ontario municipalities showing alkalinity/lime dosage requirements to attain pH 10 and 11 is shown in Figure 11.

Sludge Quantities After Anaerobic Digestion

The sludge production data summarized earlier facilitates the design of sludge handling and volume reduction facilities. When designing facilities for ultimate disposal, the sludge volume after anaerobic digestion must be known.

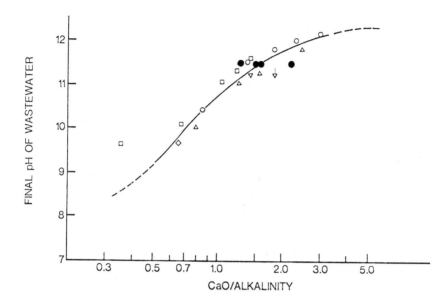

Figure 10. Ratio of lime dosage (mg/l) to initial wastewater alkalinity [7].

Such data are difficult to obtain. In many instances, this can be attributed to incomplete records concerning volume of sludge disposed of, as well as problems associated with solids concentration determinations.

The Ontario survey [6], while incomplete, provides the best currently available data base on this subject. The data relating sludge volumes disposed from standard primary plants to population served were subjected to regression analysis (Figure 12).

The equation expressing this relationship for 17 plants is shown as:

$$\text{Sludge Disposed} = 0.0169 \, (\text{Population} \times 10^{-3})^{1.131} \, (10^6 \text{ gal/yr}) \tag{1}$$

Figure 12 illustrates the fact that digester problems will result in substantial increases in sludge volumes requiring disposal.

Similarly, Figures 13 to 16 illustrate from the available Ontario data [6] various relationships between volume or mass of sludge to be disposed of from activated sludge plants after anaerobic digestion as a function of population served. The equations are summarized in Table 5.

Figure 17 summarizes all the pertinent water pollution control plant data from the Ontario survey [6].

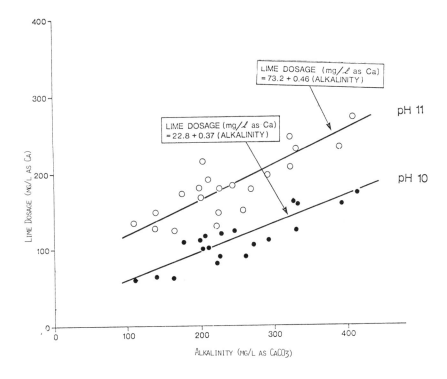

Figure 11. Lime dosage vs wastewater alkalinity [9].

Sludge Quantity Predictions for Lower Than 1 mg/l Effluent Phosphorus Targets

The aforenoted information does not address the question of "how much more sludge would be generated when imposing point source controls for effluent total phosphorus concentration of 0.5 mg/l or even 0.1 mg/l?"

A recent document [10] made a first attempt at answering this question by reporting on a computer simulation of required process modifications to meet various point source P control scenarios and the resulting sludge quantities. Figure 18 is a typical illustration of the dramatic increases in sludge mass over baseline conditions of no phosphorus removal. The example shows a simulation for 17 Canadian plants in the Lake Ontario drainage basin and represents a total flow of 582 mgd for a sewered population of 3.8 million persons. The simulation predicts a sludge mass increase from 34% over baseline conditions (no chemical addition) for an effluent total phosphorus target concentration of 1.0 mg/l.

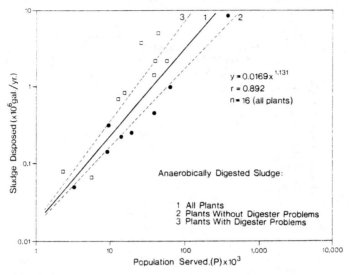

Figure 12. Sludge disposal at standard primary plants—sludge volume vs population served [6].

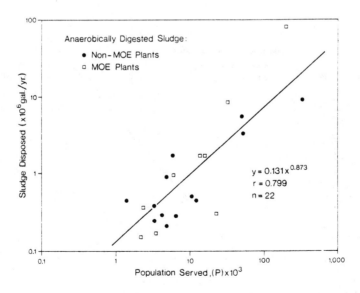

Figure 13. Sludge disposal at standard C.A.S. plants—sludge volume vs population served.

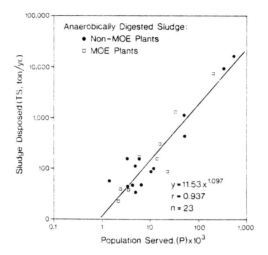

Figure 14. Sludge disposal at standard C.A.S. plants—dry weight of sludge vs population served [6].

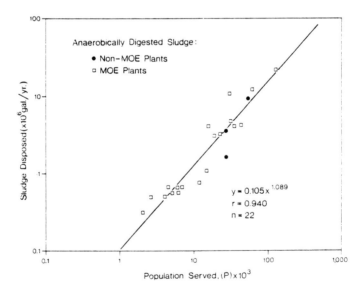

Figure 15. Sludge disposal at upgraded C.A.S. plants—sludge volume vs population served [6].

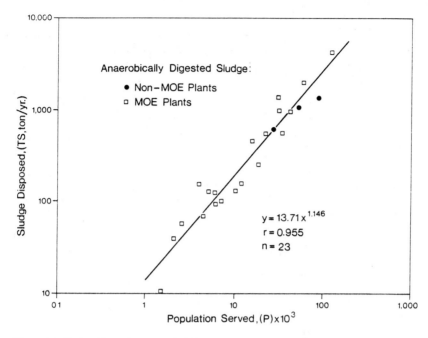

Figure 16. Sludge disposal at upgraded C.A.S. plants–dry weight of sludge vs population served [6].

Table V. Summary of Anaerobically Digested Sludge Disposal–
Relationships with Population Served [6]

Standard C.A.S. Plants

n = 22   Sludge disposed = 0.131 (Population x 10^{-3})$^{0.873}$ (2)
$(10^6$ gal/yr)

n = 23   Sludge disposed = 11.53 (Population x 10^{-3})$^{1.097}$ (3)
(TS, ton/yr)

Upgraded C.A.S. Plants

n = 22   Sludge disposed = 0.105 (Population x 10^{-3})$^{1.089}$ (4)
(x 10^6 gal/yr)

n = 23   Sludge disposed = 13.71 (Population x 10^{-3})$^{1.146}$ (5)
(TS, ton/yr)

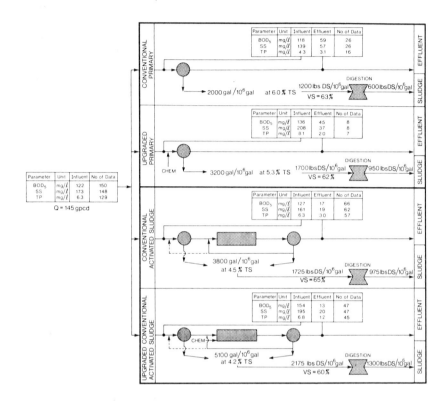

Figure 17. Summary of Ontario water pollution plant sludge production and disposal data [6].

This prediction compares favorably with the 1975 Ontario sludge survey data [6] indicating a 26% sludge mass increase. Treatment process modifications required to attain an effluent objective of 0.1 mg/l total phosphorus predict a 108% increase in sludge mass over baseline levels (no chemical addition).

It is the intent of the Phosphorus Management Strategies Task Force to update and refine the prediction model and input available data for U.S. and Canadian plants in the Great Lakes Basin. This will then enable the Task Force to identify the impact at municipal plants of various phosphorus control point source control scenarios on sludge quantities that might be generated.

The updated version of the model will also generate capital as well as O&M costs. This will allow for a relative cost comparison of the various phosphorus control scenarios.

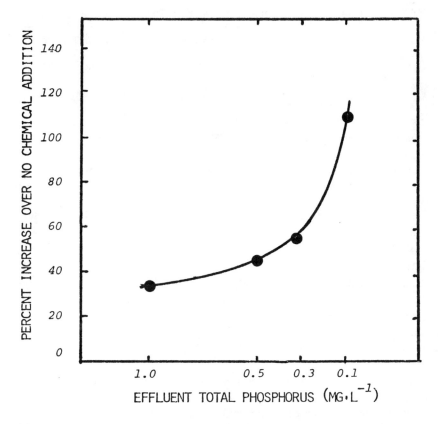

Figure 18. Increase in sludge mass due to chemical addition to meet various effluent total phosphorus targets.

SLUDGE HANDLING

Experience has demonstrated that metal salt addition to wastewater treatment processes for phosphorus removal not only results in increased sludge volumes and mass, but reduced solids concentration. The increased inorganic content due to chemical addition has the additional effect of lowering calorific values. If incineration is selected as the volume reduction process, more ash is produced.

With regard to sludge dewatering, waste-activated alum sludges, because of their gelatinous nature, are generally not dewatered by themselves but mixed with primary sludge, thickened and then dewatered. Table VI illustrates the effects of implementing phosphorus removal at two Ontario treatment plants.

Table VI. Full Scale Vacuum Filtration of Sludges From Phosphorus Removal Facilities

	West Windsor[a]			North Toronto[a]	
	None	Fe^{3+}	Al^{3+}	None	Fe^{3+}
		Primary Sludge		Digested Elutriated Sludge	
Solids Concentration (%)	11.9	8.6	7.9	8.1	7.6
Conditioning Chemicals					
(% lime)	9.9	15.9	24.0	9.5	11.8
(% ferric chloride)	1.3	0.1	1.2	0	6.6
Conditioning Cost[b]				Not	Not
($/ton of dry solids)	4.01	6.69	7.39	Recorded	Recorded
Filter Yield					
(lb/ft^2/hr)	12.4	9.6	6.7	3.8	3.9
Filter Cake Solids (%)	31	21	17	23	19
Filtrate SS (mg/l)	2830	3660	13900	5550	7690

[a]Campbell et al. [11].

[b]Cost figures as of December 1975.

The Windsor Treatment Plant is a primary facility which upon chemical addition showed reduced filter yield and cake solids concentration. Filtrate solids and conditioning costs increased. These effects were more pronounced with alum than ferric chloride. The North Toronto Sewage Treatment Plant experience using ferric chloride also showed a decrease in filter-cake solids and increases in sludge conditioning requirements. No decrease in filter yield was noted. Similar experiences are related by Farrell [7], Campbell [11] and others [12].

Lime-based sludges have invariably superior dewatering characteristics than metal, salt-based sludges. This is well-documented in the literature [7,13,14].

SLUDGE UTILIZATION/DISPOSAL

Sludge Characteristics

One of the factors impacting on potential sludge utilization schemes is that of sludge characteristics. Sludge characteristics are modulated not only by the type of waste treatment processes employed but are a function of the constituent inputs to municipal sewerage systems. More specifically, industrial discharges to municipal sewers may contain heavy metals, nitrogenous compounds, phosphates, a diversity of complex organic compounds, etc. In biological and physical/chemical treatment systems, most of these compounds are complexed, broken down and/or sorbed by sludge flocs. In most instances, additional chemicals such as lime and/or iron salts are added to enhance the dewatering characteristics of the sludge. Furthermore, as phosphorus removal is practiced, sludges not only contain appreciable amounts of phosphorus, the precipitant used to complex the phosphorus, but higher metal concentrations.

In sludge application to land, a number of factors must be considered. For instance, the heavy metal concentration in the sludge will dictate the total amount of sludge that can safely be applied over the lifetime of a site. The "total" metals are generally considered as an indicator of the likely ultimate effect and is used by many to calculate sludge application rates. If an "immediate" effect needs to be ascertained, then this is represented by the "available" fraction of the metal(s) as determined using suitable reagents for extraction. It is important to recognize the potential cumulative and thus long-term effect of metal addition to soil in excess of the small amounts taken up by plants and that lost due to leaching.

The nitrogen content of a sludge will dictate the annual application rate and should be consistent with use of nitrogen by agronomic crops. This will reduce the potential for nitrate pollution of groundwater.

Sludge as a Source of Pollutants

As noted earlier, the major problem in the area of sludge utilization on land concerns the content of potentially toxic substances. However, the level and nature of the substance(s) will also dictate the choice of land utilization alternative. More specifically, sludge may be used as a soil builder and/or organic fertilizer for land reclamation or application to agricultural land.

Sludge from wastewater treatment facilities practicing chemical phosphorus removal contains almost all of the metals which are discharged into sewers. In the case of heavy metals occurrence in Ontario sludges, Table VII summarizes this information from 40 Ontario water pollution control plants (10 primary, 30 secondary).

The concentrations of heavy metals in digested sludges from primary and secondary plants are similar except for chromium, which is three times lower in primary digested sludges.

In 1975, approximately 34% (53,000 tons dry weight) of the sludge produced in Ontario was applied to agricultural land. This resulted in annual heavy metal loadings as shown in Table VIII.

Table VII. Ontario Fluid Sludges–Heavy Metal Concentrations[6]

Component	Primary Plants[a] Anaerobically Digested Sludges			Secondary Plants[b] Anaerobically and Aerobically Digested Sludges		
	Range	Mean[c]	Stand. Dev. ±	Range	Mean	Stand. Dev. ±
	mg/l	mg/l	mg/l	mg/l	mg/l	mg/l
Zinc	2.8 - 130	74.3	48.3	4 - 225	55.5	57.4
Copper	4.6 - 150	54.5	46.9	7 - 148	34.8	30.2
Nickel	0.7 - 15	4.4	4.7	0.26 - 16.8	6.5	14.9
Chromium	2 - 68	16.5	20.9	2 - 430	41.6	51.7
Lead	11 - 86	40.9	30.3	3.7 - 60	21.8	25.1
Cadmium	0.2 - 2.6	0.7	0.7	0.1 - 8.7	1.4	2.0
Cobalt	< 0.6 - 1.4	1.0	0.3	0.3 - 3.6	0.8	0.8

[a]No. of plants = 10.
[b]No. of plants = 30.
[c]Arithmetic mean.

Table VIII. Annual Heavy Metal Loadings to Sludged Ontario Soils (1975 Estimate)[6]

Metal	Average mg/kg soil	Average kg/ha[a]	Total (metric tons)
Zn	3.4	6.9	76.7
Cu	2.2	4.4	48.9
Ni	0.4	0.8	8.5
Cr	2.4	4.8	53.1
Pb	1.4	2.8	31.4
Cd	0.1	0.2	1.8
Co	0.05	0.1	1.1

[a]Multiply by 0.8924 = lb/ac.

Sludge as a Fertilizer

Major plant nutrients, nitrogen, phosphorus and potassium are contained in sewage sludge. Typical concentrations are 3% nitrogen, 2.5% phosphorus and 0.5% potassium on a dry weight basis. The nutrients in sludge are at a level of one-fifth of the usual chemical fertilizers.

Sludge quality data from the Ontario sludge survey [6] were obtained for 43 water pollution control plants (10 primary, 33 secondary). Forty of these plants disposed of sludge in fluid form, two disposed of sludge cake, and one disposed of composted sludge. Table IX summarizes data on total solids (TS), volatile solids (VS), ammonia nitrogen, total Kjeldahl nitrogen (TKN) and phosphorus for digested sludges at primary and secondary plants.

In applying 296 x 10^6 gallons (53,000 dry wt tons) of digested sludge to agricultural land in Ontario, the amount of nutrients applied during one year is summarized in Table X. The data also indicate a relationship between TKN and total solids for anaerobically digested sludge from 23 secondary plants practicing metal salt addition for phosphorus removal (Figure 19). This relationship can be used to calculate the TKN loading to farmland as follows:

$$\% \text{ TKN} = 16.6 \ (\% \text{ TS})^{-0.799} \tag{6}$$

The survey data showed an ammonia nitrogen to TKN ratio varying between 12 and 57% (average 30%). The ammonia nitrogen loading to farmland can thus be approximated by using this average value.

By making a number of assumptions it is possible to estimate the potential nutrient value of sludge when applied to farmland in Ontario:

1. The 'available' nitrogen in fluid sludge is equal to the soluble nitrogen (NH_4-N) (a conservative estimate),
2. Only one-half of the total phosphorus in liquid sludge is potentially plant available [15] (a conservative estimate),
3. The 'available' potassium in fluid sludge is equal to the total potassium,
4. Based on commercial fertilizer prices (October, 1978) the prices for nitrogen, phosphorus and potassium are 23¢, 25¢ and 13¢ per lb respectively.

Using the amounts of nutrients applied to farmland as shown in Table X and the aforenoted assumptions, the sludge fertilizer value can be calculated as follows:

NH_4-N	=	810 tons x 2,200 lbs/ton x $0.23/lb	= $409,900
P	=	1,280 tons x 2,200 lbs/ton x 0.5 x $0.25/lb	= $352,000
K	=	270 tons x 2,200 lbs/ton x $0.13/lb	= $ 77,200
		TOTAL	= $839,100

This analysis shows that the fertilizer value of the sludge now applied to farmland is approximately $900,000 per annum ($30/ac). This excludes any other potential benefits such as the presence of calcium, magnesium, or the

Table IX. Ontario Fluid Sludges—Nutrient Characteristics [6]

Constituent	Primary Plants[a] Anaerobically Digested Sludge			Secondary Plants[b] Anaerobically Digested Sludge			Secondary Plants[c] Aerobically Digested or Waste Activated Sludge		
	Range	Mean	Stand. Dev. $\pm \sigma$	Range	Mean	Stand. Dev. $\pm \sigma$	Range	Mean	Stand. Dev. $\pm \sigma$
Total Solids - TS%	2.8 - 12.5	8.8	2.9	2.0 - 12	4.1	1.8	2.2 - 4.5	2.75	0.95
Volatile Solids - VS%	24 - 61	43.4	10.5	36 - 70	51	8.5	41 - 69	55.8	9.9
Ammonia Nitrogen - N mg/l	100 - 590	326	78	250 - 1200	628	245	20 - 180	110	24.8
Total Kjeldahl Nitrogen - N mg/l	950 - 2900	1736	913	1600 - 3000	2114	495	650 - 2300	1358	576
Total Phosphorus - P mg/l	240 - 2600	713	399	390 - 2900	975	603	440 - 1200	730	303

[a]No. of plants = 10.
[b]No. of plants = 25.
[c]No. of plants = 8.

Table X. Sludge Nutrients Applied Annually to Ontario Farmland (Estimate) [6]

Constituent	Amount (metric ton/yr)
TKN	2,800
NH$_4$-N	810
Total P as P	1,280
K	270
TS	53,000
VS	30,000

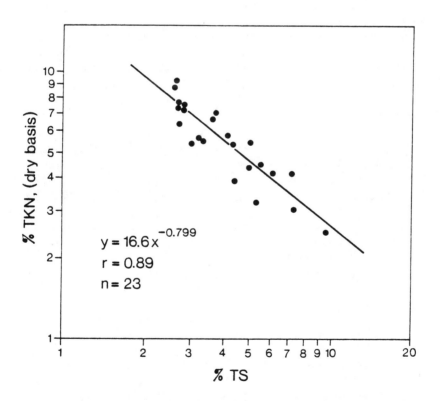

Figure 19. Total Kjeldahl nitrogen vs total solids in anaerobically digested sludge at standard C.A.S. plants [6].

considerable amount of organic matter in the sludge. If the farmer required organic matter to improve soil structure (and moisture retention capacity), sludge could have a value of approximately $20/ton. Based on the volatile solids applied to farmland, the sludge would be worth $600,000 per annum.

The farmers interviewed for the Ontario survey [6] attempted to quantify yield increases due to sludge application. The average increase in hay yield was estimated at 8 metric ton/ha and in corn yield at 1 metric ton/ha. Benefits resulting from cattle weight gain were also noted. A reasonable estimate of the benefits of sludge use on agricultural land in Ontario lies somewhere between $2,000,000 and 3,000,000 per annum [16]. At this time, the farmers receive sludge free of charge with the transportation costs charged against disposal costs, borne by the municipalities. In 1975, sludge haulage to farmland costs were approximately $2,250,000.

To date, no negative effects on crop yield were reported by farmers applying sludge for periods in excess of five years. However, long-term studies are required to assess whether heavy metals will have negative effects on plants, soil or leachate.

Field monitoring of soil, plants and leachate quality at selected sites where sludges containing high concentrations of heavy metals have been applied for extended periods would be desirable. A preliminary study in this regard is in progress at the Wastewater Technology Centre [17].

Continuing Studies

Investigations concerning sludge/soil interactions at laboratory greenhouse, field trial and lysimeter scale have been in progress since 1973 at the University of Guelph [18-22] and the Wastewater Technology Centre [23-26]. Some of the more significant conclusions from these studies are:

- Sewage sludges supplied nitrogen and phosphorus for crop production but were low in potassium. Sludges produced crop yields at least as high as were obtained with chemical fertilizers.

- Sewage sludge application did not result in marked increases in runoff of nutrients, heavy metals or bacteria on 2% and 6% slopes except when heavy rain occurred immediately after sludge application.

- Soil salinity was not a problem in the field under Ontario conditions. It might pose a problem in less humid areas. Boron levels in some sludges tested would also be expected to pose a problem in arid regions.

- Reapplication of the sludges between crop plantings did not lead to increased metal concentrations in the plant materials.

- Large amounts of metals were added to soils in some sludges, and their removal by crop uptake or leaching was very limited.

- The organic nitrogen in sludges was mineralized gradually, and the mineralization rate varied from one sludge to another. As with other sources of nitrogen, applications in excess of crop requirements lead to high levels of nitrate in the soil solution.

- The average NH_4-N content of sludges studied was 1.3% on a dry weight basis or 27% of the total nitrogen. In two experiments, 40% and 48% of the NH_4^+-N was lost by volatilization from sewage sludge applied to the soil surface. This loss occurred in five and eight days, respectively.

- Salmonellae were isolated from five of the 207 sludge samples tested. If vegetables are not grown and animals not grazed immediately following sludge application, and if reasonable care is exercised in spreading, sludge does not pose a serious health hazard.

- At least twice as much nitrogen must be applied in fluid sludge as in commercial fertilizer to obtain equivalent yields.

- The Cd, Cr, Cu, Ni, Pb and Zn concentrations in orchard grass and wheat plant materials have not exceeded suggested maximum "tolerance" or "toxic" levels.

- The maximum concentrations of Cd, Cr, Cu, Ni, Pb and Zn in leachates have not exceeded drinking water standards.

- Soluble P in the leachates from fluid sludge treatments ranged as high as 10 mg/l during summer 1977. Soluble P from the air-dried sludge treatments never exceeded 2 mg/l.

- Total organic carbon in leachates from both the fluid and air-dried sludge experiments was greater than 50 mg/l in 1976 at the highest sludge loading rate. Static bioassay toxicity tests using *Daphnia* showed no toxicity in these leachate samples.

Ontario Guidelines for Sludge Utilization on Agricultural Land

The amount of NH_4-N applied to Ontario farmland in 1975 was 810 metric ton/yr (Table X). Combining this information with the data on heavy metal application to farmland (Table 8) allows for an assessment as to whether or not Ontario sludge is generally suitable for land application if the criteria of the Provisional Guidelines for Sewage Sludge Utilization on Agricultural Land [27] are applied. This assessment is summarized in Table 11 and shows that the sludge is generally suitable for agricultural land application. The exception is the cadmium content. Sources of high cadmium content sludges are few and isolated. It is important to stress that average values are extremely misleading, and that it is imperative that each sludge source be characterized separately in order to determine its limits of suitability for application to farmland.

Table XI. Suitability of Ontario Sludge for Utilization on Farmland [6]

Constituent[a]			NH$_4$-N: Heavy Metal		
NH$_4$-N	Heavy Metal		Actual Ratio	Maximum Required Ratio[b]	Suitability
810					
	Zn	76.7	11	4	Yes
	Cu	48.9	17	10	Yes
	Ni	8.5	95	40	Yes
	Cr	53.1	15	15	Yes
	Pb	31.4	26	15	Yes
	Cd	1.8	450	500	No
	Co	1.1	736	50	Yes

[a]In metric tons/yr (Tables VIII and X).

[b]Provisional Guidelines for Sewage Sludge Utilization on Agricultural Land [27].

CLOSING REMARKS

The information presented represents a summary of Canada's exeprience in the Province of Ontario with increased sludge production due to chemical removal of phosphorus to 1.0 mg/l total phosphorus when using metal salts.

Ontario's current sludge management strategy consists of applying the most cost-effective and environmentally acceptable solution. Sludge utilization for its nutrient value on agricultural aland is one such management strategy followed by an increasing number of municipalities who, as well as the farmers, are concerned about potentially long-term harmful impacts on soils due to heavy metal addition.

While technological solutions to phosphorus point source control to 0.1 mg/l are available, the impact on sludge quantities generated, handling and disposal still remains to be more closely defined. Only with information on relative costs between alternatives to achieve these goals can an effective point source phosphorus control management strategy be proposed.

Computer simulation is one approach to assess potential management strategies. It may well turn out that point source control to levels substantially lower than currently practiced will cause more problems elsewhere. The current activities of the IJC Phosphorus Management Strategies Task Force address this subject.

REFERENCES

1. Jank, B. E., and P. H. M. Guo, "Biological Treatment of Meat and Poultry Wastewater" presented at Meat Technology Transfer Seminar on Meat and Poultry Industry Regulations and Guidelines, February, 1978.
2. Kormanik, R. A., "Estimating Solids Production for Sludge Handling", Water and Sewage Works, December, 1972.
3. Campbell, H. W., R. J. Rush and R. Tew, "Sludge Dewatering Design Manual", COA Research Report No. 72, January, 1978.
4. Sutton, P. M., K. L. Murphy and B. E. Jank, "Nitrification Systems with Integrated Phosphorus Precipitation", presented at PCAO Conference, Toronto, April, 1977 and B. C. Water and Waste Association Conference.
5. Stepko, W. E., and D. T. Vachon, "Phosphorus Removal Demonstration Studies Using Lime, Alum and Ferric Chloride at C. F. B. Borden", Environmental Protection Service, EPS 4-WP-78-2, Water Pollution Control Directorate, Ottawa, February, 1978.
6. Antonic, M., M. F., Hamoda, D. B. Cohen and N. W. Schmidtke, "A Survey of Ontario Sludge Disposal Practices", Project No. 74-3-19, COA Research Report (in press).
7. Farrell, J. B., "Design Information on Dewatering Properties of Wastewater Sludges", Sludge Handling and Disposal Seminar, COA Conference Proceedings No. 2, Toronto, 1974.
8. Brouzes, R. J. P., "The Use of Lime in the Treatment of Municipal Wastewaters", COA Research Report No. 21, Ottawa.
9. Prested, B. P., E. E. Shannon and R. J. Rush, "Development of Prediction Models for Chemical Phosphorus Removal, Volume II", COA Research Report 78, June, 1978.
10. Drynan, W. R., "Relative Costs of Achieving Various Levels of Phosphorus Control at Municipal Wastewater Treatment Plants in the Great Lakes Basin", Technical Report to the International Reference Group on Great Lakes Pollution from Land Use Activities of the International Joint Commission, July, 1978.
11. Campbell, H. W., R. Tew and B. P. Le Clair, "Some Aspects of Chemical Sludge Thickening and Dewatering", presented at the Alternatives for Nutrient Control Seminar, Kelowna (1975).
12. Moss, W. H., R. E. Schade, S. J. Sebesta, K. A. Scheutzew, P. V. Beck and D. B. Gerson, "Full-scale Use of Physical/Chemical Treatment of Domestic Wastewater at Rocky River, Ohio", J. Water Poll. Cont. Fed., 49, November, 1977.
13. Stickney, R., and B. P. Le Clair, "The Use of Physicochemical Sludge Characteristics and Bench Dewatering Tests in Predicting the Efficiency of Thickening and Dewatering Processes", Sludge Handling and Disposal Seminar, COA Conference Proceedings No. 2, Toronto, 1974.

14. EPA, "Process Design Manual for Sludge Treatment and Disposal", U.S. EPA Technology Transfer Report, EPA 625/1-74-006, October, 1974.
15. U.K. Department of the Environment, National Water Council, "Report of the Working Party on the Disposal of Sewage Sludge to Land", Standing Technical Committee Report Number 5, London, England.
16. Schmidtke, N. W., and D. B. Cohen, "Municipal Sludge Disposal on Land, A Down-to-Earth Solution",, presented at the Western Canada Water and Sewage Conference, Edmonton, Alberta, September 28-30, 1977.
17. Monteith, H., D. N. Bryant and M. D. Webber, "Assessment of PCB's and Heavy Metals at Selected Disposal Sites in Ontario", Project No. 76-3-26, COA Research Report (in preparation).
18. Bates, T. E., E. G. Beauchamp, R. A. Johnston, J. W. Ketcheson, R. Protz and Y. K. Soon, "Land Disposal of Sewage Sludge - Volume I", COA Research Report No. 16, Ottawa, 1975.
19. Bates, T. E., E. G. Beauchamp, R. A. Johnston, J. W. Ketcheson, R. Protz and Y. K. Soon, "Land Disposal of Sewage Sludge - Volume II", COA Research Report No. 24, Ottawa, 1975.
20. Bates, T. E., E. G. Beauchamp, R. A. Johnston, J. W. Ketcheson, R. Protz and Y. K. Soon, "Land Disposal of Sewage Sludge - Volume III", COA Research Report No. 35, Ottawa, 1976.
21. Bates, T. E., E. G., Beauchamp, R. A. Johnston, J. W. Ketcheson, R. Protz and Y. K. Soon, "Land Disposal of Sewage Sludge - Volume IV", COA research Report No. 60, Ottawa, 1975.
22. Bates, T. E., E. G., Beauchamp, R. A. Johnston, J. W. Ketcheson, R. Protz and Y. K. Soon, "Land Disposal of Sewage Sludge - Volume V", COA Research Report No. 73, Ottawa, 1978.
23. Chawla, V. K., J. P. Stephenson and D. Liu, "Biochemical Characteristics of Digested Chemical Sewage Sludges", COA Conference Proceedings No. 2, 1974.
24. Chawla, V. k., D. N. Bryant, D. Liu and D. B. Cohen, "Chemical Sewage Sludge Disposal on Land - (Lysimeter Studies) - Volume I", COA Research Report No. 67, Ottawa, 1977.
25. Cohen, D. B., M. D. Webber and D. N. Bryant, "Land Application of Chemical Sewage Sludge (Lysimeter Studies)", Sludge Utilization and Disposal Seminar, Toronto, Ontario, 1978.
26. Cohen, D. B. and D. N. Bryant, "Chemical Sewage Sludge Disposal on Land - (Lysimeter Studies) - Volume II", COA Research Report No. 79, Ottawa, 1978.
27. OMAF (Ontario Ministry of Agriculture and Food) and OMOE (Ontario Ministry of the Environment) Ad Hoc Joint Committee to prepare "Guidelines for Sewage Sludge Utilization on Agricultural Lands", June, 1976.

CHAPTER 16

WATER QUALITY AND WASTEWATER TREATMENT CONSIDERATIONS FOR DETERGENT SUBSTITUTES

E. E. Shannon

CH2M Hill, Inc.
Gainesville, Florida

INTRODUCTION

Laundry detergents, which replaced soaps about 20 yr ago, have been considered major contributors to water pollution. For example, early detergent formulations contained nonbiodegradable surfactants (ABS) that caused considerable foaming in receiving waters. The replacement of "hard" surfactants with biodegradable surfactants has essentially eliminated the problem. Recent investigations into the processes of lake eutrophication have shown that phosphorus is the nutrient most frequently limiting to aquatic plant production. The most expedient and economical eutrophication abatement action involves the removal of phosphorus from the wastewater effluents entering the lake (for example, as provided for in the Canada-U.S. Agreement on Great Lakes Water Quality). Phosphorus can be removed by chemical treatment (iron or aluminum salt or lime additions to raw or treated wastewater and appropriate plant modifications). Prior to major reformulation efforts by the detergent industry, domestic phosphorus levels could be reduced drastically (25-50%) by reducing or replacing the phosphate builder material of laundry detergents. However, in the past several years, the phosphate content in detergents has declined so that further reductions in phosphate builder might be anticipated to reduce wastewater phosphorus levels by 10% to 25%. The reduction-replacement of phosphate builders in detergents has been practiced by a number of countries, including Canada

391

and sections of the U.S. In Canada, legislation was passed to limit the phosphate content of laundry detergents to less than 5% as P_2O_5 as of January 1973.

The limiting of phosphate builder material in laundry detergents has been a matter of financial and technological concern to the detergent manufacturing industry. Builder alternatives presently being used to various extents include the sodium salt of nitrilotriacetic acid (NTA), sodium carbonate and sodium citrate. More recently, aluminosilicate builders such as sodium zeolite, Type A, have seen some limited usage, and organic materials such as carboxymethyloxysuccinate (CMOS) and Monsanto's Builder-M have shown promise as substitute builders. NTA is presently being used at the prime builder substitute in Canada, with the weighted average detergent content being in the order of 15% as sodium-NTA. The weighted average of phosphorus in Canadian formulations is about 1.9% as P. On the U.S. side of the Great Lakes, the average phosphorus content in detergents is about 6%. The major substitute builders in zero phosphate areas are carbonate formulations, with the zeolites having some recent impact. A thorough evaluation of the environmental implications of NTA by an IJC task force has resulted in a conclusion that NTA does not constitute an obvious environmental hazard. The detergent industry is optimistic that the Office of Toxic Substances will complete their review in the near future and reach a similar conclusion. Conceivably NTA could then enter the U.S. market as a viable substitute for phosphate builders.

BACKGROUND

Detergent builders can have a definite effect on domestic wastewater characteristics and consequently a potential effect on the eventual wastewater treatment processes and the receiving water. At the Wastewater Technology Center (WTC) of the Environmental Protection Service, Burlington, Ontario, wastewater detergents effects were under investigation as early as 1971. The subsequent program continued through 1974, covering many phases. This chapter summarizes and highlights the findings of these studies. All of the investigations have been detailed in separate reports. For specific information on procedures, these reports should be referenced. Selected results from the following projects are prescribed herein.

1. "Detergent Substitution Studies at CFS Gloucester" [1].
2. "Development of Prediction Models for Chemical Phosphorus Removal" [2].
3. "Effect of Citrate- and Carbonate-Based Detergents on Wastewater Characteristics and Treatment" [3].

4. "Activated Sludge Degradation of Nitrilotriacetic Acid (NTA)-Metal Complexes [4].
5. "A Study of NTA Degradation in a Receiving Stream" [5].
6. "Impact of Nitrilotriacetic Acid (NTA) on an Activated Sludge Plant" [6].

In addition, results from other relevant builder investigations are discussed.

DETERGENT SUBSTITUTIONS AT CFS GLOUCESTER

Detergent substitutions were carried out at a Canadian Forces Station (CFS) Gloucester, a training station that is located approximately 15 miles (24 km) southeast of Ottawa. The station has accommodation, cafeteria, training and recreational facilities for 150 enlisted men, and permanent married quarters. A schematic diagram of the station is shown in Figure 1.

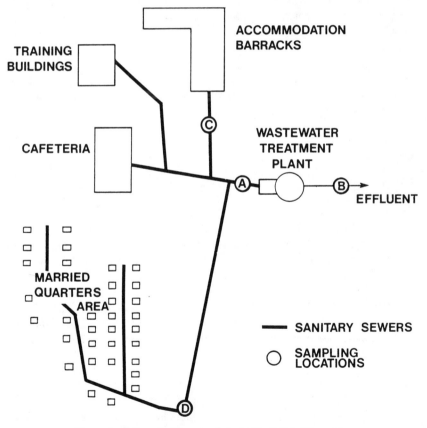

Figure 1. Schematic diagram of study site C.F.S. Gloucester.

Wastewater from the barracks, permanent married quarters, cafeteria and other areas is collected by a sanitary sewer system and conveyed to the wastewater treatment plant at the eastern side of the station (Figure 1). The average daily wastewater flow during the substitution periods was 1.64×10^5 liters. The treatment facility is an activated sludge plant equipped with mechanical aeration designed for a daily flow of 1.09×10^5 liters and an average biochemical oxygen demand (BOD) of 200 mg/l.

Five separate detergent conditions were evaluated. The first condition involved an assessment under baseline; that is, all consumers using detergents of free choice, which were at the time restricted to less than 20% P_2O_5. The four subsequent conditions involved replacement of all laundry detergents with NTA, carbonate, and phosphate- and citrate-based products, respectively. During the citrate substitution, automatic dishwashing detergents were also replaced with a citrate-based product. All detergents were products commercially available in Canada or the U.S. Their compositions and recommended usage levels are summarized in Table I. Each substitution ran for three weeks, detailed sampling being conducted during the latter two weeks of each period.

Table I. Composition of Detergents Used in the Substitution Study

Builder	Builder Content (%)	Recommended Amounts (cups/load)
NTA	19.7 as H_3NTA	1.0
Carbonate	66.9 as Na_2CO_3	0.5
Phosphate	12.1 as phosphorus	1.0
Citrate	21.0 as sodium citrate	0.5 (liquid)

One of the objectives of this study was to assess the relationships between detergent formulations and chemical phosphate removal. For this investigation, standard jar testing procedures were used with three different phosphorus removal coagulants at predetermined dosages: lime, 100-500 mg/l as $Ca(OH)_2$; ferric chloride; 5-30 mg/l as Fe; and aluminum sulfate, 5-30 mg/l as Al. A series of jar tests were conducted on grab samples from the influent and effluent streams of the wastewater treatment plant during the last six days of each substitution period.

Wastewater Effects

Average data for some of the significant changes are presented in histogram form on Figure 2. The wastewater demonstrated a marked pH and

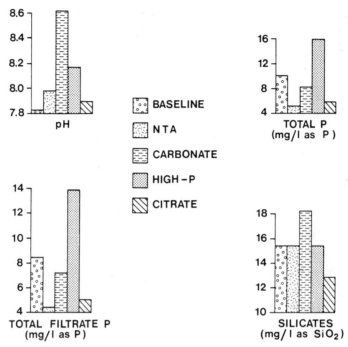

Figure 2. Comparison of married quarters wastewater characteristics for five substitution periods.

alkalinity rise during the carbonate substitution. This was not unexpected since the additional load provided by the detergent was estimated to increase the natural alkalinity of the water supply by 10% to 15%. As anticipated, the use of the phosphate-free NTA-, carbonate- and citrate-based detergents resulted in much lower total and filtrate phosphorus levels in the wastewater. The carbonate detergent caused a significant increase in the silica content of the wastewater. This had been predicted because the sodium silicate content was 7% to 8% by weight compared with that of the other detergents, which averaged 5% to 6%.

Pronounced and interesting diurnal variations in wastewater characteristics were evident during the substitutions. The effect of laundering activities on the diurnal fluctuation of total phosphorus is shown in Figure 3. Under phosphate and baseline conditions (most detergents being used were phosphate-based, although limited to less than 20% as P_2O_5), daytime total phosphorus concentrations were 2-3 times higher than during the periods of inactivity. When a phosphate-free detergent was used, these fluctuations were reduced. The variations in condensed (total filtrate minus ortho) phosphorus (Figure 4) illustrate this trend even more dramatically. Thus, not only does

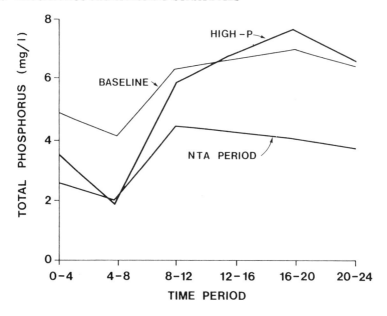

Figure 3. Diurnal variation of total phosphorus in raw wastewater.

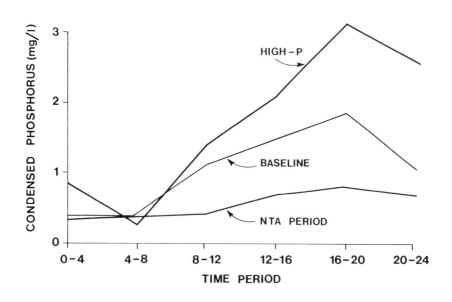

Figure 4. Diurnal variation of condensed phosphorus in raw wastewater.

the substitution of a phosphate-free detergent reduce the quantity of phosphorus in the wastewater, it also reduces the amplitude of its diurnal fluctuations. On the other hand, builder substitution can have a significant effect on other wastewater characteristics, as is demonstrated by the sodium patterns in Figure 5. Sodium concentrations rose from a background level of

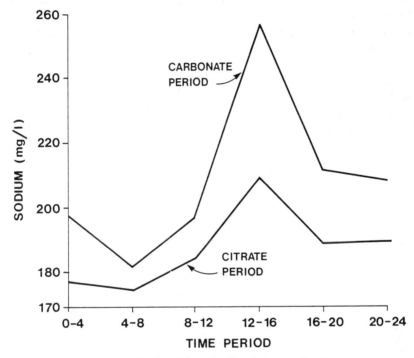

Figure 5. Diurnal variation of sodium in raw wastewater.

175 to 185 mg/l during the nighttime hours to a daytime peak of 260 mg/l for the carbonate period and 210 mg/l for the citrate period. Diurnal fluctuations during the other periods were similar, the relative amplitude of the peak being in agreement with the sodium content of the detergent (that is, carbonate > phosphate ≅ NTA ≅ baseline > citrate). Citrate and NTA wastewater levels, during their respective substitution periods, demonstrated distinct dirunal trends similar to the condensed phosphorus trends of the baseline and phosphate periods.

The per capita (equivalent station population = 293) total phosphorus loadings for each period were baseline, 2.79 g/day; NTA, 2.07 g/day; carbonate, 2.17 g/day; high-phosphate, 3.03 g/day; and citrate, 1.98 g/day. The per capita figure from the high-phosphate study compares favorably with the loading range (3.04 to 4.35 g/day) reported by Zanoni and Rutkowski [7] for domestic wastewater. Their figures were derived by monitoring the wastewater from a residential area of Milwaukee in which high-phosphate detergents were used for laundry purposes. The average per capita contribution for the three phosphate-free substitutions was 2.07 g/day. This represented a reduction of 25% from the baseline contribution. In a later study carried out at a larger Canadian forces base (Uplands), Stepko and Shannon [8] observed the impact of the Canadian detergent reformulation program on the reduction of wastewater phosphorus loadings. The average diurnal variations in influent wastewater phosphorus before reformulation (all detergents $< 20\%$ P_2O_5) are shown in Figure 6. The corresponding per capita contribution reduction was from 5.2 to 2.3 g/day, or 56%

Treatability study data collected for several Ontario municipalities presented in Table II indicated that reformulation has caused an average reduction of 20% in raw wastewater phosphorus levels and 24% in secondary effluent phosphorus levels.

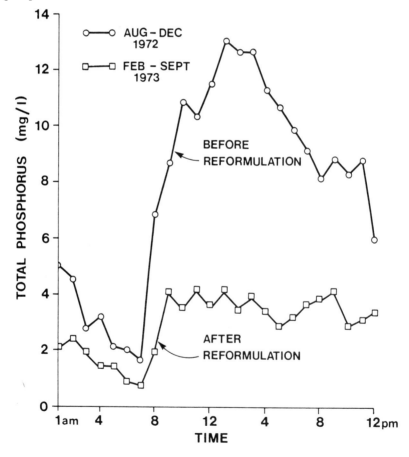

Figure 6. Diurnal variation of total phosphorus in raw wastewater at C.F.B. Uplands.

Table II. Impact of Detergent Reformulation on Wastewater Phosphorus Levels [2]

	Before Reformulation (Prior to January 1973)[a]	After Reformulation (1973 to Date)[a]	Percent Reduction
Raw wastewater total phosphorus (mg/l as P)	7.1 (n[b] = 38)	5.7 (n = 35)	20
Secondary effluent total phosphorus (mg/l as P)	4.5 (n = 31)	3.4 (n = 25)	24

[a]Average value.

[b]n = number of observations.

Detergents and Phosphorus Removal

The jar test data collected during the substitution studies indicated that the presence of NTA or citrate at levels up to 15 mg/l did not interfere with phosphorus removal by lime, ferric chloride, or alum. It has been postulated that NTA could interfere with phosphorus removal by either complexing with the active ion (Fe, Al, Ca) or forming mixed ligand (NTA-phosphate-metal) complexes.

The reduction in total phosphorus levels brought on by the use of phosphate-free detergents (NTA, citrate, and carbonate) actually resulted in lower chemical requirements for phosphorus removal to a level of 1 mg/l. This reduction in coagulant requirements far outweighed any interference effect that the builder materials may have had on the phosphorus removal process. The relationships for alum addition to raw wastewater and to secondary effluent are shown in Figures 7 and 8, respectively. Similar but not as pronounced relationships were obtained for ferric chloride. As might be expected, initial phosphorus removals did not influence phosphorus removal by lime.

The regression equations given in Figures 7 and 8 were used to estimate alum requirements for phosphorus removal for the wastewater when phosphate-free detergents were being used (that is, average total phosphorus concentrations during the carbonate, NTA, and citrate periods were substituted into the regression equations to estimate an average alum requirement) and the wastewater when phosphate-based detergents were being used (that is, average total phosphorus concentrations during the baseline and high-phosphate periods were substituted in the equations to compute an average

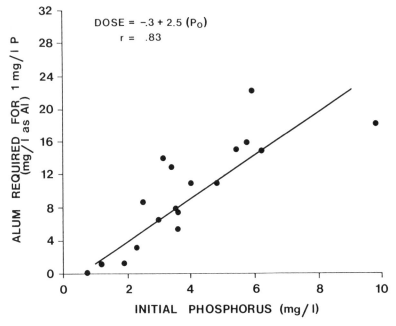

Figure 7. Relationship of alum dosage and initial phosphorus level–raw wastewater.

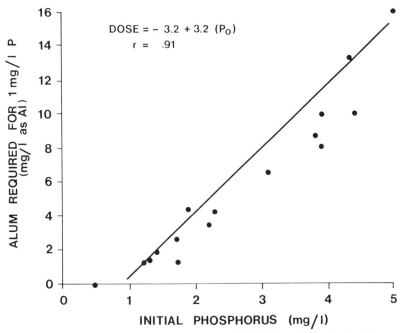

Figure 8. Relationship of alum dosage and initial phosphorus level–effluent.

alum requirement). For removal down to a 1-mg/l phosphorus residual in the raw wastewater, phosphate-free alum requirements were 9.2 mg/l as Al as compared with 13.4 mg/l for the high-phosphate conditions, or an overall reduction in alum requirement of 31%. Similarly, for phosphorus removal in the effluent, alum requirements for the phosphate-free and high-phosphate periods were, respectively, 7.5 mg/l and 12.6 mg/l, or a reduction in alum requirement of 37%. Identical calculations with regression equations derived for ferric chloride yielded chemical requirements for raw wastewater of 18.9 and 15.4 mg/l as Fe for phosphate detergent and phosphate-free detergents, respectively, or an overall reduction of 19%; for the treated effluent, the requirements were 18.0 and 12.1 mg/l, respectively, or reduction of 32% in ferric chloride requirements.

Jar test data collected as part of the Ontario treatability program for a number of different municipalities (Table III) confirmed the reductions observed at Gloucester.

Table III. Average Jar Test Dosages of Alum or Ferric Chloride Required to Achieve a 1-mg/l Total Phosphorus Residual Prior to 1973 and After 1973 [2]

	Raw Wastewater Addition			Mixed Liquor Addition		
Chemical	Pre-1973	1973-Present	Reduction (%)	Pre-1973	1973-Present	Reduction (%)
Ferric chloride (mg/l) as Fe^{3+})	30.5 ($n^a = 37$)	17.2 ($n = 36$)	44	21.5 ($n = 31$)	12.8 ($n = 26$)	41
Alum (mg/l as Al^{3+}	19.4 ($n = 38$)	13.1 ($n = 36$)	32	12.8 ($n = 31$)	6.9 ($n = 27$)	46

$^a n$ = number of observations.

DEGRADATION OF NTA-METAL COMPLEXES

Mixed ligand complexes of the form NTA-metal-phosphorus were prepared in 1:1:1 molar ratios. Stock solutions of these complexes were made up for the following metals: Al^{3+}, CD^{2+}, Ca^{2+}, Cr^{3+}, Cu^{2+}, Fe^{3+}, Hg^{2+}, Ni^{2+}, Pb^{2+} and Zn^{2+}. Batch activated sludge reactors were acclimated to each NTA-metal complex and a series of degradation studies were carried out to determine first order degradation coefficients.

The degradation constants obtained from this study are summarized in Table IV. Typical results for selected NTA-metal complexes are shown in Figures 9 and 10.

Table IV. First-Order Degradation Constants k (hr^{-1}) for Several NTA-Metal Complexes

| NTA-Complex with | Degradation Constants[a] | | | |
	Temperature 5° C	Number of Experiments	Temperature 10-15° C	Number of Experiments
Ca^{2+}	-0.067	6	-0.100	3
Fe^{3+}	-0.066	6	-0.102	3
Al^{3+}	-0.057	6	-0.075	7
Pb^{2+}	-0.072	5	-0.099	3
Cr^{3+}	-0.040	5	-0.068	3
Cu^{2+}	-0.025	4	-0.072	9
Zn^{2+}	-0.051	5	-0.055	6
Cd^{2+}	-0.008	4	-0.020	1
Ni^{2+}	-0.005	5	-0.021	2
Hg^{2+}	0	1	0	2

[a]Average value.

The NTA-metal complexes tested fall into two distinct groups.

1. Readily degradable: NTA-Ca, –Fe, –Al, –Pb, –Cr, –Cu and –Zn.
2. Resistant to biodegradation: NTA-Cd, –Ni and –Hg.

These results are in agreement with other studies which have reported limited degradation for NTA-Ni, -Cd and -Hg complexes.

As expected, the degradation rate constants were strongly influenced by temperature, i.e., considerably lower at 5°C than at 15°C. The few experiments conducted at 10°C indicated little difference from the degradation rates at 15°C. Again these data are consistent with the results of other studies in which reduced NTA degradation at wastewater temperatures of less than 10°C was observed.

Application of the rate constant data of Table IV to a conventional activated sludge plant operating under conditions of:

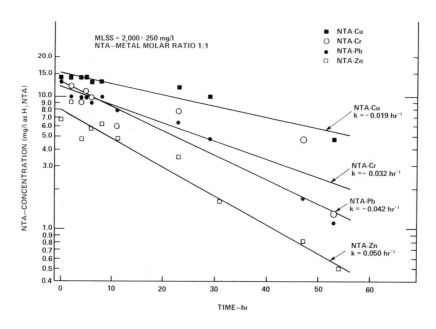

Figure 9. Typical degration of NTA-Cu, –Cr, –Pb, –Zn, complexes in activated sludge at 5°C.

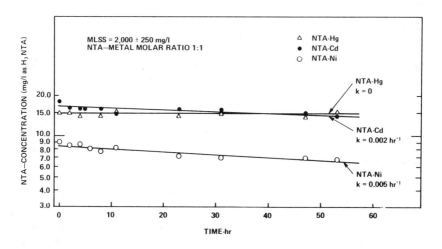

Figure 10. Typical degration of NTA-Hg, -Zn, -Ni, complexes in activated sludge at 5°C.

1. Wastewater temperatures of $10°$ to $15°C$;
2. Initial NTA levels of 10 mg/l as H_3NTA; and
3. Aeration tank residence time of 8 hr.

and assuming that all of the NTA present is complexed with the particular metal of interest suggests that overall NTA removal efficiencies for the readily degradable complexes would be in the order of 50%. Removal efficiencies for the bioresistant complexes would approximate 10%. For wastewater temperatures of $5°C$, NTA removal efficiencies for the two metal groups would drop to approximately 30% and 5%, respectively. The full-scale NTA degradation studies carried out at the Waterdown plant concurrent with these experiments [6] demonstrated higher NTA removal efficiencies, i.e., in excess of 80% for wastewater temperatures greater than $10°C$ and 40%-50% for temperatures between $8°$ and $10°C$. Although effluent NTA concentrations from treatment plants will definitely be higher during the winter months when wastewater temperatures fall below $10°C$, the fact that NTA degradation occurs in receiving streams at water temperatures as low as $2°C$ [5], and that receiving water temperatures are generally in excess of $10°C$ for several months of the year indicates that a buildup of NTA in the aquatic environment is extremely unlikely.

The results of this study show that the NTA-Al, -Fe and -Ca complexes degrade readily. Consequently, the probable interference of NTA with chemical phosphorus removal systems would be minimal. This was confirmed in the Waterdown full-scale studies by Wei et al. [6]. In the case of calcium, concentrations of calcium present in the wastewater in excess of a 1:1 NTA: metal ratio appeared to increase the degradation rate. A similar effect for the degradation of NTA-Cd and -Cu complexes has been observed by Walker [9] for wastewaters having iron levels in excess of 1:1 NTA:iron.

From this study, it would appear the NTA could drastically affect Cd, Ni and Hg transport through treatment plants because these complexes were poorly degraded. However, actual conditions which prevail in wastewaters and wastewater treatment plants reduce this possibility because Cd, Hg and Ni levels in municipal wastewaters are usually very low when compared to other metals such as Cr, Zn and Pb. For example, recent analyses by Atkins and Hawley [10] have shown that even in a heavily industrialized community such as Hamilton, Ontario, the Cd, Ni, Cr, Zn, and Pb levels in that waste-water average no more than 0.02, 0.44, 1.2, 8.1 and 1.2 mg/l, respectively. Mercury concentrations are expected to be in the same order of magnitude as cadmium. Thus, the actual levels of Cd and Hg complexes that could exist in wastewaters are very low. Nickel complexes may occur at slightly higher levels, but the majority of the NTA would probably be complexed with other dominant cations such as Ca, Fe and Al.

This study showed that NTA-metal complexes with Ca, Fe, Al, Pb, Cr, Cu, and Zn are readily biodegradable. The poor biodegradability of NTA-metal complexes with Cd, Ni and Hg appears to be a problem of little practical significance.

EFFECT OF NTA ON CHEMICAL PRECIPITATION OF PHOSPHORUS

Jar testing with raw wastewater and secondary effluents and supplemental additions of NTA up to 10 mg/l demonstrated that NTA had no effect on chemical precipitation of phosphorus when alum or ferric chloride was used. These results were confirmed during subsequent full-scale testing at the Waterdown activated sludge plant.

NTA DEGRADATION

The full-scale Waterdown investigations confirmed that NTA is biodegradable and can be removed by conventional activated sludge systems. Typical results illustrating the effect of temperatures upon the removal are shown in Figure 11.

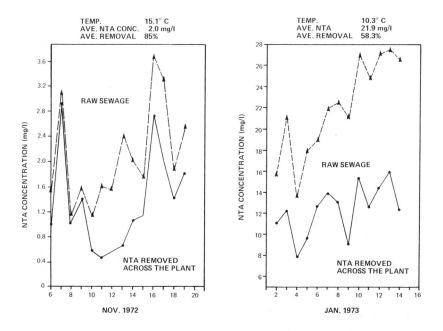

Figure 11. NTA degradation.

NTA AND METAL TRANSPORT

In the Waterdown study, the net effect of chelate formation and their biodegradation was assessed indirectly by comparing heavy metal removals efficiencies for periods of different NTA dosages. Figure 12 shows the

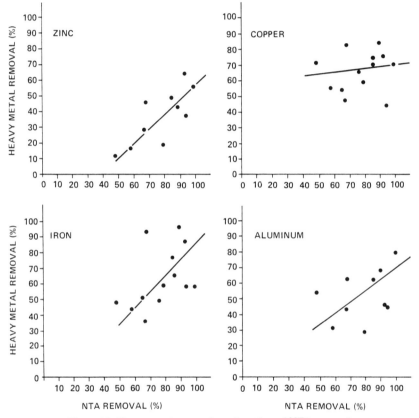

Figure 12. Heavy metal removal as a function of NTA removal.

relationships between heavy metal removal efficiencies and percent NTA removal. The heavy metals examined included Zn, Cu, Fe, and Al. Except for Cu, the general trend is that higher NTA removal results in higher heavy metal removal. This is not unexpected since a higher NTA removal efficiency means less NTA is available to form chelates with the heavy metals.

DEGRADATION OF NTA IN A RECEIVING STREAM

Grindstone Creek, the receiving stream for the Waterdown treatment plant, was monitored during the NTA investigations at the plant. The locations of the sampling stations are shown in Figure 13. In addition, laboratory degradation studies were conducted to develop reaction rate data.

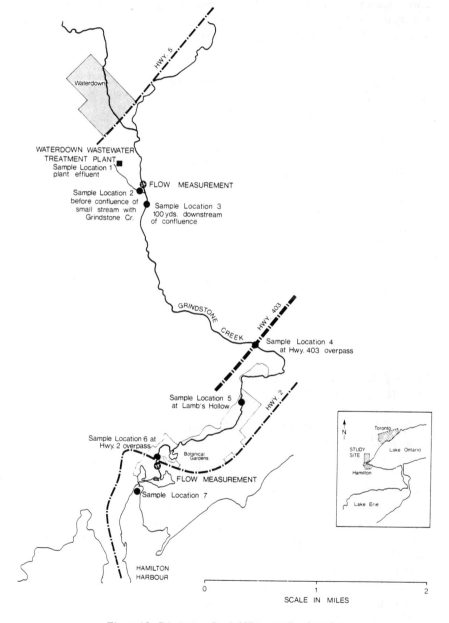

Figure 13. Grindstone Creek NTA sampling locations.

During the summer sampling period, NTA levels at all monitoring points in the stream were less than 10 mg/l. However, higher levels were observed during the winter sampling when the stream temperature fell to less than 3°C. The winter NTA profiles are shown in Figure 14. The laboratory degradation data for 18° C and 2° C are presented in Figure 15.

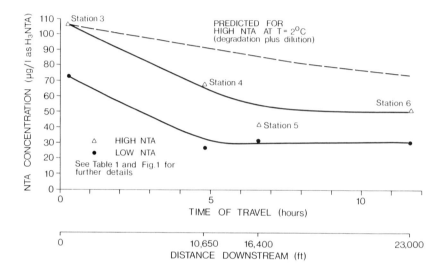

Figure 14. NTA profiles in Grindstone Creek during winter.

EFFECT OF CITRATE DETERGENT ON
CHEMICAL PHOSPHORUS REMOVAL

Typical effects of citrate detergent additions (expressed in mg/l as H_3Cit) on alum phosphorus removal jar test results using degritted raw sewage are shown in Figure 16. A similar trend was observed for ferric chloride and, to a lesser extent, for lime. A definite trend toward reduced phosphorus removal was observed as the citrate level increased above 10 mg/l.

Higher soluble iron/aluminum levels at increased citrate detergent addition levels, and a similar effect in tests using pure sodium citrate additions showed that this interference with phosphorus removal was due to complexing of the precipitant cation (Fe^{3+}, Al^{3+}, Ca^{2+}) by the citrate and not due to the surfactant interfering with the floc settling properties.

The Gloucester citrate substitution studies by Shannon and Kemp [1] showed that wastewater citrate levels averaged 7 mg/l (as H_3Cit) when citrate-based detergents were the only detergents used. This is in the lower range of

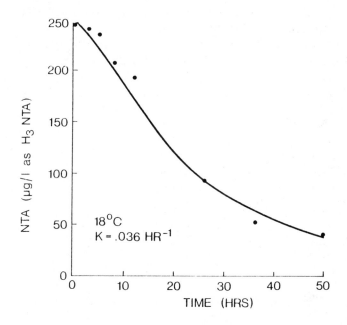

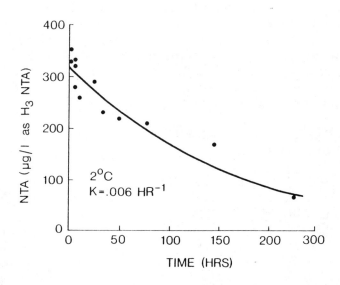

Figure 15. Degration of NTA in accumatized Grindstone Creek water at 18 and 2°C.

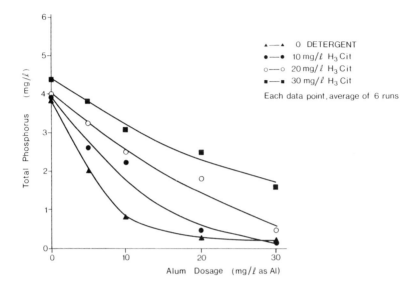

Figure 16. Effect of citrate-based detergent on phosphorus removal by alum (jar tests).

the levels investigated here. However, the jar test data suggest that if all detergents now in use were to be replaced by citrate-based detergents (20% H_3Cit by weight), the ferric chloride, alum and lime requirements to achieve a given level of phosphorus removal (i.e., 1 mg/l residual P) in primary treatment systems may be expected to increase. Subsequent citrate addition and phosphorus removal investigations at Waterdown demonstrated that on a full-scale basis, citrate did not have a quantifiable effect on phosphorus removal with alum or ferric chloride.

EFFECTS OF CARBONATE DETERGENT
ON PHOSPHORUS REMOVAL

Physical-chemical pilot plant experiments [131 m^3/day (20 gpm)] were carried out to determine the effects of carbonate detergent additions upon: (1) a low-lime [240 mg/l as Ca (OH)$_2$], and (2) a high-lime [600 mg/l as Ca (OH)$_2$] phosphorus removal system. Results for the high-lime system are summarized in Table V. The high lime phosphorus removal performance did not appear to be significantly affected at any of the carbonate addition levels. At least 90% removal was achieved in all cases. Thus, full-scale usage of carbonate detergents which increase the raw wastewater alkalinity by 10% to 20% should have no significant effect on lime phosphorus removal systems. A survey of 20 Ontario municipal wastewaters by Prested et al. [2] found

Table V. Effects of Carbonate Detergent Additions on a High-Lime
($\cong$ 600 mg/l) Phosphorus Removal System

| Parameter (mg/l) | Detergent Addition Level Percent Increase in Total Alkalinity | | | | | | | |
| | 0 | | 10 | | 25 | | 50 | |
	Inf.	Eff.	Inf.	Eff.	Inf.	Eff.	Inf.	Eff.
Total P	4.2	0.3	4.7	0.2	5.1	0.5	5.3	0.4
Filtered P	2.3	0.1	2.5	0.1	2.5	0.1	2.1	0.2
Hardness (as CaCo$_3$)	314	348	301	373	281	288	258	239
Alkalinity (as CaCo$_3$)	238	292	236	342	239	307	234	330
Dissolved silica	4.7	3.8	3.5	3.7	3.7	4.4	2.6	3.2
Sodium	99	79	92	100	105	130	85	121

that alkalinitios varied from 100 to 425 mg/l as CaCO$_3$, with an average of 240 mg/l as CaCO$_3$. Consequently, the majority of wastewaters would probably respond to carbonate detergents in a manner similar to these pilot studies.

CITRATE DEGRADATION

The full-scale activated sludge degradation of citrate averaged in excess of 90% removal over a 6-month period which included fall and winter operation. Citrate degradation was decreased at wastewater temperatures less than 11°C, as shown in Figure 17. These results are similar to those obtained for NTA. A survey of baseline citrate levels in several Ontario wastewater treatment plants (Figure 18) indicated that citrate was reduced approximately 52% by primary plants and 92% by secondary plants.

ZEOLITE TREATABILITY

A synthetic aluminosilicate, Type A zeolite, a potential partial replacement for phosphate builders, has been subjected to bench-scale and full-scale wastewater treatability investigations [10]. In the full-scale investigations, a zeolite laundry detergent formulation (20% by weight of zeolite) was supplied to families in an 81-home subdivision. The performance of the activated sludge plant serving the subdivision was monitored under the substitution condition and compared to baseline conditions (without zeolite detergents).

Zeolite removals of 80% to 90% were observed at the treatment plant with zeolite tending to accumulate in the mixed liquor solids of the plant. This

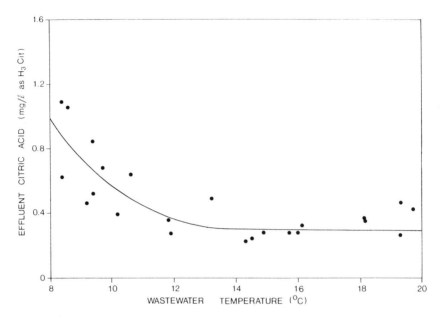

Figure 17. Relationship between effluent citric acid levels from an activated sludge plant and wastewater temperature.

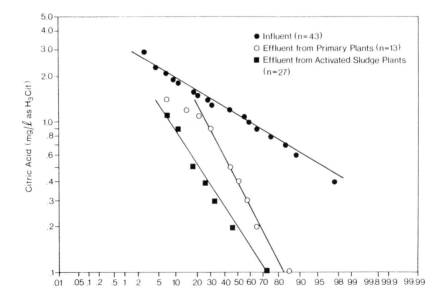

Figure 18. Influent-effluent critic acid concentration at Ontario treatment plants.

accumulation eventually leveled off at a "steady state" condition. The zeolite had no quantifiable effect on treatment plant performances from BOD and suspended solids removal standpoints.

CONCLUSIONS

In the investigations carried out at the WTC during 1971-1975, different detergent builders (NTA, carbonate and citrate) were subjected to extensive bench-scale and full-scale testing to determine potential wastewater treatment and water quality effects. More recent studies have tested zeolite builder effects. Conclusions that can be drawn from these studies are:

1. The substitution of nonphosphate detergents for existing detergents (phosphate content $< 20\%$ as P_2O_5) will usually result in at least a 25% reduction in the total phosphorus content of domestic waste-waters. With the recent decline in the overall phosphate content in detergents, it is expected that the impact of a switch to nonphosphate detergents would result in phosphorus reductions somewhat lower than this.

2. Detergent-induced increases in NTA carbonate or citrate levels did not have any quantifiable effect on phosphorus removal with alum, ferric chloride or lime; although citrate demonstrated a possibility of inter-fering with raw wastewater (primary) alum or ferric chloride phosphorus removel systems at citrate level in excess of 10 mg/l. Detergent-induced reductions in wastewater phosphorus levels were observed to result in a 25% to 30% reduction in the ferric chloride or alum require-ments to achieve a residual phosphorus level of 1 mg/l. Again recent reductions in the overall phosphate content of detergents may lessen the magnitude of these chemical reductions.

3. None of the available detergent builder materials (NTA, carbonate, citrate, and zeolite) had an adverse effect on the activated sludge process, although carbonate formulations can cause an increase in wastewater pH and alkalinity.

4. NTA, citrate and zeolite are removed by activated sludge systems, with year-round removals being in the order of 80%[+] for NTA and zeolite and 90%[+] for citrate. Both NTA and citrate exhibit lower removal efficiencies when wastewater temperatures fall below $10°C$. There is no evidence, however, of an NTA buildup in receiving waters due to these lower winter removal efficiencies. The Canadian long-term moni-toring of over 70 municipal water supplies confirms this.

5. Certain NTA-metal complexes, e.g., Cd and Ni, exhibited poor degrada-bility by laboratory activated sludge systems. However, there appears to be little full-scale significance of this finding in terms of increased metal transport.

ACKNOWLEDGMENTS

The author would like to express his appreciation to Alan Brownridge of Proctor and Gamble, Hamilton, Ontario, for providing recent information on detergent builders.

REFERENCES

1. Shannon, E. E., and L. J. Kamp. "Detergents Substitution Studies at CFS Gloucester," Environmental Protection Service Report. EPS 4-WP-73-3 (1973).
2. Prested, B. P., E. E. Shannon and R. J. Rush. "Development of Prediction Models for Phosphorus Removal, Volume I," Canada-Ontario Agreement on Great Lakes Water Quality Report #68 (1977).
3. Shannon, E. E., N. W. Schmidtke and P. J. A. Fowlie. "Effect of Citrate-and Carbonate-Based Detergents on Wastewater Characteristics and Treatment," Canada-Ontario Agreement on Great Lakes Water Quality Report No. 61 (1977).
4. Shannon, E. E., N. W. Schmidtke and B. A. Monaghan. "Activated Sludge Degradation of Nitrilotriacetic Acid (NTA) -Metal Complexes," Environmental Protection Service Report, EPA 4-WP-78-5, Water Pollution Control Directorate, Ottawa, Ontario (1978)
5. Shannon, E. E., P. J. A. Fowlie and R. J. Rush. "A Study of Nitrilotriacetic Acid (NTA) Degradation in a Receiving Stream," Environmental Protection Service Report EPS 4-WP-74-7 (1974).
6. Wei, N. R. Stickey, P. Crescuolo and B. P. LeClair. "Impact of NTA on an Activated Sludge Plant--A Field Study," Environmental Protection Service Report (in preparation).
7. Zanoni, A. E., and R. J. Rutkowski. "Per Capita Loadings of Domestic Wastewater," J. Water Poll. Control Fed. 44:1756 (1972).
8. Stepko, W. E., and E. E. Shannon. "Phosphorus Removal Demonstration Study Using Ferric Chloride and Alum at C.F.B. Uplands," Environmental Protection Service Report EPS 4-WP-74-5 (1974).
9. Walker, A. P. "Ultimate Biodegradation of NTA in the Presence of Heavy Metals," Proceedings of the 7th International Conference of Water Pollution Research, Paris, September, 1974.
10. Atkins, E. E., and T. R. Hawley. "Sources of Metals and Metal Levels in Municipal Wastewaters," Canada-Ontario Agreement Research Report No. 80, Project No. 75-1-43 (1978).
11. Hopping, W. D. "Activated Sludge Treatability of Type A Zeolite," J. Water Poll. Control Fed. 50(3):433 (1978).

CHAPTER 17

A MULTIPURPOSE PROGRAM FOR MANAGING
PHOSPHORUS IN THE GREAT LAKES

J. R. Sheaffer

Sheaffer and Roland, Inc.
Chicago, Illinois

B. C. Nagelvoort and M. A. Moser

Sheaffer and Roland, Inc.
Washington, D.C.

THE PROBLEM

This conference addresses the problem of phosphorus and its effects on the aging process of the Great Lakes. In the Great Lakes Water Quality Agreement of 1972 the United States and Canada proposed limiting phosphorus in discharges from municipal sewage treatment facilities larger than 1 mgd capacity to 1 mg/l [1]. This limit is not perceived as the standard required for clean water in the Great Lakes; rather, the standard reflects the "state of the art" for phosphorus removal by conventional sewage treatment processes with added-on facilities for advanced waste treatment (AWT).

The Water Quality Board of the International Joint Commission (IJC) reported in 1976 and 1978 that the 1-mg/l effluent phosphorus limit had not been achieved. This is due in part to delays in completion of facilities, but also in part to "the unreliability of phosphorus removal in some facilities which have been brought on line." The 1976 report also cited new studies using mathematical models showing eutrophication rates would not be stabilized even if the phosphorus limit were achieved at all municipal treatment plants [1,2].

The above conclusions led the IJC to recommend consideration of limiting phosphorus in discharges from all municipal treatment plants to 0.1 mg/l. The IJC also recommended limiting phosphorus to 0.5 mg/l in detergents manufactured for use in the Great Lakes Basin [1,3].

In a major departure from past EPA policy on phosphorus control, the June 1977 report of the Region V Phosphorus Committee also advocated a ban on phosphorus in detergents. The reason cited was the failure of conventional treatment to achieve the water quality objective of significantly decreasing and stabilizing eutrophication [4].

A phosphate in detergent ban and a stringent municipal limit of 0.1 mg/l of phosphorus would clearly reduce the level of phosphorus in the Great Lakes. For example, it is estimated that these measures would limit phosphorus to the target load level in Lake Michigan for 28 yr, if implemented by 1983. The 28-yr period would allow sufficient time in which a managment strategy to address nonpoint source of phosphorus in the Lake Michigan Basin could be formulated [4].

This chapter will identify and contrast two of the potential options for phosphorus removal from wastewater—conventional treatment plants and land treatment systems. It is assumed that most readers are familiar with conventional primary and secondary treatment. AWT options are adequately covered in other chapters in this book. Because land treatment systems are not so well known, the concepts of land treatment will be outlined prior to a discussion of the potential for land treatment to achieve desired management objectives. Multipurpose opportunities in wastewater management will be highlighted at the end of this chapter.

LAND TREATMENT

A variety of land treatment systems are reported in the literature. Common names applied to the various systems include spray irrigation, slow rate irrigation, overland flow and rapid infiltration. (See Appendix A for further details on each process.) Each of these systems is site specific and produces a different degree of wastewater renovation. Because these different systems are referred to generally as land treatment, there frequently is confusion when land treatment is being discussed and evaluated.

To avoid such confusion, land treatment, with respect to this paper, refers to what in initial studies was called a spray irrigation system and is currently referred to as slow-rate irrigation. Soil processes and a growing crop are integral parts of such land treatment systems.

A land treatment system must be designed to fit specific site characteristics. There is danger in repeating the design of a successful land treatment system at a new site. With different soil conditions, for example, certain pollutants removed at one site may not be treated adequately at another. Even outspoken opponents of land treatment agree that it is possible to design a land treatment system that would not result in the pollution of underground water resources. To illustrate, one critic stated, "the impression should not be left that no waste materials can or should be placed on, in, or under the ground surface. Given the proper hydrogeological conditions and using appropriately designed facilities, there are situations when selected wastes can be disposed of into the ground without appreciably modifying the quality of the potable ground water" [5]. Therefore, emphasis must be placed on proper design and operation. Simply piping wastewater to a site is not land treatment.

The six basic components in a complete land treatment system are:

1. collection and transport of raw sewage,
2. pre-treatment,
3. storage,
4. management of the site to be irrigated,
5. an agriculture cropping program, and
6. management of renovated water.

A network of pipes collects and conveys the wastewater to a selected location. Gravity flows are desirable, but pumping equipment is often necessary for rough terrain or long distances. Conveyance systems may transport wastewater a short distance or up to a hundred miles, depending on the treatment site location.

The second component provides appropriate pretreatment of the wastewater prior to storage or application to the land. The proper level of mechanical and biological treatment is provided to minimize health hazards and to avoid creation of nuisance odor conditions. Biological stabilization, i.e., secondary treatment, is regarded as the maximum level of pretreatment that may be required. Lagoon aeration to less than secondary treatment standards will likely meet pretreatment standards established by most states.

Storage, the third component, provides capacity in ponds for the retention of pretreated wastewater during nongrowing season and for periods of heavy rainfall or high wind when irrigation is undesirable. It also allows for the interruption of irrigation during planting and harvest periods. Storage facilities must be designed to prevent or manage leakage.

Characteristics of the fourth component, the irrigation site, must be intensively evaluated to allow selection of the most appropriate irrigation process, wastewater application rate and the crop management scheme. Detailed field testing of soil properties must be made to ascertain soil characteristics related to pollutant removal. The following group of analyses will help to establish the potential for prolonged irrigation of the soil and to evaluate its potential to purify the wastewater:

1. cation exchange capacity (CEC);
2. pH;

3. calcium carbonate ($CaCO_3$) if exceeding 0.1%,
4. particle size distribution,
5. total organic matter,
6. total organic nitrogen,
7. exchangeable Ca, Mg, K and Na,
8. total soluble salts,
9. chlorides, and
10. iron.

These analyses and other soil factors can be used to assess pollutant removal potential. The proper design to match the land treatment system to site characteristics ensures successful wastewater renovation.

The fifth component, growing of agriculture crops on irrigated land, provides wastewater renovation through the removal of nutrients from soil. The crops selected must be compatable with the soil, climate and wastewater characteristics. A soil system with a growing crop can recycle nutrients and can extract some substances that should be confined and contained in the environment. Cadmium, for example, is an element in sewage which is extracted by some vegetable crops. However, it is deposited in stems and foliage of grain crops so that the seeds may be safely utilized. Nitrogen and phosphorus are, of course, the major nutrients recycled in crop growth.

The sixth component is a management system for the renovated wastewater. This important component can be either natural (as in ground water recharge), or installed (drain tiles or wells), or a combination of the two. The purpose of an underdrainage system is to prevent waterlogging of the soils and the resultant disruption of chemical and biological processes in the soils contributing to wastewater renovation.

It is important to note that many systems referred to as land treatment systems are not designed properly and do not contain all of the components of a complete system. In some instances, the existing treatment plant does not work, so the effluent is conveyed to a nearby field and discharged. This is not a land treatment system. Similarly, an industry with a seepage bed does not have a land treatment system. The performance of such a system should not be used to prejudge the performance of a properly designed land treatment system.

POTENTIAL SOLUTIONS

The goal of 0.1 mg/l of phosphorus in municipal wastewater discharges can be achieved either by conventional sewage treatment or land treatment processes. A detergent phosphorus ban alone is not sufficient to meet Great Lakes water quality objectives. Meeting the 0.1-mg/l goal means that the typical 8-mg/l phosphorus concentration in wastewater must be reduced by 98.75%.

Considerations related to the choice of conventional or land treatment processes to control phosphorus include:

1. cost;
2. reliability;
3. the generation and disposal of sludge;
4. the degree of difficulty in design and operation of facilities;
5. the potential necessity for treating other wastewater components such as nitrogen and heavy metals in the future; and
6. potential hazards from wastewater treatment.

Cost

Removal of phosphorus in conventional sewage treatment processes is costly. While capital costs may be relatively low, operation and maintenance (O&M) costs are high due to the large quantity of chemicals and energy required to produce and process the phosphorus-containing sludge. Even if phosphorus were banned from detergents, the cost for removing the remaining phosphorus in municipal sewage would remain high due to the threshold level of chemical precipitant needed to initiate phosphorus removal.

The Upper Occoquan Sewage Authority facility in Fairfax, Virginia removes phosphorus to 0.1 mg/l at an O&M cost of $1.41/1000 gal treated, with phosphorus removal as the only AWT element [6]. The South Lake Tahoe AWT system is currently reported to cost $3.00/1000 gal treated to provide phosphorus control and nitrogen removal [6]. These two plants are described as proving "state of the art" AWT. However, the Lake Tahoe AWT seldom functions as it was designed and has caused numerous widespread fish kills in the Indian Creek Reservoir. Unreliability and excessive costs of AWT may result in a congressional cutback in the entire EPA construction grant program [6]. If such a cutback occurs, water quality improvement efforts in the Great Lakes area will be severely hampered.

Contrast these AWT implications with the low-cost, high-performance land treatment system at Muskegon County, Michigan. In 1976 the underdrainage discharge from the utilization of 28 mgd of pretreated wastewater to irrigate 5000 ac of corn contained an average of 0.07 mg/l phosphorus [7]. In 1977 the discharge averaged 0.08 mg/l. Data for July 1978 (the only data available) indicated a discharge of 0.03 mg/l [7]. Therefore, the Muskegon County phosphorus discharge is already far better than the proposed standards for the Great Lakes.

The Muskegon system costs relatively little to operate. User charges for the first five years of operation were 17¢/1000 gal [8]. In 1979, with a 56% increase in power costs in the past two years and substantial increases in hourly wage rates, the system managers were obligated to raise user charges by 3¢ or 18%, an average increase of 3% over the past 6 yr of dramatic inflation [7]. The system keeps costs low because it takes advantage of natural processes in the soils along with sunlight and wind, and because of revenue from the sale of the crops irrigated with the wastewater.

Capital costs at Muskegon were reasonable as well, amounting to less than 89¢/gal of treatment capacity when the collection system costs of $5.2 million are substracted from total costs of $42.7 million for a 42-mgd capacity system (1972 prices) [8]. More current information indicates that land treatment systems are likely to be cost competitive with conventional sewage treatment, particularly if high levels of phosphorus removal are required. Under conditions which are just moderately favorable for land treatment (i.e., land is reasonably available and soils are suitable), an authority reports land treatment to be less costly than conventional systems reducing phosphorus to the level of 0.1-0.5 mg/l [9].

Current law recognizes the viability and encourages the use of land treatment technologies. Public Law 95-217, the Clean Water Act of 1977, provides for 85% federal funding for construction of land treatment systems compared with 75% for conventional systems. It provides that no federal funds will be available for a treatment system unless land treatment is given full consideration in the planning process.

Reliability

Conventional sewage treatment with consistent phosphorus removal to 0.1 mg/l, though proven feasible, is not commonly achieved. Only "state of art" demonstration treatment facilities attempt high levels of phosphorus removal. Consequently, the reliability of AWT phosphorus removal processes under more normal circumstances is unknown.

Conventional secondary treatment plants are normally out of operation 10-30% of the time because of biological upsets, and thus do not treat phosphorus. Also, they are not designed to treat excessive stormwater flows, which means they are often overloaded and discharge highly polluted effluents to surface waters.

Properly designed and managed land treatment systems can consistently meet the IJC proposed phosphorus removal requirements. Table I shows effective phosphorus removal at nine existing long-term land treatment systems. All systems except Melbourne meet the IJC standard or reduce phosphorus to the level found in the natural ground water.

Land treatment systems must be carefully designed to meet site requirements, their pre-treatment retention times (usually three days in aerated lagoons) and storage capacity (for retention during the nongrowing season) provide buffering for protection from both biological upsets and excessive storm flows.

Table I. Phosphorus Removal By Slow Rate Irrigation

Land Treatment System	Treatment Process	Start Up Date	Present Flow (mgd)	Applied Wastewater	Total Phosphorus Concentration (mg/l)		Seepage Creek
					Groundwater		
					Treatment Area	Site Boundary	
Pleasanton, CA [10]	SR	1957	1.4	4.8	0.2 - 0.5	0.2	
Unicoi State Park, GA [10]	SR[a]	1974	0.02	12.1	0.2 - 0.5	0.2[b]	
Muskegon, MI [11]	SR	1975	28.5	1.4			0.05
Dickinson, ND [12]	SR	1959	1	6.9		0.9[b]	
Roswell, NM [13]	SR	1944	4	8.6 - 15.8		0.08 - 1.4[b]	
Lubbock, TX [14]	SR	1937	16.5		0.12		
San Angelo, TX [10]	SR	1958	5.8	5.9		0.01 - 0.02[b]	0.09
Braunschweig, W. Germany [15]	SR	1939	13	13		0.5[b]	
Melbourne, Australia [16]	SR[c]	1896	56	7.8			1.1

[a]Forest irrigation.

[b]Not significantly different from local groundwater.

[c]Pasture irrigation with animal grazing, ditch drainage.

Sludge

Conventional systems with high levels of phosphorus removal produce considerable amounts of sludge. Processing and disposal of these sludges are costly and energy intensive. Every ton of lime or alum added for phosphorus precipitation produces about three tons of sludge. Most phosphorus sludges are difficult to process. Once dewatered, transportation and disposal by land spreading, landfill or incineration must occur. Sludge processing results in additional wastewater treatment costs.

Some sludge will be produced in the treatment of wastewater prior to land application. The sludge will be mostly bacterial cell mass rather than chemical in nature. Recommended preapplication or prestorage treatment by lagoon aeration produces a relatively small volume of sludge, even when compared to conventional secondary treatment processes without phosphorus removal. In most cases, sludge digestion and storage take place in the treatment lagoons or storage ponds, with sludge removal and disposal required at infrequent intervals, perhaps as long as 15-20 yr. These aged sludges are easily dewatered and can be land spread or land filled, often within the land treatment site.

Facility Operation

Highly trained, skilled technicians are required to operate and maintain AWT facilities. The Upper Occoquan plant includes multiple computers to insure consistent levels of treatment [6]. The Lake Tahoe AWT facility has seldom functioned as designed due to mechanical breakdowns, chemical intensive elements, and loss of trained operators. These high-treatment facilities require constant, expert attention.

Land treatment systems, on the other hand, are less susceptible to short-term changes in wastewater flow and characteristics; and their treatment processes require less skilled operators. However, the preapplication and irrigation need attention by knowledgeable maintenance personnel. In addition, capable management is necessary to maintain a profitable agriculture operation.

Other Pollutants

To this point, this chapter has focused on phosphorus as the controlling nutrient in eutrophication in the Great Lakes and hence the nutrient for which treatment is required. However, future studies may show nitrogen control to be necessary in reaching water quality objectives. For example, preliminary indications from the EPA Annapolis Field Office show that high levels of available phosphorus make nitrogen the limiting nutrient for algal growth in the Potomac Estuary [17]. Years of phosphorus accumulation in Great Lakes sediments may provide similar results. The IJC has scheduled an intensive examination of Lake Ontario for 1981-82, due in part to a steady increase in the nitrogen content of the Lake over the past nine years.

The add-on costs of nitrogen removal in conventional sewage treatment would be extremely high. Nitrogen removal by AWT is all but unknown, except at bench and pilot scale. No reliable or proven year-round method of nitrogen removal by conventional AWT exists.

The natural processes of plant and microbial nutrition utilize nitrogen applied to cropped areas. In addition, substantial nitrification and denitrification takes place in storage ponds and in the soils. Utilizing these processes, land treatment removes substantial amounts of nitrogen from applied wastewater.

Land treatment removes more than phosphorus and nitrogen. The Muskegon County system reduces BOD_5 and suspended solids by 99% and 97% respectively, far higher removal levels than provided by all but the most sophisticated AWT processes. Pathogens and heavy metals are reduced to trace amounts far below levels meeting Public Health standards [8,9]. Also, preliminary research shows nearly total removal of toxic substances and chlorinated hydrocarbons in the system. Perhaps more significant, the multiple reduction of pollutants by land treatment is achieved as a first-cost investment, in contrast to costly increments of additional treatment required by conventional technologies.

Health Issues

Conventional treatment with AWT elements will remove high percentages of bacteria and viruses, but will also concentrate a portion of those pathogens along with chemicals, heavy metals and toxic organics in sludges. The substances remaining in the wastewater, including organics chlorinated in the disinfection processes, are discharged to surface waters that may provide drinking water supplies, such as the Great Lakes. In essence, conventional treatment takes one problem, that of potential health hazards in sewage, and creates two major problems—the low level discharge of hazardous materials to drinking water supplies, and the disposal of sludges which may contain potentially hazardous substances or require costly management.

Frequent critics of land treatment raise questions relating to pathogens in aerosols and possible groundwater contamination. While there is no categorical response to these questions, substantial investigations indicate that the risks to public health can be minimized with properly designed and managed systems.

Some researchers suggest that safety depends "on the origin of the waste-water, the treatment process used, the crops planted and above all the quality of the planning and care" [18]. Others choose to concentrate attention on relative risk, suggesting that embracing a goal of complete removal of all viruses of human origin from any waters that man may contact may not be the best protection for public health or in the public interest [19]. Researchers from the Food and Drug Administration recommend that food crops that are to be consumed raw by humans should not be irrigated and/or fertilized with wastes from sewage treatment plants [20]. Several researchers find that land application systems provide protection equal to or better than that provided by conventional activated sludge systems [21]. A study currently underway finds that aerosol levels downwind from an activated sludge facility to be greater by almost an order of magnitude than those found downwind from a spray irrigation facility [22]. One researcher reported that "Based on the currently available data, it is reasonable to assume that with regard to aerosols, minimal risk, not zero risk, will exist for those wastewaters that have undergone secondary treatment and effective disinfection. Under these conditions, there may be no need for buffer zones" [23]. And finally, a review of health issues related to currently operating land treatment systems led a researcher to state, "In summary, the potential health hazards from land treatment are no greater and probably less than those from conventional treatment. Land treatment, under proper management, is more effective, reliable and predictive in the removal of toxic trace metals and organics, viruses, and pathogens than conventional sewage treatment. The degree of acceptable risk has to be defined and compared with the risk associated with other alternatives of wastewater treatment and disposal" [24].

MULTIPURPOSE PHOSPHORUS MANGEMENT

Conventional treatment has only one potential benefit, clean water, from the dollars expended for concrete, steel, chemicals and energy. It might be suggested that those dollars would be better used to provide multiple benefits when an opportunity exists to do so.

Land treatment of wastewater involves a multidisciplinary, multipurpose approach to wastewater management, and therefore phosphorus management, within the Great Lakes Basin. The Muskegon land treatment system will illustrate the point. For the past several years, the system there has produced an average of almost 400,000 bushels of corn which is used for livestock feed. The availability of the corn locally means livestock can be raised locally at less cost. The Muskegon system has not only prevented sprawl of develop-ment to the east of the built-up areas adjacent to the City of Muskegon, but also is a substantial recreation area being used for bow and arrow deer hunting and rabbit hunting. The storage lagoons in the systems have become one of the 19 major resting areas for migrating waterfowl in the State of Michigan. The system is also the site of the county's sanitary landfill, and a greenhouse operation which raises much of the county's flowers and shrubs for downtown park purposes. The landfill is underdrained so that leachate may be piped for treatment into the system.

While Muskegon County was required to pay for all of the land for the system, land used for treatment is now eligible for purchase with federal and state construction grant funds. However, mutually beneficial approaches can be developed between farm and city. It is possible to lease land or to provide pretreated wastewater to farmers for irrigation purposes, thus avoiding direct ownership of land by sewer authorities and allowing farmers to retain ownership of their land.

In Northglenn, Colorado, treated wastewater will be supplied to a local irrigation district and used to irrigate thousands of acres of cropland. Controlling urban sprawl by land treatment of wastes from the urban area preserves farmland and open space and therefore keeps the cost of farm products low by retaining productive farmland close to urban centers.

Land treatment can be integrated with floodplain management. The Department of the Army observed that the authorities in the Federal Water Pollution Control Act Amendments of 1972 regarding the acquisition of sites for land treatment of wastewater when combined with the authorities of Section 73 of the Water Resources Development Act of 1974 offer an outstanding opportunity for multiple uses of floodplains while preserving green space and providing recreational opportunities. The Army spokesman inquired, "why not use our floodplains in urban areas for crop production, golf courses, forests, and other uses which can capitalize on the nutrients in our wastewater and provide tertiary waste treatment at the same time?" [25].

Land treatment influences energy use. The nitrogen in a year's flow of domestic wastewater in the United States requires the equivalent of 2.25 x 10^9 gallons of crude oil to replace as fertilizer [26].

The implementation of a land treatment system allows a community to transfer the cost of sewage treatment (a social inflationary cost) to the positive side of the ledger (an investment in the production of future food and fiber). This is the rationale behind the 10% bonus in federal construction grants for land treatment systems. When the federal government supports a land treatment system over a conventional treatment plant, the construction grant shifts from a federal subsidy to an investment.

The fact that land treatment systems are land intensive is probably the greatest obstacle to the implementation on a broad scale of this wastewater management technique. However, the land requirements, which likely range from 125 to 350 ac/mgd treatment capacity in the Great Lakes area, should be viewed as an asset in efforts to implement the process, when all of the potential benefits are considered.

Land treatment can also be incorporated into existing wastewater management plants and should not be viewed solely as a "new source control" option. For example, adding irrigation of cropland with the effluent from a conventional secondary treatment facility may be the cost-effective manner to meet required higher treatment standards. In later years, when replacement facilities are required for the conventional system, it may be reasonable to shift away from the sludge-intensive conventional system to one which is better integrated with the land treatment process. On the other hand, it may be more cost-effective to abandon the operating conventional facility and simply include its depreciated value in the cost of the new alternative, paying off any indebtedness on the old system as part of the cost of the new one. The appreciation of land values and the ability of land treatment systems to provide lower operating costs (which may not rise as much as inflation because of higher crop prices) are strong incentives for abandoning conventional facilities which are capable only of becoming increasingly costly to operate as energy and labor costs rise. It may also be reasonable to utilize conventional systems for the treatment of storm water runoff in urban areas while shifting municipal sewage flows to beneficial uses. Urban runoff has become a major element in the continued pollution of surface waters and must be managed if high levels of cleanup of our rivers and lakes are to be accomplished.

SUMMARY

Phosphorus is currently considered the nutrient controlling eutrophication of the Great Lakes. Reducing phosphorus concentration to 0.1 mg/l in municipal wastewater discharges and reducing phosphorus in detergents to 0.5 mg/l in the Great Lakes Basin would minimize the rate of eutrophication in the Great Lakes.

Conventional treatment with AWT add-ons or land treatment could meet a 0.1 mg/l discharge limit for phosphorus. Such conventional treatment is costly, has not been proven reliable, demands highly skilled operators, and results in chemical sludges requiring expensive disposal techniques.

Land treatment at Muskegon County, Michigan and elsewhere has demonstrated low cost, ease of operation and reduction of phosphorus to levels below the proposed phosphorus discharge requirement, thereby protecting water quality in Lake Michigan. A multitude of land treatment systems operating for decades show consistent high degrees of phosphorus removal. Land treatment systems also remove pathogens, nitrogen, heavy metals and toxic chemicals from wastewater.

Land treatment systems must be carefully designed, constructed and managed. While land treatment may not be possible at all locations, in most situations it can be implemented to realize multiple benefits in addition to protecting water quality. Among such benefits are protection of open space and farmland, creation of recreational opportunities, sound utilization of floodplains, preservation of wildlife habitat and reduced energy consumption.

RECOMMENDATIONS

From analysis of performance and costs, it is apparent that land treatment is a concept that matches the demands of the public for a clean environment while at the same time meeting equally strong demands for lower cost government services.

Land treatment can effectively replace conventional treatment with AWT add-ons and insure water quality protection in the Great Lakes Basin.

Existing wastewater treatment facilities and plans for new facilities in the Great Lakes basin should be carefully examined to determine the potential to upgrade treatment and provide for the enjoyment of multiple benefits from land treatment systems.

There are those who call land treatment difficult to implement. However, it is evident that until engineers and wastewater managers begin to accommodate land treatment in their pollution control plans, the Great Lakes will continue to be degraded, with little hope for the kind of improvement desired by the public.

A final note quotes the consultant who designed and operated the Lake Tahoe AWT facility, and who in May of 1978 recommended that the advanced treatment elements of the facility be abandoned. He said, "A lot of people are surprised I'd advocate a switch to land treatment but you've got to consider energy and the realities of federal funding programs for land application." He also stated, "secondary treatment with land application would reduce total annual costs (at Lake Tahoe) by about 27%, local costs by 40%, and energy consumption by 50%, as comapared to the lowest cost AWT system" [6].

APPENDIX: LAND TREATMENT PROCESSES

Several land treatment processes have been identified: slow rate, rapid infiltration and overland flow. Each process had different soil and site requirements and produces a different degree of wastewater treatment. Table II lists the design and operation features of land treatment processes, while Table III summarizes site characteristics.

Table II. Comparison of Design Features for Land Treatment Processes[a]

Feature	Slow Rate	Rapid Infiltration	Overland Flow
		Principal Processes	
Application techniques	Sprinkler or surface[a]	Usually surface	Sprinkler or surface
Annual application rate (m)	0.6 - 6.0	6.0 - 170	3 - 21
Field area required[b] (ha)	22 - 220	0.8 - 22	6.4 - 45
Typical weekly application rate (cm)	1.3 - 10	10 - 305	6 - 15[c] 15 - 40[d]
Minimum preapplication treatment provided in United States	Primary sedimentation[e]	Primary sedimentation	Screening and grit removal
Disposition of applied wastewater	Evapotranspiration and percolation	Mainly percolation	Surface runoff and evapotranspiration with some percolation
	Surface discharge if drainage recovery	Surface discharge if drainage recovery	
Need for vegetation	Required	Optional	Required

[a]Includes ridge-and-furrow and border strip.

[b]Field area in acres not including buffer area, roads, or ditches for 1 mgd (43.8 liter/sec) flow.

[c]Range for application of screened wastewater.

[d]Range for application of lagoon and secondary effluent.

[e]Depends on the use of the effluent and the type of crop.

1 in. = 2.54 cm
1 ft   = 0.305 m
1 acre = 0.405 ha

Slow rate irrigation is designed in a manner similar to an agricultural irrigation system. However, the goal is wastewater treatment rather than optimum crop growth. Slow rate is by far the most commonly used land treatment process and can be expected to remove oxygen demanding materials, nutrients, heavy metals and pathogenic microorganisms. Figure 1 diagrams natural processes functioning to treat wastewater during slow rate irrigation.

Table III. Comparison of Site Characteristics for Land Treatment Processes [10]

Characteristics	Principal Processes		
	Slow Rate	Rapid Infiltration	Overland Flow
Slope	Less than 20% on cultivated land; less than 40% on noncultivated land	Not critical; excessive slopes require much earthwork	Finish slopes 2 - 8%
Soil permeability	Moderately slow to moderately rapid	Rapid (sands, loamy sands)	Slow (clays, silts, and soils with impermeable barriers
Depth to groundwater	0.6 - 0.9 m (minimum)	3 m (lesser depths are acceptable where underdrainage is provided)	Not critical
Climatic restrictions	Storage often needed for cold weather and precipitation	None (possibly modify operation in cold weather)	Storage often needed for cold weather

1 ft = 0.305 m

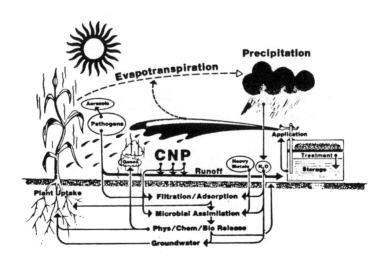

Figure 1. Natural processes associated with slow rate irrigation [27].

Rapid infiltration removes oxygen demanding materials, suspended solids, phosphorus and heavy metals and is used to treat large flows of wastewater in small areas of highly permeable soils. Application schedules must be defined to control nitrogen. Figure 2 is a schematic cross section of a rapid infiltration basin.

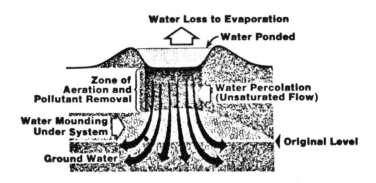

Figure 2. Rapid infiltration [27].

Raw or partially treated wastewater is applied at the head of carefully prepared overland flow slopes. Figure 3 schematically shows the overland flow process. Impermeable soil minimizes percolation. Overland flow removes suspended solids, nitrogen and heavy metals to low concentrations but does not completely remove phosphorus.

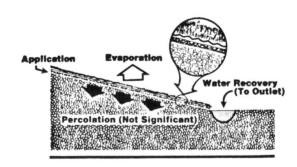

Figure 3. Overland flow [27].

REFERENCES

1. "Great Lakes Water Quality Board Sixth Annual Report," International Joint Commission, Washington, DC, (1978).
2. 1975 International Joint Commission, "Water Quality Board Report," International Joint Commission, Washington, DC, (1978).
3. "Great Lakes Focus on Water Quality," International Joint Commission, Windsor, Ontario, 5:1 (March 1979).
4. "Detergent Phosphorus Ban," Position paper prepared by the Region V Phosphorus Committee, U.S. EPA, Chicago, IL, EPA 905/2-77-003 (June 1977).
5. Johnson, C. C., Jr. "Drinking Water Policy Problems: Background of the Current Situation," paper presented at the National Conference on Drinking Water Policy Problems, March 6, 1978.
6. "A Report to the Committee on Appropriations," U.S. House of Representatives, Surveys and Investigations Staff (March 1979).
7. Demirjian, A., Manager, Muskegon County Land Treatment System. Personal Communication. (1978, 1979).
8. "Wastewater: Is Muskegon County's Solution Your Solution?" U.S. Environmental Protection Agency Region V, Chicago, IL (1978).
9. Culp, G. et al. "Costs of Land Application Competitive with Conventional Systems," *Water Sewer Works* 125(10) (1978).
10. "Process Design Manual for Land Treatment of Municipal Wastewater," U.S. Environmental Protection Agency, Washington, DC (1978).
11. Nutter, W. L. et al. "Land Treatment of Municipal Wastewater on Steep Forest Slopes in the Humid Southeastern United States," proceedings of "State of Knowledge in Land Treatment of Wastewater," International Symposium, August 1978. U.S. Army Corps of Engineers, Cold Regions Research and Engineering Laboratory, Hanover, NH (1978).
12. "Long Term Effects of Applying Domestic Wastewater to the Land–Dickinson, North Dakota Slow Rate Irrigation Site," R. S. Kerr Environmental Research Lab., U.S. Environmental Protection Agency, Ada, OK (1978).
13. "Long Term Effects of Applying Domestic Wastewater to the Land–Roswell, New Mexico Slow Rate Irrigation Site," R. S. Kerr Environmental Research Lab., U.S. Environmental Protection Agency, Ada, OK (1978).
14. Wells, D. M. et al. "Effluent Reuse in Lubbock," in *Land as a Waste Management Alternative,* R. C. Loehr, Ed. (Ann Arbor, MI: Ann Arbor Science Publishers, 1976), pp. 451-466.
15. C. Tietjen et al. "Land Treatment of Wastewater in Braunschweig and in Wolfsburg, Germany," proceedings of "State of Knowledge in Land Treatment of Wastewater," International Symposium, U.S. Army Corps of Engineers, Cold Regions Research and Engineering Laboratory. Hanover, NH (1978).

16. McPherson, J. B. "Renovation of Wastewater by Land Treatment at Melbourne Board of Works Farm Werribee, Victoria, Australia," proceedings of "State of Knowledge in Land Treatment of Wastewater," International Symposium, U.S. Army Corps of Engineers, Cold Regions Research and Engineering Laboratory, Hanover, NH (1978).

17. Clark, L. J. et al. "Assessment of 1977 Water Quality Conditions in the Upper Potomac Estuary," (July 1978).

18. Braude, G. L. et al. "Use of Wastewater on Land—Food Chain Concerns," proceedings of "State of Knowledge in Land Treatment of Wastewater," International Symposium, U. S. Army Corps of Engineers, Cold Regions Research and Engineering Laboratory, Hanover, NH (1978).

19. Lenette, E. H., and D. P. Spath. "Overview—Health Considerations Associated with Land Treatment of Wastewater Systems Compared with Other Human Activities," proceedings of "State of Knowledge in Land Treatment of Wastewater," International Symposium, U.S. Army Corps of Engineers, Cold Regions Research and Engineering Laboratory, Hanover, NH (1978).

20. Larkin, E. P. "Land Application of Sewage Wastes: Potential for Contamination of Foodstuffs and Agricultural Soils by Viruses, Bacterial Pathogens and Parasites," proceedings of "State of Knowledge in Land Treatment of Wastewater," International Symposium, U.S. Army Corps of Engineers, Cold Regions Research and Engineering Laboratory, Hanover, NH (1978).

21. Uiga, A. et al. "Relative Health Factors Comparing Activated Sludge Systems to Land Application Systems," proceedings of "State of Knowledge in Land Treatment of Wastewater," International Symposium, U.S. Army Corps of Engineers, Cold Regions Research and Engineering Laboratory, Hanover, NH (1978).

22. Clark, C. S. et al. "A Seroepidemiologic Study of Workers Engaged in Wastewater Collection and Treatment," proceedings of "State of Knowledge in Land Treatment of Wastewater," International Symposium, U.S. Army Corps of Engineers, Cold Regions Research and Engineering Laboratory, Hanover, NH (1978).

23. Sorber, C. A. et al. "Discussion Summary," in *Risk Assessment and Health Effects on Land Application of Municipal Wastewater and Sludges,* B. P. Sagik and C. A. Sorber (Eds.), Center for Applied Research and Technology, University of Texas, San Antonio, TX, December 12-14, 1977.

24. Iskandar, I. K. "Overview of Existing Land Treatment Systems," proceedings of "State of Knowledge in Land Treatment of Wastewater," International Symposium, U.S. Army Corps of Engineers, Cold Regions Research and Engineering Laboratory, Hanover, NH (1978).

25. Ford, C. R. "Effect of New Legislation on Management of River Systems," Transactions of the 40th North American Wildlife and National Resources Conference, Washington, DC. 1975.

26.   Sheaffer, J. R. "Land Application of Waste—Important Alternative," *J. of Groundwater Div. Nat. Water Well Assoc.* Vol. 1. (1979).

27.   Urban, N. W., and M. A. Moser, "Overview of Land Treatment of Wastewater," presented at Symposium on Regionalization and the Technical Response to PL 92-500, Hillsboro, OR, Sept. 1976.

CHAPTER 18

PHOSPHORUS IN STORMWATER: SOURCES AND TREATABILITY

W. G. Lynard

Metcalf & Eddy, Inc.
Palo Alto, California

R. Field

U.S. Environmental Protection Agency
Storm and Combined Sewer Section
Edison, New Jersey

INTRODUCTION

Urban stormwater runoff and combined sewer overflows can contribute significant loads of phosphorus to receiving waters. Several characterization studies and a number of full-scale prototype and pilot plant treatment systems have been implemented where phosphorus data were collected. These data give an indication of the sources, the magnitude of the problem, and the treatability of phosphorus by stormwater treatment systems. However, the primary purpose of most storm and combined sewer overflow countermeasures is volume control and suspended solids and biochemical oxygen demand (BOD) removal. A comparison of phosphorus in municipal sewage (dry weather) is required to place the magnitude of the problem in perspective.

PHOSPHORUS

Phosphorus is a major nutrient, stimulating algal and aquatic growth in receiving waters. Typical phosphorus concentrations in receiving waters in excess of 0.015 mg/l are considered critical and can trigger biological growth [1]. Phosphorus in wastewaters can also interfere with treatment processes that use chemically assisted unit operations.

Forms of Phosphorus

Several forms of phosphorus occur in wastewaters, each having distinct characteristics of bioavailability and treatability. The common forms include orthophosphate, polyphosphate and organic phosphorus. Orthophosphates are the most readily usable for biological growth and the most easily treated. Orthophosphates can occur in equilibrium as PO_4^{3+}, HOP_4^{2-}, $H_2PO_4^-$, and H_3PO_4; but, PO_4^{3+} usually predominates at pH values normally found in wastewater. Polyphosphates slowly undergo hydrolysis to orthophosphate forms; and as organic materials decompose, the phosphorus is also converted to orthophosphate [2].

Phosphorus Loads

The principal sources of phosphorus loads result from municipal sewage and agricultural runoff. However, on a storm event basis, potential loads from urban wet-weather discharges can contribute a substantial portion of the total load. Total phosphorus concentrations in municipal sewage of 8-10 mg/l are typical. Representative phosphorus concentrations in combined sewage runoff are about 2 mg/l and 0.6 mg/l, respectively, as shown in Figure 1 [3].

Phosphorus concentrations in combined sewage and stormwater runoff are small in comparison with the concentrations in municipal sewage, or even municipal effluent after secondary treatment. Wet-weather sources contribute only about 15%, or less, of the total annual phosphorus load to receiving waters, indicating that the annual phosphorus load may potentially be best controlled by municipal sewage treatment (chemically assisted phosphorus removal).

During storms, however, over 60% of the phosphorus load can come from combined sewage and stormwater runoff sources, as shown in Figure 2. Representative phosphorus loads for the major Great Lakes cities are also presented, are are slightly less than the national average. Discrete and intermittent storms actually discharge the corresponding 15% of the total annual load in a series of high volume events over a relatively short period of time. Further narrowing of the time frame could result in over 90% of the load from wet-weather sources occurring during the peak hourly flow.

These high volume shock discharges, while relatively minor on an annual basis, can cause short-term water quality impacts on sensitive receiving waters.

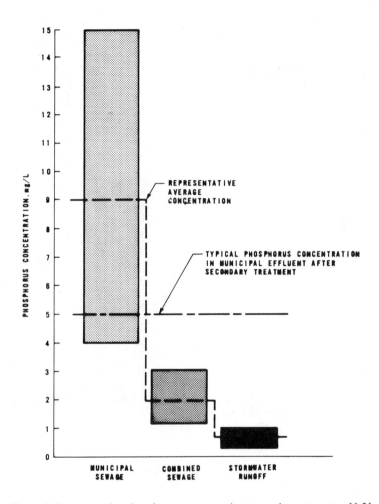

Figure 1. Representative phosphorus concentration ranges in wastewaters [1,3].

Even with a high degree of continuous treatment of dry-weather sources, phosphorus problems may still persist whenever it rains.

URBAN PHOSPHORUS SOURCES

Phosphorus in urban stormwater and combined sewage originates from numerous sources, many of which are difficult to isolate and quantify in

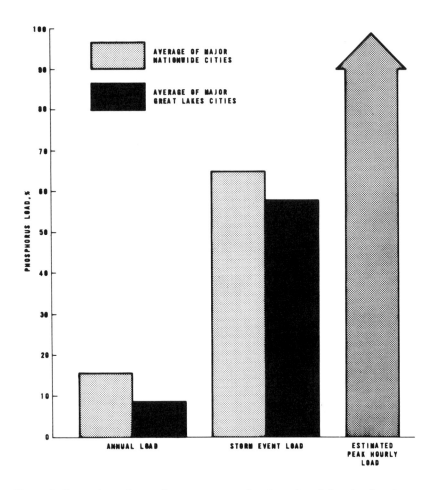

Figure 2. Comparison of annual, storm event and estimated peak hourly phosphorus load from wet-weather sources [4].

terms of actual loads. Many of the sources contribute phosphorus collectively to the runoff from various land uses and land activities.

Sources in Urban Runoff

Suspected major sources of phosphorus in urban runoff include eroding soils, leachate from leaves, and atmospheric loads. Other sources such as fertilizer application and animal populations in urban areas can also contribute to the phosphorus load. Fuel additives and some street deicing compounds also contain phosphorus, but loads from these sources are minor [5].

Urban Soil Erosion

Eroding soil from new construction sites, road cuts and fills, and stream banks is a major contributor of sediment in urban runoff, and is potentially the major source of phosphorus. Studies have indicated that the correlation of total phosphorus to total suspended solids in urban runoff from several urban land uses is the highest compared with the correlations for total nitrogen, BOD_5, and chemical oxygen demand (COD) [6]. Thus, phosphorus is probably associated with suspended solids or held in complex with eroding soils.

Representative sediment yields for urban and natural land areas are shown in Table I. Sediment yields from developed urban areas average about 3400 lb/ac-yr, and can range from less than 300 to over 7000 lb/ac-yr. Sediment yields from developing areas and construction sites can range from 40 to over 90 times the urban rate, and from 200 to 3000 times the natural sediment yield rates.

Table I. Sediment Yields from Developed, Developing and Natural Areas [7-9]

Location	Area Description	Sediment yield (lb/ac-yr)
Urban		
Lake Tahoe, CA	Poorly planned residential development	3200
	Well planned residential development	300
Detroit, MI	Area under development	138,000
	Urban area	5400 - 6300
Kensington, MD	Area under extensive construction	156,000
	General small urban construction	3100 - 310,000
Washington, DC	Avg. of 750 mi^2 of urban area	3100
	Range from underdeveloped to urbanized	400 - 7200
Montgomery Co., MD	Potomac watersheds	4700
	Watts Branch watershed	260
Natural		
Potomac River Basin	Native cover	46 - 600
Pennsylvania and Virginia	Natural drainage basins	600
Mississippi River Basin	Throughout geologic history	1000

Using an estimated representative phosphorus content in surface soil of about 0.1% by weight, average phosphorus loads of about 5 lb/ac-yr were estimated for developed urban areas [8]. The expected range for developed urban areas is 0.5-10.5 lb/ac-yr, depending on land uses in the drainage area.

Comparing these estimates with actual sampling data for general urban, commercial, and residential land uses (4.9, 2.2 and 0.7 lb/ac-yr, respectively), phosphorus sources from eroding soil sediment potentially contribute the largest fraction of the total phosphorus load [10].

Leachate from Leaves

Leachate from leaves on streets, in gutters, and on impervious surfaces can also contribute substantial phosphorus loads in urban runoff. Laboratory simulation of the rain leaching process on oak and poplar leaves yielded total soluble phosphorus loads of 54 and 140 μg/g of leaves, respectively. Soaking of the leaves for extended periods, up to about one day, yielded about 270 μg/g of leaves [11].

Rainfall and Dustfall

Atmospheric phosphorus loads from rainfall and dustfall have also been identified as potentially significant. Wind-blown soil and dust are believed to be the major sources of the atmospheric load.

An estimated 16% of the total phosphorus load entering Lake Michigan has been attributed to atmospheric sources, based on loading rates ranging from 0.2 to 0.3 lb/ac-yr [12]. Weighted average total phosphorus concentrations found in the rainfall ranged from 0.05 to over 0.06 mg/l.

Rainfall sampling in the Washington, DC area also found high levels of phosphorus originating from atmospheric sources. The range of total, total soluble and orthophosphate concentrations is presented in Table II [13]. Phosphorus loads for individual storms ranged from 0 to 0.03 lb/ac for orthophosphate, and 0.002 to 0.23 lb/ac for total phosphorus. Other studies have shown orthophosphate concentrations rates ranging from 0 to 0.9 mg/l and averaging as high as 0.24 mg/l [8].

Dustfall loadings determined for several urban land uses averaged about 0.002 lb/ac-day for orthophosphate and 0.005 lb/ac-day for total phosphorus, indicating that the deposition of atmospheric phosphorus by dustfall is less significant than during rainfall [13]. Other findings in this study include:

- Rainfall pollutants tend to wash out in relatively uniform amounts throughout the urban area, irrespective of the original sources.

- Atmospheric pollutants wash out during the first stages of the storm.

Table II. Phosphorus Concentrations in Rainfall, Washington, DC Area [13]

Parameter	No. of samples	Concentration Range[a] (mg/l)	Weighted Average Concentration (mg/l)
Total phosphorus	77	0.04 - 0.62	0.14
Total soluble phosphorus	42	0.02 - 0.62	0.11[b]
Orthophosphate	77	0 - 0.17	0.02

[a]Individual samples.

[b]Estimated using the average of the difference between the soluble and the total phosphorus for corresponding samples.

- Factors affecting atmospheric pollutant loads include turbulent or stagnant air and the number of antecedent dry days.

Other Sources

Although no definitive numbers are available for contributions from lawn areas, excessive application of phosphorus-based fertilizers can create a potential loading source. Similarly, animals and pets can contribute significant nutrient loads in urban areas.

Characterization by Land Use

Since identifying and quantifying the sources of phosphorus loads is often difficult, characterizing phosphorus in runoff by land use may be more easily obtained. Factors such as imperviousness, antecedent conditions, rainfall, and soil characteristics all affect the phosphorus level from a particular land use. Ranges of expected concentrations and mean annual loads for several land uses are presented in Table III.

Phosphorus loads on a storm basis can produce different and potentially higher unit loadings for the different land use categories. The values in Table IV represent the total storm load from a drainage area divided by the duration of runoff. Since most runoff durations are less than one day, these values may be more representative of peak loading rates to receiving waters. Reporting values on a daily basis may more closely reflect the effects of antecedent dry conditions, catchment area, rainfall characteristics, and runoff duration. A high intensity short duration rainfall would produce a higher pollutant yield than the same volume at a lower intensity (long duration) [14].

The effects of rainfall/runoff volume are shown in Figure 3 in a basinwide correlation of phosphorus load and runoff volume for two watersheds in

Table III. Characteristic Phosphorus Concentrations and Loads
for Several Urban Land Uses [14,15]

Land Use	Phosphorus Concentration (mg/l)	Annual Phosphorus Load (lb/ac/yr)
Low density residential	0.3 - 0.4	0.38 - 0.46
Medium density residential	0.4 - 1.0	0.78 - 1.13
High density residential	0.3 - 0.5	1.16 - 1.29
Multifamily residential	0.2 - 0.3	1.25 - 1.36
Commercial	0.2 - 0.3	1.88 - 2.57

Table IV. Range of Average Storm Phosphorus Yields
for Several Land Uses [6,14]

Land use	Range of Storm Phosphorus Yield (lb/ac-day)
Low density residential	0.03 - 0.04
Medium density residential	0.09 - 0.12
High density residential	0.11 - 0.13
Multifamily residential	0.07 - 0.10
Commercial	0.12 - 0.13

northern Virginia. The relationships indicate a log-linear increase in phosphorus load with increasing runoff volume for a 118,400-ac mixed urban-rural watershed and a 219,400-ac watershed with predominantly agricultural land use. The correlation coefficients for these two regression lines are 0.94 and 0.93, respectively [17]. Although the difference in phosphorus yields between the two basins is about fivefold, the rate of increase in yields is roughly parallel.

Phosphorus concentrations in runoff from streets, parking lots, bare areas, corporation yards and construction sites can range from about 0.5 to over 3 mg/l [16]. Over 56% of the phosphorus found on streets is associated with very fine silt-like particulates (< 43 μ), and over 85% is associated with particle sizes less than 104 μ [8].

Combined Sewage Sources

Phosphorus in combined sewage comes from both domestic and stormwater/urban runoff sources. Phosphorus in municipal sewage comes from

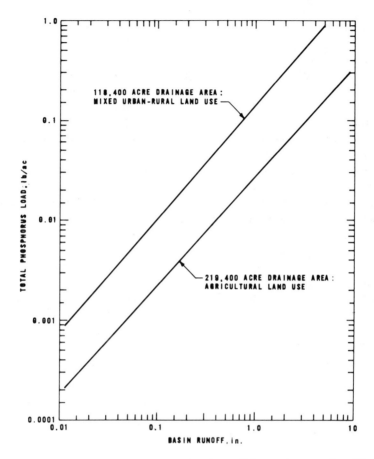

Figure 3. Relationship of phosphorus yield and runoff in two large northern Virginia watersheds [17].

human and food wastes, 30 to 50%, and from household and commercial compounds such as detergents, 50 to 70% [2]. The high phosphorus concentrations in municipal sewage (8 to 10 mg/l) are diluted by the high volumes of stormwater which have phosphorus concentrations of about 0.3 to 1.0 mg/l. The resulting characteristic phosphorus concentrations in combined sewage (1.0 to 3.0 mg/l) are significantly less than the concentrations in municipal sewage but slightly more than those found in urban runoff.

Combined sewage flows of about five times the average dry-weather flowrate could discharge equal mass loadings of phosphorus as untreated municipal sewage. This estimate may be conservative, as typical overflow volumes can exceed 10 times the dry-weather flow.

TREATABILITY

Most storm and combined sewer overflow treatment facilities have been designed to reduce the solids and organic loads occurring during storm events. Several unit processes using chemical addition have been demonstrated at the pilot scale [3]. Chemical coagulants are used primarily to increase solids capture, particularly where high-rate devices are used or where the varying flow and loading necessitate chemical addition to improve performance at nonoptimum operating conditions.

High levels of treatability are achieved by physical unit processes with chemical addition, usually alum, ferric chloride, or lime, to precipitate and settle the soluble phosphorus. Without chemical addition, only the insoluble fraction of the total phosphorus is readily treatable, resulting in removals of between 10 to 30%.

A study on the treatability of phosphates in stormwater showed that ferric chloride performed best. However, higher dosages were required than that of alum. Lime dosages averaging 800 mg/l were required to raise the pH of stormwater (4-5.5) to 11.0 or higher to obtain reasonable phosphorus removal. Dosages of alum and ferric chloride to achieve the optimum phosphorus removal are shown in Table V for nonoptimum and optimum pH ranges. Phosphorus removals using ferric chloride ranged from 80 to 87% for dissolved phosphate and orthophosphate in quiescent settling tests [18].

Table V. Alum and Ferric Chloride Dosages
for Phosphorus Removal in Stormwater [18]

Coagulant	Dosage (mg/l)	
	optimum pH	nonoptimum pH
Alum	45 - 60	80 - 90
Ferric Chloride	70 - 80	110 - 140

Source Control Measures

Source control measures, or best management practices (BMPs), are designed to prevent stormwater pollution problems form developing at the source. These measures are equally applicable to stormwater runoff or combined sewage flows. BMPs can include (1) regulatory action to limit the

use of polluting chemicals or control development so that pollutant generation is minimized; (2) maintenance measures such as street sweeping; and
(3) low structural soruce controls such as detention facilities. Street cleaning,
for example, can reduce pollutants entering both combined sewer systems or
being washed off in urban runoff.

The discussion of treatability or removal effectiveness of BMPs is limited
to the "most promising" technical approaches—those that are being or have
been implemented most widely. These measures include street cleaning and
source detention/treatment.

Street Cleaning

A demonstration project in San Jose, California, evaluating improved
street cleaning practices has shown street cleaning effective, or moderately
effective in removing litter, oil, grease and heavy metals; but it has a low
suitability for removing phosphorus [19]. Accumulation rates of orthophosphate ranged from 0.37 to 1.1 lb/curb-mi-yr and removal rates ranged from
0.01 to 0.08 lb/curb-mi per equipment pass. The estimated effectiveness of
street cleaning is summarized as a function of the number of passes of street
cleaning equipment:

1 - 2	passes per weekday:	1 - 20%
1	pass per week:	$<$ 1%
1 - 3	passes per 3 months:	$<$ 1%

Street cleaning was shown to be effective only on the larger ranges of
street particulate sizes, and since over 85% of the phosphorus has been associated with the very fine fraction of street solids, low phosphorus removals
would be expected.

Urban Runoff Detention

Onsite detention of urban runoff is used to control runoff volumes and
rates in downstream drainage systems. The reduction of flowrate may
directly affect the erosion potential and thus the phosphorus load potential
from the eroding soils.

Source detention facilities can also provide treatment. Where flows are
detained and released, 10 to 30% of the phosphorus (insoluble fraction) can
be removed. Where detention or retention is followed by treatment such as
infiltration-percolation, removals over 90% are possible.

In Orange County, Florida, source detention/treatment facilities were
found to have high phosphorus load removal effectiveness, as shown in

Table VI. The principal mechanism for orthophosphate removal, by infiltration, is adsorption by the clay minerals. The potential exists for contamination of the groundwater. However, in one example, after 88 years of rapid infiltration of wastewater in Calumet, Michigan, groundwater phosphorus concentrations only range from 0.1 to 0.4 mg/l [1].

Table VI. Phosphorus Removal from Stormwater Detention/Treatment
Facilities in Orange County, Florida [20]

Stormwater control measure	Drainage area (ac)	No. of storms sampled	Influent phosphorus load (lb/yr)	Effluent phosphorus load (lb/yr)	Average removal (%)
Diversion/percolation[a]	4.6	6	8.7	0.05	99[+]
Retention/percolation	13.8	6	14.7	0	100
Swales[b]/percolation	4.9	4	21.4	1.7	92
Underdrains[c]	9.0	4	14.4	1.1	93
Detention/sedimentation	4.4	6	48.0	11.6	76

[a]Diversion of first flush materials to a storage/percolation basin.

[b]Detention in natural swales.

[c]Percolation and collection by underdrains.

Maintenance Operations

Maintenance operations and general city housekeeping can reduce the phosphorus loads available for transport to receiving waters. Information on the effectiveness of these measures is limited; however, flushing tests on a 15-in, 136-ft combined sewer near Boston revealed that approximately 40-60% of the organics and nutrients are transported considerable distances during flushing. Mass phosphorus removals of about 0.03 lb/flush from this pipe segment were reported [21]. The effectiveness of sewer flushing, however, may vary considerably depending on the characteristics of the combined sewer and the dry-weather pollutant deposition and buildup potential.

City housekeeping activities such as effective leaf removal from streets can also significantly reduce the phosphorus load.

Combined Sewage Treatment Facilities

Combined sewage treatment facilities are usually "end of pipe" solutions with most full-scale prototype facilities using combinations of storage and treatment. Storage is considered an essential component to reduce the peak volumes of flow in the system and reduce the size of the treatment capacity required. Most combined sewage treatment facilities use physical treatment processes because they can handle high and variable influent concentrations and flow rates and are particularly applicable for removal of settleable and suspended solids. Physical unit processes usually have short startup and shutdown times, compared with biological systems. Biological systems treating combined flows have used lagoons, usually aerated facultative lagoons; and one facility in Kenosha, Wisconsin used a contact stabilization process (modified activated sludge process) to treat combined flows, netting a phosphorus removal of about 58%.

A summary of the expected effectiveness for several physical treatment unit processes used to treat combined flows is presented in Table VII. These processes have been demonstrated in either pilot-scale or are full-scale facilities used for areawide combined sewage treatment. Phosphorus removals from physical processes not using chemical coagulant aids range from 10 to 30%, indicating that only the insoluble fractions of the total phosphorus are

Table VII. Comparison of Expected Physical Treatment Efficiencies
for Combined Sewer Treatment Processes [3]

| | Pollutant Reduction (%) | | | |
Physical Unit Process	Suspended Solids	Settleable Solids	Total Nitrogen	Total Phosphorus
Sedimentation				
Without chemicals	20 - 60	30 - 90	38	20
Chemically assisted	40 - 80	50 - 95		50[a]
Swirl concentrator/	40 - 60	50 - 90		10 - 15[a]
Flow regulator				
Screening				
Microscreens	50 - 95[b]		30	20
Drum screens	30 - 55	60	17	10
Rotary screens	20 - 35	70 - 95	10	12
Disk screens	10 - 45			
Static screens	5 - 25	10 - 60	8	10
Dissolved air flotation[c]	45 - 85	30 - 80	35	55
High rate filtration[d]	50 - 80	55 - 95	21	50

[a]Estimated phosphorus reductions.

[b]Approaching 95% with chemical additions.

[c]Includes prescreening and flotation with chemical additions.

[d]With chemical addition.

readily removed. With chemical addition, the removals increase to about 50-55%. Most of the chemically assisted physical processes are high-rate processes operated at high pollutant loading and hydraulic loading rates. Phosphorus removals in high-rate, chemically assisted processes are usually less than those achieved in dry-weather phosphorus removal facilities. Adequate mixing and flocculation conditions are usually more difficult to control in stormwater treatment systems and result in lower removals.

Experienced phosphorus removal capabilities are evaluated in the following selected case studies of pilot and full-scale treatment systems.

Saginaw, Michigan–Storage/Sedimentation

Saginaw's Hancock Street storage/sedimentation system serves a combined sewer area of about 10,200 ac and consists of an integrated system of inline storage, storage/sedimentation basins, and the dry-weather treatment facilities. Inline storage is used to reduce the peak wet-weather flowrates in the interceptor. Small storms may be totally contained in the inline system and released to the interceptor to the dry-weather facilities as capacity becomes available.

Larger storms may activate the storage/sedimentation basins where storms that are totally contained are released to the interceptor. If the capacity of the basins is exceeded, the facilities provide treatment by sedimentation. During a brief monitoring program conducted in the facilities, three storms causing overflow from the basins were sampled, showing an average phosphorus reduction of about 35%. The phosphorus reductions by sedimentation for each storm event are shown in Table VIII. Phosphorus removal by storage and sedimentation may vary depending on: (1) the volume of overflow stored and treated by sedimentation, (2) the hydraulic overflow rates during overflow from the basin, and (3) the timing of the influent loads (first flush, if any).

Table VIII. Phosphorus Removal by Sedimentation, Saginaw, Michigan[a]

	Hydraulic Loading Rate (gal/ft^2 · day)		Average phosphorus concentration (mg/l)		
Storm Date	Average	Peak	Influent	Effluent	Removal (%)
8/19/78	970	1500	2.6	1.3	50
9/13/78	1235	6500	2.8	2.4	14
9/20/78	2270	6300	2.2	1.3	41

[a]No chemical addition.

Similar results were obtained at New York City's Spring Creek storage/ sedimentation facilities where a 22% removal of total phosphorus and a 7% reduction of orthophosphate was achieved by sedimentation [3].

Fort Wayne, Indiana—Pilot Evaluation of Screening Devices

An evaluation of the effectiveness of a static screen, drumscreen, and rotary screen on combined sewage flows was conducted in Fort Wayne [22]. Screens are used to achieve various levels of suspended solids removal corresponding to:

- Main treatment—screening used at the primary treatment process.
- Pretreatment—screening used to remove suspended and coarse solids before further treatment.

Static screens have a stationary, inclined slotted screening surface with slot spacings ranging from 250 to 1600 μ. The combined sewage flows over a control weir at the top of the screen and flows down and through the inclined screening surface. The coarse solids are retained on the screen and move down the surface of the screen to collection.

The drumscreen is a horizontally mounted cylindrical cage covered with a screen fabric having an aperture range from 100 to 800 μ. Combined sewage enters the partially submerged drum and flows radially out through the screen fabric, trapping solids on the inside of the screen. The collected solids are backwashed from the screen into a collection trough.

The rotary screen is a vertically aligned drum with screen fabric apertures ranging from 74 to 167 μ. The drum rotates about a vertical axis, and the influent is introduced into the center of the rotating drum and directed along horizontal baffles to the screen surface. Solids are retained on the inside of the screen and are backwashed when the flow split becomes excessive, usually 70:30 [3].

Phosphorus removals obtained by the Fort Wayne facilities are presented in Table IX. The expected removals for all three screens are between 10 and 15%, which indicates that as with other physical treatment processes without chemical addition, only insoluble phosphorus is readily removable and is associated with the level of suspended solids removal.

Tests in Philadelphia, using a 23-μ microscreen, resulted in an average 9% increase in phosphorus removal, from 15 to 24%, with the addition of moderately charged, high molecular weight, cationic polyelectrolytes. Polyelectrolyte concentrations ranged from 0.25 to 1.5 mg/l [23].

Table IX. Experienced Phosphorus Removal from Screening Processes
Fort Wayne, Indiana [3,22]

Screening Device	Projected Average Suspended Solids Removal[a] (%)	Phosphorus Removal (%)	
		Range	Average
Static screen	12	0 - 28	14
Drum screen	40	0 - 22	11
Rotary screen	28	0 - 24	11

[a]Estimated value at an influent solids concentration of 400 mg/l [19].

Milwaukee, Wisconsin—Screening/Dissolved Air Flotation

Pilot plant studies were conducted in Milwaukee to determine the effectiveness of screening and dissolved air flotation (DAF) in treating combined sewage. The 5-mgd pilot plant included a 297-μ drumscreen before the DAF unit. The DAF tank, 61 x 18 x 8.5 ft, received the screen effluent and a pressurized flow charged with air. Once in the tank, the pressurized flow releases the air in solution and forms tiny bubbles that attach themselves to the solids and rise to the surface. Tests were also conducted using a 22-μ microscreen on raw combined sewage, and as a final treatment unit for flotation tank effluent.

Phosphorus removals, using the screening/DAF combination with a chemical addition averaging about 25 mg/l ferric chloride and about 4 mg/l cationic polymer, ranged from 30 to over 60% removal, as shown in Table X. The system was operated at both high and low overflow rates on the DAF unit with resulting average phosphorus removals of 42 and 48%, respectively. Phosphorus removal by screening/DAF appears to follow a similar trend as suspended solids removals in physical processes with increased removal rates at higher influent pollutants concentrations, as shown in Figure 4. Orthophosphate removals ranged from about 85 to over 95% [24].

Tests were also conducted using alum ranging in concentration from 20 to 60 mg/l, powdered carbon averaging about 40 mg/l, and a cationic polymer at about 1.0 mg/l. Phosphorus removals for these tests averaged about 63%, and with the higher dosage of alum ranged up to about 80-85%.

The effectiveness of screening alone on phosphorus removal was minimal; however, tests using a microscreen with an effective aperture size of 22 μ, resulted in phosphorus removals of about 22% [24]. These results are representative of the results for both microscreening of raw combined sewage and polishing of screening/DAF effluent.

Table X. Phosphorous Removal by Screening Dissolved Air
Flotation with Chemical Addition [24][a]

Event	Influent P Concentration (mg/l)	Effluent P Concentration (mg/l)		P Removal (%)	
		High Rate[b]	Low Rate[c]	High Rate	Low Rate
1	0.70	0.44	0.45	37	36
2	2.47	1.27	0.77	49	69
3	0.75	0.25	0.32	67	57
4	0.98	0.59	0.51	40	48
5	1.52	0.98	0.81	36	47
6	0.43	0.28	0.25	35	42
7	0.95	0.53	0.52	44	45
8	0.25	0.17	0.16	32	36
Average removal				42	48

[a]Chemical addition: about 25 mg/l $FeCl_3$ and 4 mg/l polymer.

[b]About 2.7-3.9 gal/ft^2-min.

[c]About 4.0-7.8 gal/ft^2-min.

Similar results were obtained in Racine, Wisconsin where 50-60% phosphorus removals were recorded using screening/DAF with chemical addition [3].

Cleveland, Ohio—High Rate Filtration

High rate, dual-medial filtration of combined sewage was piloted in Cleveland. The high rate filters consisted of a column of No. 3 (effective size 4.0 mm) anthracite (48 to 60 in) over No. 612 (effective size 2.0 mm) sand (36 in.). The combined flows received pretreatment with a drumscreen to remove coarse solids before entering the filter. Chemical coagulant aids, alum and anionic polymer, at about 30 mg/l and 1 mg/l, respectively, were added to improve filter performance. Anionic polymers proved to be the most effective at this test site. For comparison, several test runs were made with no chemical addition.

Phosphorus removal using alum and anionic polymer averaged about 53% with specific averages of 63, 52, and 44% at hydraulic loading rates of 8, 16, and 24 gal/ft^2-min, respectively [25]. Without chemical addition, the average phosphorus removal was approximately 32%. As the loading rate on the

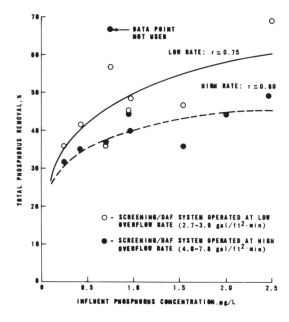

Figure 4. Phosphorus removal by dissolved air flotation, Milwaukee, WI [24].

filters increased from 8 to 24 gal/ft^2-min, the phosphorus removal efficiency dropped to about 45% with chemical addition, and about 30% without chemical addition, as shown in Figure 5.

In addition to the apparent influence of hydraulic loading rate, influent concentrations also affect the filter performance. The effects of reduced efficiency resulting from higher hydraulic loading rates are combined with the effects of increased influent loads, as shown in Figure 6.

Phosphorus removal results by high-rate, dual-media filtration in pilot plant studies in Rochester had similar results, with an average removal of 57%. Removals of total inorganic phosphorus ranged from 0 to 34% without chemical addition and from 20 to 89% with chemical addition [26].

SUMMARY

Annually, phosphorus loads from dry-weather municipal sewage represent about 85% of the total load, and wet-weather sources contribute the remaining 15%. However, during storms, nonpoint sources can contribute up to 65% of the total phosphorus load.

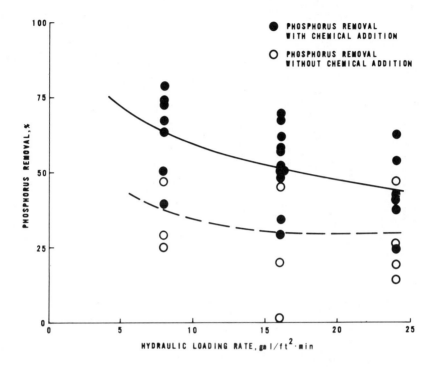

Figure 5. Phosphorus removal by high rate, dual-media filtration [25].

The principal source of phosphorus in urban runoff can be linked to land uses that increase runoff and erosion. No single source control measure is considered a total control approach; rather, a combination of source control strategies (nonstructural-maintenance, low structural-source detention, and regulatory actions) may be required to control many sources, each of which contributes to the phosphorus load. Source controls may also be practiced in combined sewered areas to reduce potentially large, identified sources.

Combined sewage treatement systems have demonstrated various phosphorus removal capabilities—with and without the use of chemical aids. The unit capital costs of combined sewage treatment systems, capable of 50 to 60% phosphorus removal with chemical addition, are significantly higher than primary treatment removal in the range of 10 to 30%, as shown in Figure 7. The wide range in costs for these processes depends on auxiliary process requirements such as pumping and/or pretreatment and scales of economy. These facilities may also sit idle for long periods between storms.

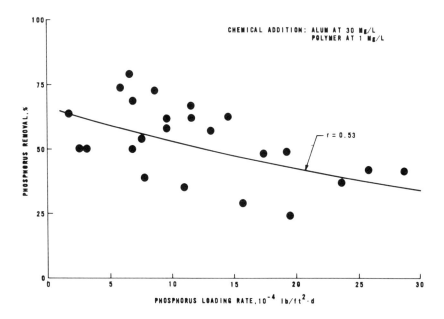

Figure 6. Effects of mass loading rates on high-rate filter phosphorus removal performance.

Use of high-rate or high-performance processes may, however, be an attractive alternative, particularly where land is limited or land costs are very high. In most full-scale, areawide applications, however, combined sewage treatment by storage and sedimentation represents the most cost-effective technology in terms of $/lb of suspended solids removed.

Control of phosphorus in combined sewage may be best achieved by integrating the wet- and dry-weather treatment facilities. Most large combined sewage treatment systems store and treat excess combined sewage and release the stored volume to the dry-weather treatment facilities.

Stored combined sewage volumes released to the dry-weather facilities may contain a large portion of the total combined phosphorus load, particularly if a first flush is captured. Dry-weather treatment potentially offers the most cost-effective approach for phosphorus control (chemical phosphorus removal), since higher continuous loads and higher concentrations are experienced annually. The control of phosphorus removal may be more easily maintained in the dry-weather facilities since they do not have the extreme transient flow and quality conditions of combined sewage treatment processes.

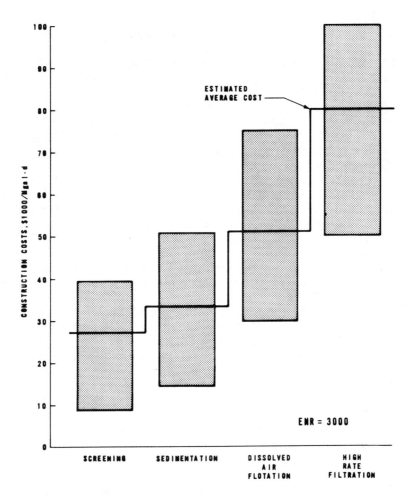

Figure 7. Representative ranges of construction costs for combined sewer overflow treatment processes 3.

REFERENCES

1. Metcalf & Eddy, Inc. and G. Tchobanoglous. *Wastewater Engineering: Treatment Disposal Resuse,* 2nd ed. (New York: McGraw-Hill, Inc., 1979).
2. "Process Design Manual for Phosphorus Removal," U.S. Environmental Protection Agency Technology Transfer, EPA-625/1-76-001a (April 1976).
3. Lager, J. A., W. G. Smith, W. G. Lynard, R. M. Finn and E. J. Finnemore. "Urban Stormwater Management and Technology: Update and User's Guide," EPA-600/8-77-014 (September 1977).

4.   "Report to Congress on Control of Combined Sewer Overflow in the United States," EPA-430/9-78-006 (October 1978).

5.   University of Wisconsin Water Resources Center. "Estimating Nutrient Loadings of Lakes from Non-Point Sources," EPA-660/3-74-020 (August 1975).

6.   Northern Virginia Planning District Commission and Virginia Polytechnic Institute and State University. "Planning for Nonpoint Pollution Management," prepared for EPA Conference on Watershed Management R & D, Athens, Georgia, October 18-20, 1977.

7.   White, C. A., and A. L. Franks. "Demonstration of Erosion and Sediment Control Technology—Lake Tahoe Region of California," EPA-600/2-78-208 (December 1978).

8.   Manning, M. J., R. H. Sullivan, and T. M. Kipp. "Nationwide Evaluation of Combined Sewer Overflows and Urban Stormwater Discharges, Volume III: Characterization of Discharges," EPA-600/2-77-064c (August 1977).

9.   Montgomery County Department of Environmental Protection. "Walts Branch Watershed Storm Water Management Concept Plan," draft (July 1977).

10. Randall, C. W., T. J. Gizzard and R. C. Hoen. "Impact of Urban Runoff on Water Quality in the Occoquan Watershed," Virginia Water Resources Research Center, Bulletin 80. (May 1978).

11. Cowen, W. F., and G. F. Lee. "Leaves as a Source of Phosphorus," *Envir. Sci. Technol.* 7(9) (1973).

12. Eisenreich, S. J., P. J. Emmuling and A. M. Beeton. "Atmospheric Loading of Phosphorus and Other Chemicals to Lake Michigan," *J. Great Lakes Res.* 3 (3-4) (1977).

13. Randall, C. W., D. R. Helsel, T. J. Grizzard and R. C. Hoehn. "The Impact of Atmospheric Contaminants on Storm Water Quality in an Urban Area," Department of Civil Engineering, Virginia Polytechnic Institute and State University (1978).

14. Griffin, Jr., D. M., T. J. Grizzard, C. W. Randall, and J. P. Hartigan. An Examination of Nonpoint Pollution Export From Various Land Use Types. International Symposium on Urban Storm Water Management, Lexington, Kentucky. July 24-27, 1978.

15. Smullen, J. T., J. P. Hartigan (Northern Virginia Planning District Commission, Falls Church, VA), and T. J. Grizzard. "Assessment of Runoff Pollution in Coastal Watersheds."

16. Tahoe Regional Planning Agency. "Lake Tahoe Basin Water Quality Management Plan. Volume III—Assessment of Water Quality and Environmental Impacts," (1978).

17. Grizzard, T. J., C. W. Randall, R. C. Hoehn and K. G. Saunders. "The Significance of Plant Nutrient Yields in Runoff from a Mixed Land Use Watershed," *Prog. Water Technol.* 10(5/6) (1978).

18. Alexander, S. B. "The Treatability of Stormwater Runoff From an Urban Commercial Catchment by Settling and Chemical Coagulation,"

M.S. Thesis, Virginia Polytechnic Institute and State University, (December 1978).

19. Pitt, R. "Demonstration of Nonpoint Pollution Abatement Through Improved Street Cleaning Practices," USEPA Grant No. S-804432, draft (January 1979).

20. Finnemore, E. J., W. G. Lynard and J. A. Loop. "Urban Stormwater Management and Technology: Case Histories," USEPA Contract No. 68-03-2617 (in progress).

21. Pisano, W. C. "Useful Technological Information on Sewer Flushing," presented at the EPA Technology Transfer Seminar Series on Combined Sewer Overflow Assessment and Control Procedures, Seattle, WA, June 28-29, 1978.

22. Prah, D. H., and P. L. Brunner. "Combined Sewer Stormwater Overflow Treatment by Screening and Terminal Ponding at Fort Wayne, Indiana," EPA Project No. 11020 GYU, Draft (June 1976).

23. Maher, M. B. "Microstraining and Disinfection of Combined Sewer Overflows—Phase III," EPA-670/2-74-049 (August 1974).

24. Gupta, M. K., D. G. Mason, M. J. Clark, T. L. Meinholz, C. A. Hansen and A. Geinopolos. "Screening /Flotation Treatment of Combined Sewer Overflows: Volume I—Bench Scale and Pilot Plant Investigations," EPA-600/2-77-069a (August 1977).

25. Nebolsine, R., P. J. Harvey and C. Fan. "High Rate Filtration of Combined Sewer Overflows," EPA 11023 EYI (April 1972).

26. Drehwing, F. J., C. B. Murphy, S. R. Garver, D. F. Geisser, D. Bhargava and G. C. McDonal. "Pilot Plant Studies—Combined Sewer Overflow Abatement Program," EPA Grant No. Y005141, Draft (November 1976).

CHAPTER 19

CONTROL OF PHOSPHORUS FROM AGRICULTURAL LAND IN THE GREAT LAKES BASIN

N. A. Berg

Soil Conservation Service
U.S. Department of Agriculture
Washington, DC

INTRODUCTION

When modern man discovered the Great Lakes he found their basins covered with forest. Since that time the land has undergone much change. Today, more than 3 million ha, about 5% of the basin, are in residential, commercial and industrial uses. Almost one-third of the basin, 18.6 million ha, is used for agriculture (cropland and pasture). Both Canada and the United States have profited from these land use changes. The Great Lakes basin includes the heart of North America's industrial establishment and produces a substantial portion of its food and livestock feed. Yet, these changes in land use also have brought a tremendous cost in reduced water quality.

The International Reference Group on Great Lakes Pollution from Land Use Activities (PLUARG) was established by the Canada-U.S. Great Lakes Water Quality Agreement of 1972 [1] under the aegis of the International Joint Commission to assess the extent and sources of pollution of the Great Lakes from land use activities and to recommend practicable remedial measures.

Accelerated eutrophication in the lower Great Lakes, and in certain near-shore areas of the upper lakes, results from higher than natural phosphorus loadings. PLUARG concluded that, even with point-source phosphorus control, reductions in phosphorus from diffuse sources, including agricultural lands, will be needed in several areas in the Great Lakes basin to meet target loads. The Phosphorus Management Strategies Task Force will be working on an assessment of the tradeoffs among alternative control strategies. This chapter will summarize the findings and recommendations of PLUARG [2] and others for the control of phosphorus from the basin's agricultural lands, and outline some related strategies for implementing remedial measures mandated by the Great Lakes Water Quality Agreement of 1978. It will also describe some significant efforts already underway that are helping to reduce phosphorus loadings from agricultural land as well as yielding other benefits to the Great Lakes System.

PHOSPHORUS LOADINGS FROM AGRICULTURE

There are 18.6 million ha of agricultural land in the Great Lakes basin, of which 8.1 million ha are cropland (Table I) [3]. Some 24% (10,850 metric tons) of all phosphorus (excluding shore erosion) entering the Great Lakes in 1976 came from the basin's agricultural land [2]. Agricultural land contributions vary significantly from lake to lake as is shown in Table II. Within Ontario agricultural watersheds studied by PLUARG, an estimated 70% of total phosphorus came from cropland. For individual watersheds, this ranged from less than 50 to 98%. Predicted contributions from livestock in these watersheds ranged from less than 1 to 60%, with an average of 20% [4]. Table III summarizes intensive livestock operations in the basin. About 60% of the total phosphorus in the Ontario agricultural watersheds was found to be associated with sediment [4]. These findings are comparable to those in the Maumee River Basin where about 80% of the land area is in intensive row-crop agriculture [5].

Unit area phosphorus loads from agricultural lands were calculated in the PLUARG pilot watersheds (Table IV). There were very wide ranges of unit area loads within a given type of agricultural land use (Table V). Further analyses led to the conclusion that the principal determinants involved were the clay content of soil and the proportion of the area in row-crops. Soils with high clay content were found to be high in phosphorus content, susceptible to erosion, and more readily delivered to the stream network. Figure 1 shows the general location of these soils [6].

Physiography, soil erodibility, drainage area, livestock population and—to a lesser extent—fertilizer phosphorus application also contributed to the ranges of unit area loads [4,5,7]. Phosphorus losses from tile drainage in mineral soils were found, in the PLUARG agricultural watersheds, to be insignificant, but in organic soils such losses may be high [4].

It was the great variation in unit area loads, as well as the need to account for hydrologic variability [8], which led PLUARG to the adoption of the

Table I. Agricultural Land Use in the Great Lakes Basin (ha) [3]

Lake Basin	Cropland	Pasture	Total Agriculture
Superior			
U.S.	10,604	389,924	400,528
Canada	2,241	51,154	53,395
Total	12,845	441,078	453,923
Michigan			
U.S.	2,747,744	3,374,524	6,122,268
Canada	0	0	0
Total	2,747,744	3,374,524	6,122,268
Huron			
U.S.	686,096	683,900	1,369,996
Canada	511,949	1,303,933	1,815,882
Total	1,198,045	1,987,833	3,185,878
Erie			
U.S.	2,189,144	1,794,940	3,984,084
Canada	1,182,228	670,031	1,852,259
Total	3,371.372	2,464,971	5,836,343
Ontario			
U.S.	425,188	1,124,016	1,549,204
Canada	387,729	1,056,468	1,444,197
Total	812,917	2,180,484	2,993,401
Great Lakes			
U.S.	6,058,776	7,367,304	13,426,080
Canada	2,084,147	3,081,586	5,165,733
Total	8,142,923	10,448,890	18,591,813

Table II. Phosphorus Loadings from Agricultural Land [3]

Lake	Total P From Agricultural Land (metric tons/yr)	Percent of Total P Load[a]
Superior	168	4
Michigan	1334	21
Huron	1649	34
Erie	5584	32
Ontario	2115	18

[a]Excluding P from shore erosion.

Table III. Number of Intensive livestock Operations in the Great Lakes Basin[a] [3]

Lake Basin	Poultry			Cattle			Swine			All Livestock
	U.S.	Canada	Total	U.S.	Canada	Total	U.S.	Canada	Total	
Superior	3		3	84		84	3		3	90
Michigan	280		280	5,094		5,094	1,477		1,477	6,851
Huron	51	27	78	704		863	98	111	209	1,150
Erie	206	78	284	1,816	711	2,527	1,145	546	1,691	4,502
Ontario	98	60	158	1,696	132	1,828	44	177	221	2,207
Great Lakes	638	165	803	9,394	1,002	10,396	2,767	834	36,001	14,800

[a]For the United States portion of the basin, intensive livestock operations were defined as follows: 10,000 or more poultry; 100 or more cattle, and 200 or more swine. For the Canadian portion, the definitions were: 30,000 or more poultry; 75 or more daily cattle; 150 or more beef cattle, and 300 or more swine.

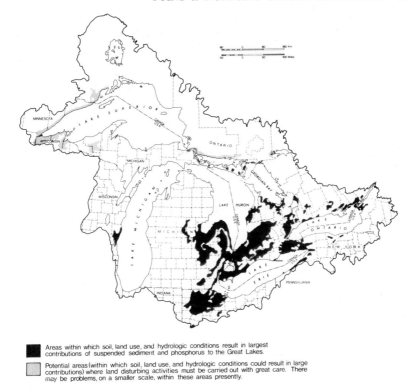

◼ Areas within which soil, land use, and hydrologic conditions result in largest contributions of suspended sediment and phosphorus to the Great Lakes.

▢ Potential areas (within which soil, land use, and hydrologic conditions could result in large contributions) where land disturbing activities must be carried out with great care. There may be problems, on a smaller scale, within these areas presently.

Figure 1. Regions of fine textured soils requiring, or potentially requiring treatment for erosion control.

Hydrologically Active Area concept. A hydrologically active area is one which produces significant amounts of runoff to stream systems, even during relatively minor rainfall and snowmelt events. More descriptive information relating to hydrologically active areas may be found in PLUARG's Final Report [2]. In a given watershed, 80-90% of the sediment may be contributed by only 15-20% of the land area [7]. This clearly has important implications for remedial strategies [9].

The preceding discussion relates only to the transport of phosphorus from agricultural land by land drainage (runoff and subsurface flow). Another component, which further research may find to be highly significant, is the phosphorus carried by wind-blown soil particles from cropland within the basin—and from areas far removed—which fall directly onto the surfaces of the Great Lakes. This was suggested in the Upper Lakes Reference Group's report to the International Joint Commission in 1977 [10].

Studies at the University of Illinois Urbana-Champaign showed large contributions of phosphorus from airborne soil particles [11]. Southwestern

Table IV. Unit-Area Loads of Phosphorus for Rural Land Uses,
from Pilot Watershed Studies [3]

Rural Land Use	Annual Unit Area Loads (kg/ha/yr)		
	Suspended Sediment	Total Phosphorus	Filtered Reactive Phosphorus
General Agriculture			
Genesee	30 -900	0.1 - 1.1	0.01 - 0.16
Grand/Saugeen	3 - 2200	0.1 - 2.3	0.01 - 0.5
Maumee	500 - 5600	1.4 - 6.9	0.2 - 0.5
Menomonee	230 - 410	0.3 - 0.6	0.2
Agricultural watersheds	30 - 800	0.1 - 1.5	0.02 - 0.6
Cropland			
Maumee River	80 - 5100	0.8 - 4.6	0.05 - 0.3
Mill Creek	20 - 70	0.2 - 0.6	0.1 - 0.3
Agricultural watersheds	400 - 800	0.9 - 1.5	0.3 - 0.4
Improved Pasture			
Agricultural watersheds	30 - 80	0.1 - 0.5	0.02 - 0.2
Forested/Wooded			
Genesee	7 - 820	0.02 - 0.67	0.01 - 0.03
Grand/Saugeen	30 - 50	0.1	0.01
Forested watersheds	1 - 5	0.04 - 0.2	0.03-0.1[a]
Idle/Perennial			
Felton-Herron	10 - 30	0.1 - 0.2	0.02 - 0.07
Genesee	7 - 820	0.02 - 0.67	0.01 - 0.03
Grand/Saugeen	30 - 50	0.1	0.01
Sewage Sludge			
Grand/Saugeen		0.2	0.01
Wastewater Spray Irrigation			
Felton-Herron		0.4 - 1.4	0.1 - 1.3
Grand/Saugeen		0.2	

[a]Total dissolved phosphorus.

Ontario's agricultural areas receive an average of about 1000 g/ha/yr of total phosphorus loadings from the atmosphere [12]. Sonzogni [13] calculated that over 4.4 million metric tons of fine-grained sediment, the type most likely to carry phosphorus, fall on the Great Lakes each year.

PLUARG estimates of total phosphorus from loads from all atmospheric inputs, as a percentage of total phosphorus from all sources, except shore erosion, are as follows:

Lake	% Atmospheric
Superior	37
Michigan	26
Huron	23
Erie	4
Ontario	4

Table V. Summary of Ranges of Unit Area Loads of Phosphorus
for Rural Land Uses from Pilot Watershed Studies [3]

Rural Land Use[a]	Annual Unit Area Loads (kg/ha/yr)		
	Suspended Solids	Total Phosphorus	Filtered Reactive Phosphorus
General Agricultural	3 - 5600	0.1 - 6.9	0.01 - 0.6
Cropland	20 - 5100	0.2 - 4.6	0.05 - 0.4
Improved Pasture	30 - 80	0.1 - 0.5	0.02 - 0.2
Forest/Wooded	1 - 820	0.02 - 0.67	0.01 - 0.10[b]
Idle/Perennial	7 - 820	0.02 - 0.67	0.01 - 0.07
Sewage Sludge		.02	0.01
Wastewater Spray Irrigation		0.2 - 1.4	0.1 - 1.3

[a]Dominant land uses, as defined in PLUARG pilot watershed study reports.
[b]Total dissolved phosphorus.

The question of the relative biological availability of various forms of phosphorus is an important one that is addressed elsewhere in this book. Bioavailability is highly dependent upon soil and other characteristics of contributing areas. The Maumee River Basin Summary Pilot Watershed Report cited substantial differences for filtered and filtered reactive phosphorus between the Cattaraugus Creek, New York and the Maumee River [5].

PRACTICES TO REDUCE PHOSPHORUS
FROM AGRICULTURAL LAND*

The location, selection and application of practices should fit within the framework of an overall management strategy. A recommended management strategy is described later in this paper under the heading, "Implementing an Agricultural Phosphorus Management Strategy."

To be effective, water quality management practices must be selected on a site-specific basis [14]. A given practice, for example, no-till farming, is very effective in reducing phosphorus losses on a wide range of soil types [15]. However, on some soils this practice must be accompanied by improved subsurface drainage in order to be effective. Furthermore, there are soils, such as the Paulding soil in Ohio, on which this practice simply will not work [5]. Some practices, which are marginal in most cases, may be the best recommendation in a given situation [16]. It is important to consider the effects of reducing both soil erosion and sediment yield in developing farm plans for improving water quality [17].

To ensure that practices are compatible with site conditions and with each other and commensurate with the farmer's ability to sustain an economically viable operation, it is recommended that farmers in hydrologically active areas develop water quality management plans for their operating units [2].

Practices for reducing phosphorus losses from agricultural land fall into three general categories: (1) those that reduce erosion; (2) those that reduce surface runoff; and (3) those that manage or control fertilizer and animal waste [16].

Practices to Control Erosion

Erosion usually can be controlled by practices, such as conservation tillage, which minimize raindrop impact on the soil surface and weaken the erosive forces of the runoff by reducing its velocity and channelization. Various investigators have concluded that, to control pollution from phosphorus and other sediment-associated contaminants, reducing detachment of soil particles is the best approach [18,5].

In many situations, erosion can be reduced with agronomic practices that improve crop residue management, cropping sequences, seeding methods,

---

*Much of this section has been adapted from the following publication, to which the author is indebted and the interested reader's attention is directed: Stewart, B. A., D. A. Woolhiser, W. H. Wischmeier, H. H. Caro and M. H. Frere. "Control of Water Pollution from Cropland. Vol. 1. A Manual for Guideline Development," U.S. EPA Report No. EPA-600-2/75-026a, U.S. Dept. of Agr. Report No. ARA-H-5-1, Washington, DC (1975), 111 pp.

soil treatments, tillage methods and timing of field operations. Generally, farming parallel to field contours will further reduce erosion. Contouring and some agronomic practices are not effective, however, where slope length or the area from which runoff concentrates is excessive. They then must be supported by practices such as terraces, diversions, contour furrows, contour listing, contour stripcropping, grassed waterways or water control structures. Table VI lists the principal types of erosion-control practices and some of their features.

Practices to Control Direct Runoff

Surface runoff from cropland rarely can be eliminated. It can be substantially affected, however, through agronomic and engineering practices by changing the volume of runoff or changing the peak rate of runoff. A change in runoff volume generally will change the peak runoff rate in the same direction; however, peak runoff rates sometimes can be changed without affecting the volume. Direct surface runoff volumes can be reduced by practices that increase infiltration rates, increase surface retention or detention storage or increase interception of rainfall by growing plants or residues. Runoff control practices are listed in Table VII.

Fertilizer and Animal Waste Management

These practices (Table VIII) reduce phosphorus losses by limiting fertilizer application to that recommended by soil test, reducing effects of erosion and runoff on the transport of phosphorus, and carefully managing livestock wastes.

EFFECTS AND COSTS OF IMPLEMENTING PRACTICES

No one wants to embark on an expensive undertaking without knowing, with some degree of certainty, what the effects will be and how much it will cost.

In the Great Lakes Basin most cultivated areas contain soils with high natural fertility. For this reason, measures which reduce erosion on cropland (Table VI) should be very effective in reducing phosphorus losses [7]. Investigations in PLUARG's eleven agricultural watersheds in Ontario showed that remedial measures which reduce erosion and sediment yield would be the most effective since an average of 60% of the total phosphorus was found to be associated with sediment. In four of these watersheds, total phosphorus reductions achievable with remedial practices were estimated to range from 36% in the Little Ausable River to 50% in Holiday Creek [4].

Table VI. Principal Types of Cropland Erosion Control Practices and Their Highlights [16]

No.	Erosion Control Practice	Practice Highlights
E 1	No-till plants in prior-crop residues	Most effective in dormant grass or small grain; highly effective in crop residues; minimizes spring sediment surges and provides year-round control; reduces man, machine and fuel requirements; delays soil warming and drying; requires more pesticides and nitrogen; limits fertilizer and pesticide placement options; some climatic and soil restrictions; especially very poorly drained soil.
E 2	Conservation tillage	Includes a variety of no-plow systems that retain some of the residues on the surface; more widely adaptable but somewhat less effective than E 1; advantages and disadvantages generally same as E 1 but to lesser degree.
E 3	Sod-based rotations	Good meadows lose virtually no soil and reduce erosion from succeeding crops; total soil loss greatly reduced but losses unequally distributed over rotation cycle; aid in control of some diseases and pests; more fertilizer-placement options; less realized income from hay years; greater potential transport of water soluble P; some climatic restrictions.
E 4	Meadowless rotations	Aid in disease and pest control; may provide more continuous soil protection than one-crop systems; much less effective than E 3.
E 5	Winter cover crops	Reduce winter erosion where corn stover has been removed and after low-residue crops; provide good base for slot-planting next crop; usually no advantage over heavy cover of chopped stalks or straw; may reduce leaching of nitrate; water use by winter cover may reduce yield of cash crop.
E 6	Improved soil fertility	Can substantially reduce erosion hazards as well as increase crop yields.
E 7	Timing of field operations	Fall plowing facilitates more timely planting in wet springs, but it greatly increases winter and early spring erosion hazards; optimum timing of spring operations can reduce erosion and increase yields.
E 8	Plow-plant systems	Rough, cloddy surface increases infiltration and reduces erosion; much less effective than E 1 and E 2 when long rain periods occur; seedling stands may be poor when moisture conditions are less than optimum. Mulch effect is lost by plowing.

E 9	Contouring	Can reduce average soil loss by 50% on moderate slopes, but less on steep slopes; loses effectiveness if rows break over; must be supported by terraces on long slopes; soil, climatic, and topographic limitations; not compatible with use of large farming equipment on many topographies. Does not affect fertilizer and pesticide rates.
E 10	Graded rows	Similar to contouring but less susceptible to row breakovers.
E 11	Contour strip cropping	Rowcrop and hay in alternate 50- to 100-foot strips reduce soil loss to about 50% of that with the same rotation contoured only; fall seeded grain in lieu of meadow about half as effective; alternating corn and spring grain not effective; area must be suitable for across-slope farming and establishment of rotation meadows; favorable and unfavorable features similar to E 3 and E 9.
E 12	Terraces	Support contouring and agronomic practices by reducing effective slope length and runoff concentration; reduce erosion and conserve soil moisture; facilitate more intensive cropping; conventional gradient terraces often incompatible with use of large equipment, but new designs have alleviated this problem; substantial initial cost and some maintenance costs.
E 13	Grassed outlets	Facilitate drainage of graded rows and terrace channels with minimal erosion; involve establishment and maintenance costs and may interfere with use of large implements.
E 14	Ridge planting	Earlier warming and drying of row zone; reduces erosion by concentrating runoff flow in mulch-covered furrows; most effective when rows are across slope.
E 15	Contour listing	Minimizes row breakover; can reduce annual soil loss by 50%; loses effectiveness with post-emergence corn cultivation; disadvantages same as E 9.
E 16	Change in land use	Sometimes the only solution. Well managed permanent grass or woodland effective where other control practices are inadequate; lost acreage can be compensated for by more intensive use of less erodible land.
E 17	Other practices	Contour furrows, diversions, subsurface drainage, land forming, closer row spacing, etc.

Table VII. Practices for Controlling Direct Runoff and Their Highlights [16]

No.	Runoff Control Practice	Practice Highlights
R 1	No-till plants in prior crop residues	Variable effect on direct runoff from substantial reductions to increases on soils subject to compaction.
R 2	Conservation tillage	Slight to substantial runoff reduction.
R 3	Sod-based rotations	Substantial runoff reduction in sod year; slight to moderate reduction in rowcrop year.
R 4	Meadowless rotations	None to slight runoff reduction.
R 5	Winter cover crop	Slight runoff increase to moderate reduction.
R 6	Improved soil fertility	Slight to substantial runoff reduction depending on existing fertility level.
R 7	Timing of field operations	Slight runoff reduction.
R 8	Plow plant systems	Moderate runoff reduction.
R 9	Contouring	Slight to moderate runoff reduction.
R 10	Graded rows	Slight to moderate runoff reduction.
R 11	Contour strip cropping	Moderate to substantial runoff reduction.
R 12	Terraces	Slight increase to substantial runoff reduction.
R 13	Grassed outlets	Slight runoff reduction.
R 14	Ridge planting	Slight to substantial runoff reduction.
R 15	Contour listing	Moderate to substantial runoff reduction.
R 16	Change in land use	Moderate to substantial runoff reduction.
R 17	Other practices	
	Contour furrows	Moderate to substantial reduction.
	Diversions	No runoff reduction.
	Drainage	Increase to substantial decrease in surface runoff.[b]
	Landforming	Increase to slight runoff reduction.
R 18	Construction of ponds[c]	None to substantial runoff reduction. Relatively expensive. Good pond sites must be available. May be considered as a treatment device

[a] Erosion control practices with same number are identical. Limitations and interactions shown in Table VI also apply to runoff control practices.

[b] Through the use of improved tile drainage, no-till planting can be practiced on a greater number of soils which, at present, contribute significant amounts of phosphorus to the Great Lakes [5].

[c] Small watershed reservoirs, constructed for flood prevention or other purposes, have been found to reduce downstream phosphorus loadings [19].

Table VIII. Practices for the Control of Nutrient Loss from Fertilizer
Applications and Animal Wastes [16]

Nutrient Control Practice	Practice Highlights
Eliminating excessive fertilization	Reduces available phosphorus losses; reduces fertilizer costs; has no effect on yield.
Incorporating surface applications	Decreases nutrients in runoff; no yield effects; not always possible; adds costs in some cases.
Timing fertilizer plow-down	Reduces erosion and nutrient loss; may be less convenient.
Livestock exclusion	Usually accomplished by fencing streambanks; often very expensive.
Livestock waste management systems	Components may include, but are not limited to: debris basins, dikes, diversions, fencing, grassed waterways or outlets, filter strips, irrigation systems, irrigation water conveyance, subsurface drains, surface drains, storage ponds, storage structures, waste treatment lagoons, and waste utilization (including the timing of manure application.

The Maumee River basin studies in Ohio suggested that practices which reduce erosion could reduce phosphorus loading significantly, by about 60-90% of the percentage reduction in gross erosion [5].

Since most of the phosphorus reduction from agricultural land would be accomplished through erosion and sediment control practices, it is important to consider other effects of these practices. One very important effect is that the soil resource, as a vital factor in the production of food, would be protected. The soils in the Great Lakes Basin are some of the most productive soils in North America.

Soil is a valuable asset on the farm, yet it becomes a liability if it is permitted to leave the farm in large quantities in the form of sediment. Sediment creates many negative externalities as a pollutant in the Great Lakes System. It has detrimental effects on the physical aquatic environment and affects fish populations in both upstream and lake habitat [13,20]. Accumulation of sediment decreases the effective life of reservoirs, increases the costs of bridge and stream channel maintenance, and requires expensive (sometimes environmentally degrading) lake harbor and channel dredging. In the Rochester, NY harbor, for example, an average of 220,000 metric tons of sediment must be dredged each year [13].

Suspended sediment detracts from the aesthetic value of the Lakes as a recreation resource and causes municipal water intake problems. At Cloquet, Minnesota, turbidity has exceeded drinking water standards 53% of the time [13].

Sediment originating from agricultural land does not seem to contribute significant quantities of heavy metals to the Great Lakes. Presently used pesticides, because of their low persistence, are not carried far by sediment. Yet residues of previously used organochlorine pesticides, such as DDT, still are carried by sediment to the Great Lakes [14]. Therefore, reduction of erosion and sediment from agricultural land not only will help meet phosphorus target loads in the Great Lakes, but will also alleviate other detrimental effects of soil loss and sedimentation.

The potential benefits of a program to reduce phosphorus loadings from agricultural land would be very significant. These benefits would be shared by many millions of people in the Great Lakes basin and by many millions more who live outside the basin but depend on it for food, commerce, or recreation.

It was emphasized earlier in this chapter that phosphorus reduction practices on agricultural land must be site-specific. It was also mentioned that, in some cases, a combination of practices is needed to attain desired effects. For these reasons and others, costs of remedial treatment will vary from lake basin to lake basin, from farm to farm, and even from field to field. Estimated costs for such practices, in four PLUARG agricultural watersheds in Ontario, range from $15 to $58 *annually* per watershed hectare [4]. In the Black Creek, Indiana project, costs for the *initial application* of practices were $146 per watershed hectare [21].

Recent estimates of costs in the U.S. Lake Erie basin present a more optimistic picture. An analysis of the economic impacts of changing tillage practices in that area indicates that a significant portion of the U.S. Lake Erie basin can be treated at minimal or no long-term cost to farmers through the use of minimum tillage or no-till planting [22]. Initial capital outlay for equipment, and for improved subsurface drainage on some soils, may be a deterrent.

IMPLEMENTING AN AGRICULTURAL
PHOSPHORUS MANAGEMENT STRATEGY

There is a major difference between pollution from municipalities and industries and pollution from agriculture. Most pollutants from municipalities and industries are regarded as *unwanted wastes* to be disposed of as economically as possible. Farmers, on the other hand, do not look upon their soil, and the fertilizers they have added, as wastes. They regard these as *assets* and would rather keep them on the land. Any strategy devised should help them to do so.

PLUARG's recommendations lay out a strategy for control of many pollutants from all nonpoint sources [2]. In this paper these recommendations are drawn upon principally as they relate to the control of phosphorus from agricultural lands. There are three major components of this strategy:

1. development of management plans;
2. implementation of management plans; and
3. review and evaluation.

In addition to these components, PLUARG made recommendations relating to the role of the publci in future studies done under the auspices of the International Joint Commission and suggested needs for future studies.

Development of Management Plan

The management plants should be prepared by appropriate jurisdictions and include a timetable with program priorities, designation of agencies responsible for implementation, formal arrangements for inter- and intra-governmental cooperation, identification of implementation programs and sources of funding, estimated reduction in loading to be achieved, estimated costs, and provision for public consultation and review.

In the development of management plans, PLUARG recommended that use be made of existing planning meachanism; that a complete review be made of presently available fiscal incentives; that greater emphasis be given to the development and implementation of information, education, and technical assistance programs; that the adequacy of legislation be assessed to insure that there is a suitable legal basis for enforcement should voluntary approaches prove ineffective; and that greater emphasis be placed on preventive aspects of laws and regulations.

PLUARG recommended very strongly that greater emphasis be given to information and education programs. Every one of the 17 public consultation panels suggested it. Every PLUARG member supported it—as a basis for wise legislation, for public acceptance of programs, and for private actions that will meet private and public goals. Their emphasis mirrored an earlier finding of the Great Plains Council in the U.S. [23] .

Implementation of Management Plans

Regional priorities for implementing management plans should be based on: water quality conditions in each Great Lake, the potential contributing areas identified by PLUARG (as further refined during the preparation of management plans) and the most hydrologically active areas found within the potential contributing areas.

PLUARG recommended that phosphorus loads be reduced to achieve individual lake target loads, and developed suggested target loads based on the best information available. It was further recommended that additional reductions be achieved in each of the Great Lakes to improve nearshore water quality and to prevent degradation.

In order to accomplish these phosphorus load reductions, PLUARG recommended that erosion and sediment control programs be improved and expanded to reduce the movement of fine-grained soil particles to the Lakes. Practices to reduce phosphorus from agricultural land were described earlier in this chapter.

With particular respect to agricultural land use, PLUARG recommended that agencies assisting farmers should help them to develop and implement water quality plants for each farm in the most hydrologically active areas. It is not a land use, per se, that affects water quality, but rather how the land is managed. These plants should consider all potential nonpoint source problems and be commensurate with the farmers' ability to sustain an economically viable operation.

PLUARG further recommended that those farmlands which have the least natural limitations for agricultural use be retained for this purpose. PLUARG suggested that water quality management programs probably will need to include both voluntary and regulatory components. Regulation should be used only when voluntary approaches do not achieve desired results. The emphasis on "voluntarism first" was supported by the PLUARG agricultural survey. Some 56% of the Canadian farmers [24] and 71% of the U.S. farmers [25] said the best policy for reducing water pollution was to rely solely on the voluntary cooperation of farmers. PLUARG felt that all levels of government should review the adequacy of their present voluntary programs first to determine if more specific guidelines or incentives may be needed. Governments should maximize the use of existing programs before creating new ones.

The U.S. Environmental Protection Agency, through its Great Lakes National Program office, and the Great Lakes Basin Commission have strong roles to play as well as the "Section 208" agencies within the Basin.

Canada has a less well-defined mechanism for onsite assistance to landowners, although there is excellent potential capability within the conservation authorities; Agriculture Canada, the Ontario Ministry of Agriculture and Food, and other agencies in the Province of Ontario; and universities.

Review and Evaluation

PLUARG recommended that the International Joint Commission ensure regular review of programs arising from its recommendations and that nonpoint source interests be represented duirng these reviews.

Tributary monitoring should be expanded to improve the accuracy of stream loading estimates of sediment and phosphorus, as well as lead and PCBs. Sampling should be based on stream response characteristics, with intensive sampling of runoff events where necessary.

The role of atmospheric inputs should be evaluated with a special effort to determine the sources of phosphorus and contaminants carried in the atmosphere.

In addition, better coordination of data is needed and the adequacy of nearshore and offshore water surveillance efforts should be examined.

The foregoing recommendations were arrived at only after careful consideration of the findings and after full exposure through an unprecedented public consultation program.

Approaches toward pollution abatement programs may need to be somewhat different in the two nations, due to differing constitutional development and perceptions of the proper role of public agencies [26]. Furthermore, U.S. programs will vary from state to state.

PLUARG did not recommend a rigid scheme for achieving target loads. Rather, it suggested that each concerned jurisdiction compare alternatives for reaching cost-effective and politically acceptable solutions. It will not be necessary to invent new planning agencies, or new implementing agencies for that matter. PLUARG suggested that governments make better use of existing agencies and programs.

A prime example of such agencies are the soil and water conservation districts in the U.S. portion of the basin. These districts are legal subdivisions of state government, responsible for soil and water conservation work within their boundaries, just as townships and counties are responsible for roads and other services, and school districts are responsible for education. There is a soil and water conservation district covering each of the 190 counties in the U.S. Great Lakes basin. A state soil conservation agency assists each of these districts. Each district is governed by a locally elected board of supervisors. While many districts in the basin have their own technicians, they also rely on personnel and facilities of several feeder and state agencies for trained manpower. Districts have been actively involved in the Section 208 water quality management planning activities in many parts of the basin.

Control of phosphorus will not require that every farm in the Great Lakes basin have a water quality plan and that every field be adequately protected in order to realize beneficial effects in the Lakes. What PLUARG has said is that by working primarily through existing agencies and by giving special attention to the cropland within the most hydrologically active areas— perhaps less than 25% of the basin's agricultural land—total phosphorus loads to the Great Lakes can be reduced by more than 1000 metric tons/yr [6]. There would be further benefits in decreased sediment damage to fish habitat, reduced harbor and channel dredging costs, reduced loadings of other sediment-associated contaminants such as pecticides, improved quality of tributary waters, and maintenance of the productive capacity of the land.

CURRENT EFFORTS THAT WILL REDUCE
PHOSPHORUS FROM AGRICULTURAL LAND

There have been many recent developments which, if continued, will reduce phosphorus loadings from agricultural land in the Great Lakes. Much of this effort is a redirection of long-standing soil and water conservation progams or mandated by the U.S. (Federal Water Pollution Control Act and amendments) but a great deal has been inspired by the Great Lakes Water Quality Agreement of 1972 and the outstanding information and public consultation work sponsored by the International Joint Commission.

Water quality management planning is well underway in the U.S. portion of the basin. This is being done by 21 areawide and 8 state "Section 208" planning agencies with strong public participation as mandated by PL 92-500. The Great Lakes Basin Commission, with a grant from the U.S. Environmental Protection Agency (EPA), has held workshops to bring the Great Lakes Water Quality Agreement and PLUARG's findings to the attention of these planning agencies. The U.S. Army Corps of Engineers, as a result of the ongoing Lake Erie Wastewater Management Study, and the U.S. Department of Agriculture Soil Conservation Service (USDA-SCS) have assisted at these well-attended workshops. SCS has provided professional conservationists on detail to EPA regional and national offices to advise on the water quality aspects of soil conservation practices [27]. In addition, the USDA Science and Education Administration-Extension (SEA-Ext) has provided an extension specialist to EPA to aid information and education efforts.

In Ontario, workshops have been held with leaders of farm organizations, and PLUARG recommendations are receiving very positive consideration from conservation authorities [25,29]. The Conservation Authority Act requires that the initiative for the formation of a conservation authority come from the local municipality—and ensures, once an authority is formed, that resource management decisions reflect local needs and priorities. The member municipalities also are required to contribute funds toward the completion of approved resource management projects in accordance with the degree of benefit to be received. On the basis of this demonstrated local commitment, the province agrees to enter into a partnership with the local municipality through the conservation authority. The province provides technical leadership in the form of comprehensive watershed services and provision of field staff and final assistance through grants for approved projects. The conservation authority maintains its corporate autonomy while providing an effective link between its member municipalities and the province [30].

Through the years the conservation authorities have proven to be organizations which can readily adapt to changing needs and priorities at both the provincial and local level. This ability to adapt, through organizational and program adjustment, has contributed to the steady expansion of the conservation authority program in the province. Since 1946, thirty-eight conservation authorities, embracing some 400 municipalities and approximately 90% of Ontario's population, have been established [30].

Implementation has already begun to some degree. For example, practice application in the Black Creek Project in Indiana has been completed. Work on this 4800-ha agricultural watershed in the Maumee River drainage to Lake Erie, administered by the Allen County Soil and Water Conservation District, serves as an example for similar projects in the Great Lakes. Five districts (four in Wisconsin and one in Minnesota) have led the Red Clay project, which borders Lake Superior. The project is scheduled for completion this year. Application of practices will begin next year in an 80,000-ha watershed in two Michigan counties draining into the Saginaw Bay, with $400,000 allocated for cost sharing during Fiscal Year 1979. This special Agricultural Conservation Program (ACP) project is located within a potential contributing area identified by PLUARG.

During 1978, throughout the U.S. portion of the Great Lakes basin, farmers—through conservation districts and with SCS assistance—developed more than 5300 conservation plans covering more than 340,000 ha. Most of the practices called for in these plans will reduce erosion and sediment yield, thereby reducing phosphorus loadings in the Great Lakes System. Among the practices actually applied by farmers during 1978 were almost 57,000 ha of minimum tillage and 300 agricultural waste management systems. These systems provide for storage and land application of wastes from 18,000 animal units, which amounts to about 650,000 kg/day, the equivalent to the volume of waste produced by a city of 360,000 people.

Soil surveys have been completed on almost 90% of the potential contributing areas within the U.S. Great Lakes basin lands and are an essential tool in the location of lands needing treatment within specific hydrologically active areas and the selection of appropriate practices.

Every SCS field office is equipped with a technical guide, based on current research and tailored to local soil and climatic conditions, which helps assure the proper selection and design of practices. An Agricultural Waste Management Field Manual [31] provides guidance for waste management practices in the United States. The Canada Animal Waste Management Guide [32] serves a comparable purpose.

A program to subsidize erosion control and manure management practices in Ontario is scheduled to begin this spring. This program, administered by the Ontario Ministry of Agriculture and Food, is expected to cost $6 million per year for the next five years. While the entire province is eligible, a high proportion of the expenditures, and the benefits, probably will be in the Great Lakes basin. There will be a strong role for conservation authorities to play in this most important program, and there will be demonstration projects and other information and education efforts.

Potentially, the most important implementing authority in the United States is the Rural Clean Water Program (RCWP) (Section 208 j. of the Clean Water Act of 1977) [34]. This cost-sharing program, influenced to a large degree by PLUARG findings, applies only in critical contributing areas identified in EPA-approved Section 208 plans. Farmers will receive technical assistance in the development of water quality plans tailored to their farm and cost-sharing for required practices under 5- to 10-year contracts. RCWP projects will be administered by soil and water conservation districts, by state soil and water conservation agencies, or by state water quality agencies wherever practicable. In other cases, USDA will retain administrative responsibility [34].

The Agricultural Conservation Program also provides federal cost-share assistance to farmers for the application of conservation practices, many of which help to reduce erosion and improve water quality practices and programs.

In the past, a keystone of soil and water conservation district programs has been that farmers voluntarily came to districts asking for assistance. This is changing somewhat. Pennsylvania and New York State laws now require conservation plans. In Ohio, a farmer is assumed to be meeting state pollution abatement performance standards if he is following a conservation plan approved by a district. In Michigan, farmers are not required to have a permit for earth-disturbing activities if they are district cooperators [26].

In aid of many local and state initiatives toward land and water quality, and particularly related to the PLUARG recommendation on important farmlands, the U.S. Secretary of Agriculture has issued an official Statement on Land Use Policy which, among other things, calls for Department of Agriculture agencies to help identify and to advocate the retention of such lands [35].

A LOOK INTO THE FUTURE

The goals for phosphorus control are stated in Annex 3 of the 1978 Great Lakes Water Quality Agreement [36]. These goals appear to be consistent with those of the PLUARG public consultation panels [37,38]. The goals have been restated many times by private citizens at public hearings conducted by the International Joint Commission. Many of the world's leading scientists, such as those participating in this conference, have dedicated themselves to helping attain these goals.

As far as agriculture is concerned, the principal planning and implementing mechanisms are in place to make its contributions toward meeting the goals. What seems to be needed is the commitment—on the part of federal, provincial, state and local governments—to direct the needed technical, financial and educational resources toward meeting the phosphorus goals described in Annex 3 of the 1978 Great Lakes Water Quality Agreement and (hoped for) by the public. Will this happen? The long-term implications of diffuse source control programs are well recognized, but we are admonished that this should not be used as an excuse for delay [9]. We now have a management strategy to follow which is adaptable to physical, social and political conditions throughout the Great Lakes basin.

Will strategies for phosphorus control be meshed with overall strategies for achieving water quality in the Great Lakes? We do not want to unnecessarily and unwisely polarize pollution abatement into single-pollutant pigeonholes. We will be most effective by taking a holistic view that permits taking advantage of "piggyback" benefits of programs as well as avoiding contradictory results from separate programs.

Will these things happen? They *can*, I think they *will*, but I think we must resolve to move a little faster.

We have not yet fully addressed the recommendations of the Lower Lakes Reference Group to the International Joint Commission ten years ago [39].

We have not yet fully addressed the early-action suggestions of PLUARG in 1974 [40]. Note, for example, that those proposals included the use of land for recycling wastes. As land use pressures within the Great Lakes Basin become heavier to meet many needs, one such pressure will likely be for locating land suitable for recycling wastes.

As of December 1977, the U.S. and Canada had obligated more than $4.5 billion for sewage treatment in the Great Lakes in just six years; expenditures for controlling nonpoint source pollution are—to be kind about the comparison—minimal, even in 1979.

It will soon be a year since PLUARG gave its report to IJC on an "Environmental Management Strategy for the Great Lakes System." Its recommendations—and the nonpoint source water pollution that it addressed —need a higher profile than they have yet received. This conference will help.

Meanwhile, as mentioned earlier in this paper, not everybody in the several states and Ontario is waiting for final reports and assignments. We see some very positive efforts underway, with grassroots public support and local leadership. I might add that not one of these efforts resulted in any new agencies being formed!

I am also greatly encouraged by a meeting I attended recently of agricultural experts from Ohio, Indiana, Michigan, New York, Pennsylvania and Ontario. There was dedication, a tremendous sharing of experience and knowledge, open and frank discussion of successes *and* failures, a repeated challenge to get the best possible information to farmers in the best possible way, and discussion of great differences in areas, soil type and its response or behavior, and in climatic conditions and patterns.

Out of this session I became more convinced than ever that we need to use all of our tools in the toolbox, in whatever mix is called for in a specific situation—focusing, say on erosion control in Ohio, point-source controls in New York. No one strategy, no one practice will do it all. No-till is great where it fits, but we should *not* "throw away the plow."

I am likewise convinced that water quality is more likely to improve if final land use and treatment decisions are made in the field. By far the best way to achieve an effective and acceptable program would be through one-on-one assistance to landowners. With the small number of employees and volunteers that we will have to do the job, however, we will have to augment the one-on-one with some other ways of expanding our knowledge and the understanding of landowners.

As PLUARG suggests in its final report [2], we will need to expand our knowledge by:

- Looking more closely at near-shore areas, which are the most affected by man's activities;
- Studying the biological availability of pollutants;
- Measuring in-lake contamination;
- Quantifying the loadings from various lands;
- Better defining pollution and better identifying hydrologically active areas; and
- Studying the short- and long-term effectiveness of remedial measures.

I am very optimistic about the future *if*, while we seek to expand our knowledge, we will move quickly to help landowners use what we already know. In the United States, we have learned much through working with our Canadian friends in the PLUARG studies. These efforts will help our nations, separately and jointly, proceed with action programs to accomplish our goals for the Great Lakes.

SUMMARY

About 24% (10,850 metric tons) of all P, exclusive of shoreline erosion, entering the Great Lakes comes from the basin's agricultural land. Most P is carried by clay-sized sediment particles eroded from fertile topsoil. Much of the soluble P from agricultural areas is from dairies and other livestock enterprises. An appreciable amount is also found in cropland runoff.

In a given watershed, 80-90% of the sediment may be contributed by only 15-20% of the drainage area (the most hydrologically active area). Unit area loads from agricultural land vary. The principal determinants are the clay content of soil and the proportion of the area in row crops. Physiography, soil erodibility, drainage area, livestock population, and—to a lesser extent— fertilizer P application also contribute the wide ranges of unit area loads for total P.

Practices that can reduce P losses from agricultural land include those that reduce erosion, surface runoff, and nutrient loss from fertilizer applications and animal wastes. To be effective, these practices must be selected to fit a specific site, properly designed and installed, and carefully maintained.

Use of these practices in the contributing areas could reduce P loading to the Great Lakes from agriculture by more than 10% (about 20% for Lake Erie), and would also decrease sediment damage to fish, reduce harbor and channel dredging costs, reduce loadings of other sediment-associated contaminants, improve the quality of tributary water, and maintain the productive capacity of some of the best agricultural land in North America.

Remedial programs should be based on management plans developed by appropriate jurisdictions such as Section 208 water quality planning agencies in the United States. These implementation plans should include a timetable and estimates of load reductions and costs and should identify the most serious contributing areas, the most suitable institutional arrangements, and available sources of funding and technical assistance. Planning and implementation also must involve the public and provide for accelerated information and education efforts. Conservation authorities in Ontario and local soil and water conservation districts in the United States can provide leadership. Program review and evaluation likewise will be important, including adequate tributary monitoring.

No rigid scheme is recommended for achieving target loads. Instead it is suggested that each concerned jurisdiction compare alternatives for reaching cost-effective and politically acceptable solutions. Better use should be made of existing agencies and programs before inventing any new agencies.

REFERENCES

1. The Governments of Canada and the United States. "Great Lakes Water Quality Agreement," Signed at Ottawa, Ontario (1972).
2. International Reference Group on Great Lakes Pollution from Land Use Activities. "Environmental Management Strategy for the Great Lakes System," final report submitted to the International Joint Commission, Windsor, Ontario (1978).
3. Task Group B of the International Reference Group on Great Lakes Pollution from Land Use Activities. "Land Use and Land Use Practices in the Great Lakes Basin," Joint Summary Report, United States and Canada, Windsor, Ontario (1977).
4. MIller, M. H., and A. C. Spires. "Contribution of Phosphorus to the Great Lakes from Agricultural Land in the Canadian Great Lakes Basin," Phosphorus Integration Report, Task Group C (Canadian Section) International Reference Group on Great Lakes Pollution from Land Use Activities (1978).
5. Logan, T. J. "Maumee River Basin Summary Pilot Watershed Report," Task Group C (U.S. Section) International Reference Group on Great Lakes Pollution from Land Use Activities (1978).
6. Johnson, M. G., J. Comeau, W. C. Sonzogni, T. Heidtke and B. Stahlbaum. "Management Information Base and Overview Modelling," for the International Reference Group on Great Lakes Pollution from Land Use Activities, Windsor, Ontario (1978).
7. Chesters, G., J. Robinson, R. Stiefel, R. Ostry, T. Bahr, R. Coote and D. M. White. "Pilot Watershed Studies; Summary Report," to the International Reference Group on Great Lakes Pollution from Land Use Activities (1978).
8. Walling, D. E. "Natural Sheet and Channel Erosion of Unconsolidated Source Material," in *Fluvial Transport of Sediment-Associated Nutrients and Contaminants",* 1976 Workshop Proceedings, International Joint Commission, Research Advisory Board and International Reference Group on Great Lakes Pollution from Land Use Activities, Windsor, Ontario (1977) pp. 11-33.
9. Vallentyne, J. R. "Implications for the Great Lakes," in *Fluvial Transport of Sediment-Associated Nutrients and Contaminants,* 1976 Workshop Proc., International Joint Commission, Research Advisory Board and International Reference Group on Great Lakes Pollution from Land Use Activities. Windsor, Ontario (1977) pp. 275-277.

10. Upper Lakes Reference Group. "The Waters of Lake Huron and Lake Superior," final report submitted to the International Joint Commission, Windsor, Ontario (1977).

11. Murphy, T. J. "Sources of Phosphorus Inputs from the Atmosphere and Their Significance to Oligotrophic Lakes," University of Illinois at Urbana-Champaign (1974).

12. Andren, A., S. Eisenreigh, F. Elder, T. Murphy, M. Sanderson and R. Vet. "Atmospheric Loadings to the Great Lakes," report to Task C, International Reference Group on Great Lakes Pollution from Land Use Activities. Unpublished.

13. Sonzogni, W. C. "Impact of U.S. Land Derived Pollutant Loadings on the Great Lakes; Literature Review," for Task D (U.S. Section) International Reference Group on Great Lakes Pollution from Land Use Activities, in draft, December 1978.

14. Coote, D. R., E. M. MacDonald and W. T. Dickinson. "Agricultural Watershed Studies in the Canadian Great Lakes Basin; Final Summary Report," to Task C (Canadian Section) International Reference Group on Great Lakes Pollution from Land Use Activities, Windsor, Ontario (1978).

15. Adams, J., U.S. Corps of Engineers, Buffalo District. Personal communication (1979).

16. Stewart, B. A. et al. "Control of Water Pollution from Cropland, Volume I: A Manual for Guideline Development," Agricultural Research Service, USDA, Washington, D.C., prepared under Interagency Agreement with the U.S. Environmental Protection Agency, Athens, GA, Publication No. EPA-600/2-75-026a (1975).

17. Haith, D. A., and R. C. Loehr (Eds.) "Effectiveness of Soil and Water Conservation Practices for Pollution control," Section 2 "Conclusions." Draft report to the U.S. EPA, Athens, GA (1978).

18. Walter, M. F., T. S. Steenhuis and H. P. DeLancey. "The Effects of Soil and Water Conservation Practices on Sediment," in *Effectiveness of Soil and Water Conservation Practices for Pollution Control,* Haith, D. A. and R. C. Loehr (Eds.). Draft report to U.S. EPA, Athens, GA (1978).

19. Olness, A. and D. Rausch. "Callahan Reservoir: III. Bottom Sediment-Water-Phosphorus Relationships," in *Trans. Am. Soc. Agric. Eng.* 29(2) (1977).

20. Bartsch, A. F., and R. A. Taft. "Settleable Solids, Turbidity, and Light Penetration as Factors Affecting Water Quality," in *Transactions of a Seminar in Biological Problems in Water Pollution.* U.S. Department of Health, Education, and Welfare (1969).

21. Lake, J., and J. Morrison. "Environmental Impact of Land Use on Water Quality," Final Report on the Black Creek Project, EPA-905/5-77-077-B (1977).

22. Forster, D. L. "Economic Impacts of Changing Tillage Practices in the Lake Erie Basin," for the Lake Erie Wastewater Management Study, U.S. Army Engineer District, Buffalo, NY (1978).

23. Great Plains Council. "The Future of the Great Plains," Report to the President (1936).

24. Statistics Canada. "Survey of Canadian Great Lakes Basin Farmers Regarding Water Pollution from Agricultural Activities," for International Reference Group on Great Lakes Pollution from Land Use Activities, Windsor, Ontario (1977).

25. Powers, E. C. and E. A. Jarecki, Eds. "Survey of U.S. Great Lakes Basin Farmers Regarding Water Pollution From Agricultural Activities," U.S.D.A. Statistical Reporting Service, for International Reference Group on Great Lakes Pollution From Land Use Activities, Windsor, Ontario (1977).

26. Castrilli, J. F., and A. J. Dines. "Control of Water Pollution From Land Use Activities in the Great Lakes Basin: A Review of Regulatory Programs in Canada and the United States," for the International Reference Group on Great Lakes Pollution From Land Use Activities, Windsor, Ontario (1978).

27. Berg, N. A. "Unmuddying the Waters," *Great Lakes Communicator,* 9(1) (1978).

28. Pleva, E. G. "The Conservation Authorities Connection and the Thames Drainage Systems," in proceedings of the Twenty-fifth National Watershed Congress, Toronto (1978).

29. Gardner, F. E., Islington, Ontario. Personal communication (1979).

30. Johnson, M. G., Canada Centre for Inland Waters, Burlington, Ontario. Personal communication (1979).

31. USDA Soil Conservation Service. "Agricultural Waste Management Field Manual," Washington, DC (1975).

32. Canada Animal Waste Management Guide Committee. "Canada Animal Waste Management Guide," under authority of Canada Committee on Agricultural Engineering (1972).

33. Brubaker, J. E., Ontario Ministry of Agriculture and Food, Guelph, Ontario. Personal communication (1979).

34. "Rural Clean Water Program," Federal Register 43(212):50845-50866. (1978), (to be codified in 7 CFR, Sec. 634.27).

35. U.S. Department of Agriculture. "Statement on Land Use Policy," Secretary's Memorandum No. 1827, revised (1978).

36. The Governments of Canada and the United States. "Great Lakes Water Quality Agreement (Revised)," Signed at Ottawa, Ontario (1978).

37. Leppard, S. "Reports of the Canadian Public Consultation Panels to the International Reference Group on Great Lakes Pollution from Land Use Activities," Windsor, Ontario (1978).

38. Clark, M. "Reports of the United States Public Consultation Panels to the International Reference Group on Great Lakes Pollution from Land Use Activities," Windsor, Ontario (1978).

39. International Joint Commission. "Pollution of Lake Erie, Lake Ontario, and International Section of the St. Lawrence River," Report of the

International Lake Erie Water Pollution Board and the International Lake Ontario-St. Lawrence River Water Pollution Board, Windsor, Ontario (1969).

40. Pollution from Land Use Activities Reference Group. "Early Action Program Report," submitted to the Great Lakes Water Quality Board, Windsor, Ontario (1974).

acetylene reduction 301
advanced waste treatment 415
advection 196
aerated lagoons 346
agricultural conservation program 477
agricultural land, guidelines for sludge use 387
agriculture 460
algae 23,64,127
algal-available phosphorus 261
alum 364
aluminosilicate builders 392
aluminum sulfate 330
ammonia 107,181
anaerobic 250
anion-exchangeable phosphorus 303
anoxia 202,249
apatite 271
atmospheric inputs 40,278
atmospheric phosphorus loads 42,53
available phosphorus 285
availability 259

Bardenpho 333
benthic organisms 78
benthos 64
best available treatment (BAT) 297
best management practices (BMP) 444
Bierman 75,155,238,240
biochemical oxygen demand (BOD) 197,333,423

biological availability 44,50,84, 466
biological removal of phosphorus 332
biological sludge production 362
biological treatment 417
Black Creek project 477
blue-green algae 64,127
BOD$_5$
 See biochemical oxygen demand
Boundary Waters Treaty 14

cadmium 418
capital costs 358
carboxymethyloxysuccinate (CMOS) 392
cation exchange capacity (CEC) 417
Cayuga Lake 4
Chapra 71,238,239
chemical costs 349,356
chemical fertilizers 382
chemical precipitation 405
chloride 94
chlorophyll 64,68,73,78,93,180, 181,188,216,223,239,250, 298
citrate-dithionite-bicarbonate 272
Cladophora 64,70,87,237
Clean Water Act 420,476
combined sewer overflows 435
conservation tillage 466
conventional treatment 416